Compendium of Corn Diseases

FOURTH EDITION

Edited by

Gary P. Munkvold
Department of Plant Pathology and Microbiology
Iowa State University, Ames

Donald G. White
Department of Crop Sciences
University of Illinois, Urbana

APS PRESS
The American Phytopathological Society
St. Paul, Minnesota, U.S.A.

Front and back cover images: Courtesy DuPont Pioneer

Library of Congress Control Number: 2015960969
International Standard Book Numbers:
Print: 978-0-89054-492-1
Online: 978-0-89054-494-5
Mobi: 978-0-89054-493-8
ePub: 978-0-89054-566-9

First edition published 1973
Second edition published 1980
Third edition published 1999
Fourth edition published 2016
Second printing, 2017
(The second printing includes substantial taxonomic updates.)

Printed in the United States of America on acid-free paper.

The American Phytopathological Society
3340 Pilot Knob Road
St. Paul, Minnesota 55121, U.S.A.

Preface

Malcolm C. Shurtleff

The American Phytopathological Society (APS) published the *Compendium of Corn Diseases* in 1973 amid an eagerness within the agricultural community for more information about diseases of corn following the southern corn leaf blight epidemics of 1970 and 1971. The corn compendium was the initial publication of APS PRESS. The overwhelming success of this publication led directly to the entire series of APS disease compendia and launched APS PRESS. Malcolm C. Shurtleff, the extension pathologist at the University of Illinois, conceived of the concept of the corn compendium and secured funding to compile the compendium before approaching APS business manager Ray Tarleton with the proposal that APS publish his manuscript. Shurtleff then edited the first two editions. We are indebted to Dr. Shurtleff for his insightful recognition of the need for the disease compendia and for his diligent efforts in compiling the first corn and soybean compendia. Forty-three years after publication of the first edition, we are excited to complete the fourth edition of the *Compendium of Corn Diseases*.

As with the first three editions, the fourth edition has been written to serve as a practical reference for both plant pathologists and general audiences. It is intended for the education of and use by crop scouts, agribusiness representatives, extension agents, area crop specialists, educators, students, growers, and researchers. The selected references are a starting point for individuals who wish to learn more about specific corn diseases.

New features of the fourth edition include a diagnostic key to aid in the identification of disease symptoms, a new section on the use of fungicides for foliar disease management in corn, a section on storage molds, and expanded entries for several increasingly important diseases, such as Diplodia leaf streak, stolbur mycoplasma, seedling blights, and nematodes that feed on corn. The section on diseases caused by viruses has been expanded to include more detail on some of the viruses that are not widely distributed. The number of color figures has increased drastically from 177 in the third edition to more than 300 in this fourth edition. Two-thirds of all the figures are new. As in other recent APS compendia, color figures are found within their appropriate section, which facilitates relating the text to the photos. We hope that these additional figures will add to the value of the compendium as a guide for identifying corn diseases. We attempted to illustrate not just the "textbook" symptom appearance for each disease but also to show some of the range in symptoms that can occur over time and across host genotypes.

As with the first three editions, this edition is international in scope but emphasizes diseases in North America. Each section of the compendium has been revised and updated by a plant pathologist; in most cases, section authors for the third edition revised their sections for this edition. The organization of the fourth edition follows closely that of the third edition. Part I is composed of infectious diseases and includes major sections for prokaryotic pathogens (Bacteria and Mollicutes), eukaryotic pathogens (Fungi and Oomycetes), viruses, nematodes, and parasitic plants. Eukaryotic pathogens are arranged according to disease type (e.g., root rots, foliar diseases, downy mildews, smuts, rusts, stalk rots, and ear rots). Part II is composed of noninfectious or abiotic disorders, including injuries caused by fertilizers, herbicides, and air pollutants, and of abnormalities caused by environmental factors.

This edition of the compendium includes an appendix that identifies common names of diseases of corn, along with their pathogens. The same content is available on the website of The American Phytopathological Society in the section "Common Names of Plant Diseases." Any updates that are subsequently made to the taxonomy and nomenclature will be reflected on the website. Throughout the text, the common names of diseases are generally provided, along with the currently accepted scientific names. Authorities for the scientific names of pathogens are included in the Appendix. Within the text, plants are mostly referred to by their common names; scientific names are also provided for those plants whose common names may be unfamiliar to readers.

The authors are grateful to all the section authors and to those who contributed figures, as well as to the financial sponsors, listed on the inside front cover. We also are grateful to Greg Tylka (Iowa State University), Doug Jardine (Kansas State University), and Snook Pataky (Monsanto Co., formerly University of Illinois) for reviewing portions of the manuscript. Special thanks also go to the editors and staff of APS PRESS for their contributions.

Gary P. Munkvold
Donald G. White

Contributors

Z. Gloria Abad
USDA APHIS PPQ CPHST
Beltsville, MD

E. Alves
DuPont do Brasil S.A.—Divisão Pioneer Sementes
Itumbiara, Brazil

Brian M. Anderson
DuPont Pioneer, Inc.
Johnston, IA

Deane C. Arny
University of Wisconsin-Madison
Department of Plant Pathology
Madison, WI

G. Avramov
Pioneer Hi-Bred Srb
Novi Sad, Serbia

Gary C. Bergstrom
Cornell University
Plant Pathology & Plant-Microbe Biology Section
Ithaca, NY

M. Bolarin
International Institute for Tropical Agriculture
Ibadan, Nigeria

M. R. Bonde
USDA ARS (retired)
Middletown, MD

Carl A. Bradley
University of Illinois
Department of Crop Sciences
Urbana, IL

Edward A. Brown
University of Georgia
Department of Plant Pathology
Colbert, GA

Robert B. Carroll
University of Delaware (retired)
Elkton, MD

Martin L. Carson
USDA ARS (retired)
University of Minnesota - Cereal Disease Laboratory
St. Paul, MN

Keith Chamberlain
Rothamsted Research
Harpenden, Hertfordshire, United Kingdom

Wedong Chen
USDA ARS
Washington State University
Pullman, WA

Larry E. Claflin
Kansas State University
Department of Plant Pathology
Manhattan, KS

Carol Collar
University of California-Davis
Department of Animal Science
Hanford, CA

Pedro W. Crous
Centraalbureau Voor Schimmelcultures
Utrecht, Netherlands

M. P. da Silva
University of Arkansas
Department of Crop, Soil, and Environmental Sciences
Fayetteville, AR

Carlos De Leon
CIMMYT (retired)
(International Maize and Wheat Center)
Mexico City, Mexico

James L. Dodd
Professional Seed Research, Inc.
Sugar Grove, IL

Anne E. Dorrance
Ohio State University
Department of Plant Pathology
Wooster, OH

Larry D. Dunkle
Purdue University
Department of Botany and Plant Pathology
West Lafayette, IN

Darin M. Eastburn
University of Illinois
Department of Crop Sciences
Urbana, IL

Dale I. Edwards
University of Illinois
Department of Crop Sciences
Urbana, IL

H. A. El-Shafey
Plant Pathology Research Institute
Agricultural Research Council
Giza, Egypt

Roger Elmore
University of Nebraska-Lincoln
Department of Agronomy and Horticulture
Lincoln, NE

Aaron G. Hager
University of Illinois
Department of Crop Sciences
Urbana, IL

E. Scott Hagood, Jr.
Virginia Polytechnic Institute and State University
Department of Plant Pathology, Physiology, and Weed Science
Blacksburg, VA

Robert Hammon
Colorado State University Extension
Mesa, CO

R. G. Hartzler
Iowa State University
Department of Agronomy
Ames, IA

Robert G. Hoeft
University of Illinois
Department of Crop Sciences
Urbana, IL

Richard S. Hussey
University of Georgia
Department of Plant Pathology
Athens, GA

Tamra A. Jackson-Ziems
University of Nebraska
Plant Pathology Department
Lincoln, NE

Douglas J. Jardine
Kansas State University
Department of Plant Pathology
Manhattan, KS

Kyle Jensen
DuPont Pioneer, Inc.
Johnston, IA

Kevin Korus
University of Nebraska
Plant Pathology Department
Lincoln, NE

Lorin R. Krusberg
University of Maryland
Department of Botany
College Park, MD

Nancy Lambrinos
Nancy Lambrinos Design
Eden Prairie, MN

Frances M. Latterell
USDA
Plant Disease Research Laboratory
Frederick, MD

Meyer LeRoux
Sensako
Delmas, South Africa

John F. Leslie
Kansas State University
Department of Plant Pathology
Manhattan, KS

Xiao Li
Institute of Plant Protection
Sichuan Academy of Agricultural Sciences
Chengdu, PRC China

Patrick E. Lipps
Ohio State University
Department of Plant Pathology
Wooster, OH

Raymond Louie
USDA ARS
Plant Pathology Department
Wooster, OH

Mark K. Mancl
DuPont Pioneer, Inc.
Woodland, CA

Charles A. Martinson
Iowa State University (retired)
Department of Plant Pathology and Microbiology
Ames, IA

Derrick Mayfield
Iowa State University
Department of Plant Pathology and Microbiology
Ames, IA

Scott Monfort
University of Georgia
Tifton, GA

Daren S. Mueller
Iowa State University
Department of Plant Pathology and Microbiology
Ames, IA

Gary P. Munkvold
Iowa State University
Department of Plant Pathology and Microbiology
Ames, IA

Chris Nacci
USDA ARS
Ohio State University
Wooster, OH

Robert L. Nielsen
Purdue University
Department of Agronomy
West Lafayette, IN

Dave C. Nowell
United Nations Food and Agriculture Organization
Rome, Italy

Gary N. Odvody
Texas A&M University
Texas AgriLife Research & Extension Center
Corpus Christi, TX

Jerald K. Pataky
Monsanto Co.
Thomasboro, IL

Timothy C. Paulitz
USDA ARS
Root Disease and Biological Control Research Unit
Pullman, WA

Gary A. Payne
North Carolina State University
Department of Plant Pathology
Raleigh, NC

Dianne Pedersen
University of Illinois
Urbana, IL

David R. Pike
University of Illinois
Department of Crop Sciences
Urbana, IL

Brent Pringnitz
Iowa State University
Agriculture and Natural Resources Extension
Ames, IA

Margaret (Peg) G. Redinbaugh
USDA ARS
Ohio State University
Department of Plant Pathology
Wooster, OH

Robert Reeder
CABI UK
Egham, United Kingdom

Bobby L. Renfro
Bartlesville, OK

Alison E. Robertson
Iowa State University
Department of Plant Pathology and Microbiology
Ames, IA

Martha Patricia Romero Luna
Purdue University
Department of Botany and Plant Pathology
West Lafayette, IN

Laura Ruttman
DuPont Pioneer, Inc.
Johnston, IA

Erika Saalau-Rojas
University of Massachusetts
Cranberry Station
East Wareham, MA

John Sawyer
Iowa State University
Department of Agronomy
Ames, IA

Keith Seifert
Agriculture and Agri-Food Canada
Eastern Cereal and Oilseed Research Centre
Ottawa, Ontario, Canada

David R. Smith
Monsanto Co.
DeKalb, IL

Lucy R. Stewart
USDA ARS
Wooster, OH

Erik L. Stromberg
Virginia Polytechnic Institute and State University (retired)
Department of Plant Pathology, Physiology, and Weed Science
Blacksburg, VA

Brett A. Summerell
Royal Botanic Gardens
Sydney, NSW, Australia

Donald R. Sumner
University of Georgia
Coastal Plain Experiment Station
Tifton, GA

Timothy C. Todd
Kansas State University
Department of Plant Pathology
Manhattan, KS

Ivo Tosevski
Institute for Plant Protection and Environment
Department of Plant Pests
Banatska, Zemun, Serbia

Donald G. White
University of Illinois
Department of Crop Sciences
Urbana, IL

Brent Wilson
DuPont Pioneer, Inc.
Johnston, IA

Carol E. Windels
University of Minnesota (retired)
Alexandria, MN

Gary L. Windham
USDA ARS
Mississippi State University
Crop Science Research Laboratory
Mississippi State, MS

Gayle L. Worf
University of Wisconsin (retired)
Madison, WI

Xiude Xu
Liaoning Academy of Agricultural Science
Plant Protection Institute
Senyang, China

Contents

Introduction

Production and Use of Corn

Corn, also known as maize, is one of the world's three most important cereal crops, along with rice and wheat. Corn originated in the Americas prior to 5000 B.C. Selection by indigenous people in many different geographical areas over several thousand years has resulted in many different races of corn. These races, along with undomesticated relatives, constitute a very diverse genetic base (Fig. 1).

Corn is a tall, monoecious determinant annual plant in the grass family (*Poaceae*) (Fig. 2). Like most grasses, it has conspicuous nodes in the stem with a single leaf at each node in a distichous pattern (two ranks of leaves in alternating positions). Leaves have a hairy upper surface and glabrous underside with numerous stomata. Corn usually has a single dominant stem, but branches can form at the nodes, and basal tillers are common. Adventitious (brace) roots usually develop from the lowermost aboveground nodes. Unlike most grasses, the corn stem (stalk) does not become hollow but remains solid. The stalk terminates with the male inflorescence (tassel), which consists of a central spike with many ranks of paired spikelets and several lateral branches, each with two ranks of spikelets. Each spikelet in turn consists of two florets, each with three anthers. An average tassel produces about 25,000,000 windborne pollen grains from about 10,000 anthers. Female inflorescences (ears) are borne on lateral branches (shanks) with compressed internodes, so that the ear is enveloped in layers of husk leaves. Spikelets are arranged in paired parallel ranks along a thick lignified axis (cob), so that the number of kernel rows is always even. Styles (silks) elongate from the florets to extend beyond the husk leaves and receive pollen. In an open-pollinated field, about 95% of the ovaries are cross-pollinated. After fertilization, the spikelet matures to become a single-seeded caryopsis (kernel), consisting of the embryo and a starchy endosperm surrounded by an aleurone layer and a hard pericarp, and is attached to the cob by a short pedicel. The monoecious nature of corn, with physically separate male and female inflorescences, has greatly facilitated controlled pollinations and the development of high-yielding hybrids.

Corn grain can be divided into six different groups according to endosperm type. Flint corn has kernels with a hard endosperm that does not grind well when dry. Floury types have an endosperm that is soft and malleable throughout. Popcorn, probably the original domesticated type, is essentially a small-kernelled flint type, with kernels that burst when the moisture trapped in the endosperm expands during heating. Dent corn is the most common type, with an endosperm intermediate between floury and flint types: flinty along the sides of the kernel and floury at the core. The endosperm core contracts during drying, forming an indentation in the crown of the grain. Sweet corns have mutations in the starch biosynthetic pathway, resulting in a high proportion of sucrose in the endosperm. Waxy types have starch composed entirely of amylopectin.

Corn plant development has typically been described using two different systems. The most commonly used scale, described by J. J. Hanway in the 1960s, divides corn development into vegetative (V) and reproductive (R) stages. This scale commences with emergence (VE) and continues with stages V1 through V*n*, with the numerals designating the number of leaves with fully visible leaf collars. The last V stage is VT, when the tassel is fully visible. The reproductive stages are R1 (silk emergence), R2 (blister stage), R3 (milk stage), R4 (dough stage), R5 (dent stage), and R6 (physiological maturity). The BBCH scale has eight stages, each with more specific substages: stage 0, germination through emergence; stage 1, leaf development (early vegetative stages); stage 3, stem elongation (late vegetative stages); stage 5, tassel emergence; stage 6, anthesis; stage 7, development of fruit (early reproductive stages); stage 8, ripening (late reproductive stages); stage 9, senescence.

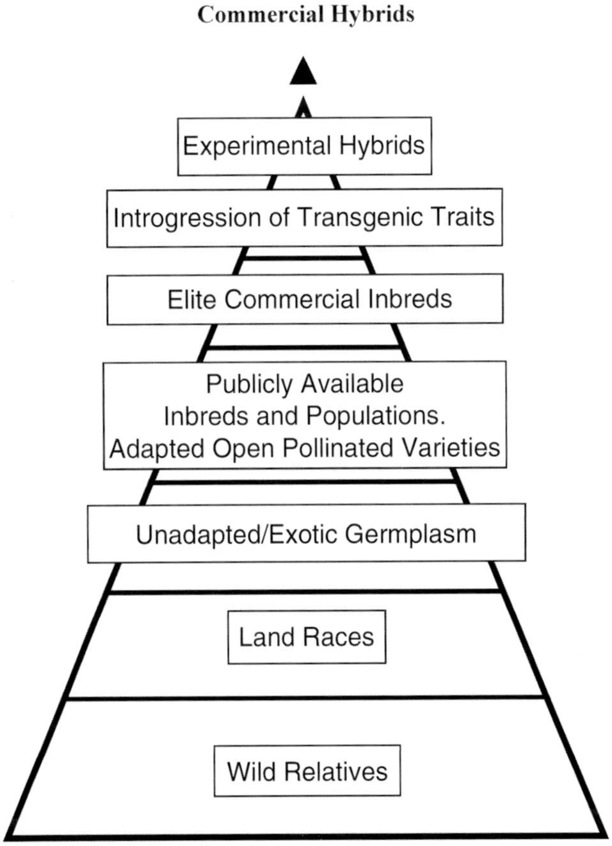

Fig. 1. Genetic basis of modern corn germplasm. (Courtesy D. G. White)

Corn currently is grown on all continents except Antarctica and is most productive where rainfall or irrigation is adequate. It is relatively easy to cultivate, and it produces large yields of grain high in starch, which can be readily metabolized into energy. Corn is not a complete food because it is deficient in the amino acid lysine, which is essential in the diet of humans and animals. When lysine is available from other sources, corn can be used as a high percentage of the diet.

Many different products can be made from corn. More than 50% of the crop produced in the United States that is not exported is processed for food, alcohol, and industrial uses. Most processed foods in grocery stores in the United States contain

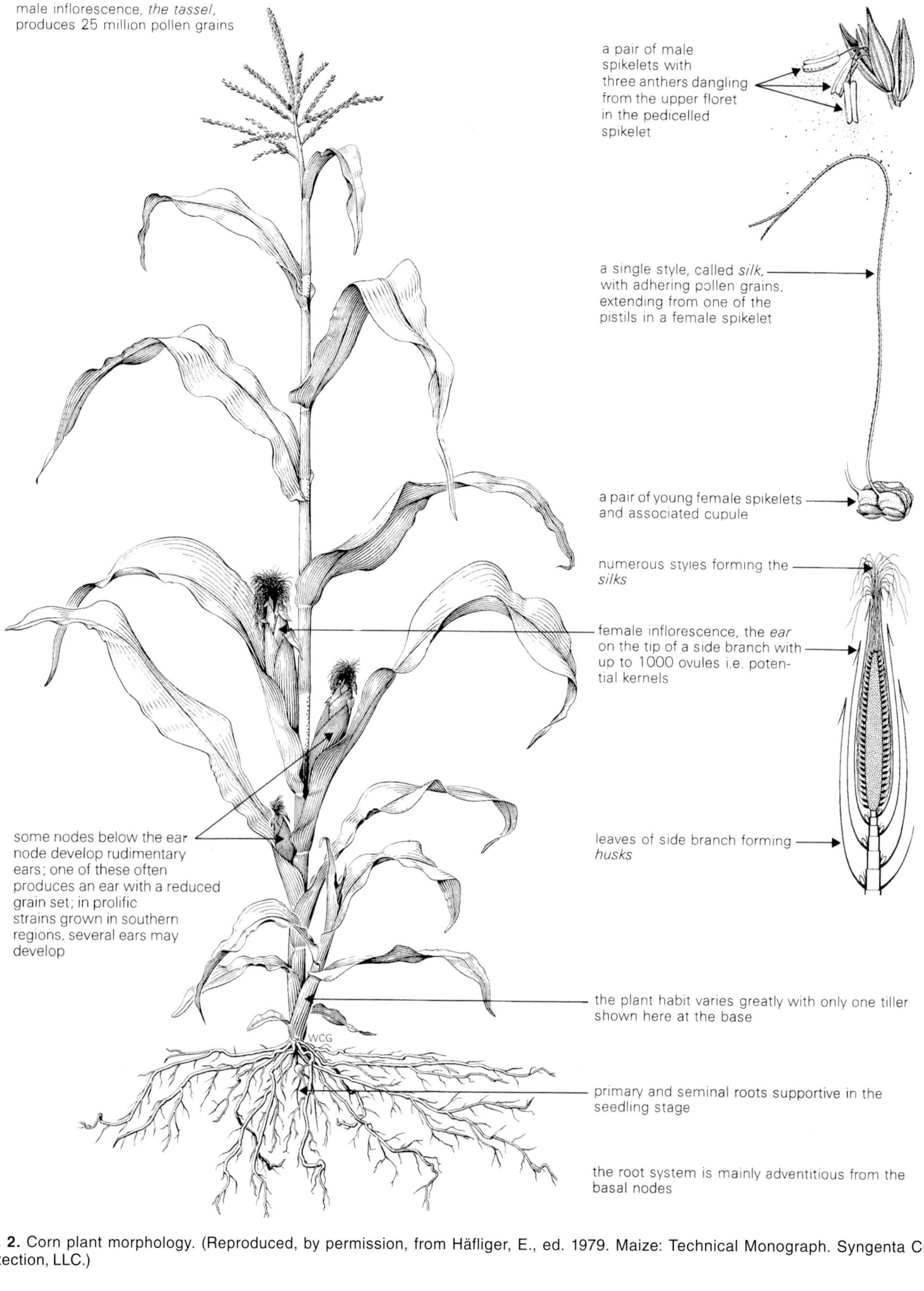

Fig. 2. Corn plant morphology. (Reproduced, by permission, from Häfliger, E., ed. 1979. Maize: Technical Monograph. Syngenta Crop Protection, LLC.)

some form of corn-based product, including high-fructose corn syrup, corn starch, modified corn starch, and corn oil. Fuel ethanol has been the fastest growing segment of corn use in the United States, rising more than sevenfold between 2000 and 2014 and consuming nearly 40% of the annual crop as of 2014 (although approximately 8% is returned to the livestock feed market as dried distillers' grain and solubles).

The United States produces nearly 40% of the world's corn crop. Corn is the most valuable crop grown in the United States, annually valued at $60–75 billion. Other major producers include the Peoples' Republic of China, Brazil, the European Union (including leading producers France, Italy, Romania, and Hungary), Mexico, Indonesia, Argentina, India, South Africa, Ukraine, and Canada. Approximately 15–20% of the corn produced in the United States is exported to other countries, making corn very important in the balance of trade. The United States is responsible for more than 50% of all corn exports globally, followed by Argentina (15–16%) and Brazil (8–10%).

The success of corn as a major food and feed crop can be largely attributed to the fact that it is grown as a hybrid. The first corn cultivars grown were open pollinated, and the seed for the following year was selected from the current year's harvest. The first commercial corn hybrids (crosses between inbreds) were produced and sold as early as the 1920s. The first hybrids were adapted primarily in the central United States, the area referred to as the "Corn Belt." Farmers accepted these hybrids because of their higher yields and more vigorous growth. By the 1940s, most farmers in the midwestern United States were using hybrids. The first hybrids produced and sold commercially were double crosses that involved four inbred lines. Starting in the 1950s and continuing through the late 1970s, a transition from double crosses to single crosses occurred; most corn hybrids used now are single crosses produced from two inbred parents. Hybrids are also used in many of the corn-producing areas of the world, except where low-input subsistence farming is practiced.

Because these corn cultivars are hybrids, the farmer must buy or produce new seed each year. The demand for seed in the United States and much of the world is met by corn seed companies. These companies may be very large with their own research and development programs or may be relatively small and depend on research and development by other companies and universities. The hybrid corn seed industry is the engine of growth for corn production. There is a competitive advantage for a corn seed company to have the highest yielding, agronomically superior cultivars with superior seed quality. Companies that do not provide good hybrids will lose customers and may not be profitable. Therefore, it is cost-effective for companies to do research and development to constantly improve the crop.

Hundreds of thousands of potential new hybrids, which are products of breeding programs in private companies, are tested each year in the United States. Many are evaluated in multiple locations for various characteristics, including disease resistance. Of the large number of potential commercial hybrids evaluated in a given year, only a few will ever be marketed in any significant geographical area. There is no other crop with such a large effort devoted to its continued improvement. Research expenditures have accelerated rapidly with the advent of biotechnology in corn hybrid development. Molecular markers have been identified for disease resistance and other traits, and high-throughput marker-assisted selection is widely used. In addition, a major part of the hybrid development effort is currently directed toward the discovery of new transgenes and their integration into high-yielding, stress-tolerant commercial hybrids.

Corn production in the United States is now predominantly accomplished with hybrids that express transgenic traits for insect resistance or herbicide tolerance, alone or in combination. Approximately 90% of the area planted to corn in the United States in 2013 was planted with transgenic hybrids. Seventy-one percent of the area was planted with "stacked trait" hybrids that include both insect resistance and herbicide tolerance; 14% was planted with herbicide-tolerant hybrids; and 5% was planted with insect-resistant hybrids. The primary herbicide tolerance trait was for glyphosate, and the majority of insect-resistant hybrids include genes for both lepidopteran insects (such as European corn borer) and corn rootworm (*Diabrotica* spp., order Coleoptera). Hundreds of potential new transgenes are tested each year, and traits for improved end-use characteristics will likely be commercially available in the near future. Globally, transgenic corn hybrids were grown on more than 57 million hectares in 17 countries, accounting for more than 32% of global corn planting as of 2013. After the United States, countries with significant area planted to transgenic corn include Argentina, Brazil, Canada, and South Africa.

Selected References

Hanway, J. J. 1963. Growth stages of corn (*Zea mays* L.). Agron. J. 55:487-492.

James, C. 2013. Global Status of Commercialized Biotech/GM Crops: 2013. ISAAA Brief No. 46. International Service for the Acquisition of Agri-Biotech Applications, Ithaca, NY.

Lancashire, P. D., Bleiholder, H., Langeluddecke, P., Stauss, R., van den Boom, T., Weber, E., and Witzen-Berger, A. 1991. A uniform decimal code for growth stages of crops and weeds. Ann. Appl. Biol. 119:561-601.

National Corn Growers' Association. 2013. World of Corn 2013. National Corn Growers' Association, Chesterfield, MO. www.ncga.com

Sprague, G. F., and Dudley, J. W., eds. 1988. Corn and Corn Improvement, 3rd ed. American Society of Agronomy, Madison, WI.

(Prepared by D. G. White, M. L. Carson, and G. P. Munkvold)

Diseases of Corn

All parts of the corn plant are susceptible to a number of diseases that reduce the ability of the plant to grow in a normal manner and may ultimately result in the loss of grain yield or quality. The actual annual loss caused by disease is impossible to measure because at least some disease is present in every field every year. Therefore, yield in the complete absence of disease is not known. Factors such as environmental conditions, yield potential, soil fertility, genetics of particular hybrids, pathogen population size and composition, and the synergistic action of different diseases and insects all impact yield loss and cannot be evaluated with any precision, especially over a large geographical area. Best-guess estimates of disease loss for corn in the United States range from 2 to 15% annually. The most devastating disease epidemic of corn to occur in the United States in modern times was the southern corn leaf blight in 1970, when approximately 20% of the corn crop was lost. Losses in some areas were 100%, while in drier and more northern areas, where conditions did not favor disease, there were no losses at all. Perhaps the greatest losses caused by diseases are from those that occur annually with little or no variation and whose effect on the crop is unrecognized.

Corn diseases may be managed by cultural practices (e.g., burying the residue of diseased crops by tillage and rotation with nonhost crops), by the use of chemicals or biocontrol agents, and by the use of plants with genetic resistance. Implementation of management practices can be constrained by logistical considerations and the balancing of multiple priorities. Tillage of plant debris and crop rotation may or may not be practical depending upon economic and environmental considerations. Burying diseased debris results in some control of leaf diseases but does not have much effect on many root and stem diseases. Crop rotation has value for disease management, but economic conditions in some areas result in the continuous cultivation of corn and this can be done successfully with the effective use of other crop management practices. Chemical seed treatment is routinely used for the control of seedling blights and, in some areas, downy mildew diseases. Foliar fungicides are routinely used to protect inbreds in hybrid seed production fields and sweet corn. They are also used to protect corn grown for grain against several leaf diseases. However, the frequency of their use in grain crops is generally connected with grain prices. Chemical control of nematodes with soil or seed application of nematicides can be useful but may not be cost-effective in all cases and can be subject to environmental concerns.

Disease resistance is undoubtedly the most effective and widely used means of managing corn diseases. Although high levels of resistance to the important diseases may be found among commercial hybrids, no hybrid is highly resistant to all diseases. However, hybrids often are available with adequate resistance to prevent significant losses caused by potentially important diseases in a geographical area, except under optimum conditions for disease development. The types and levels of disease resistance needed to protect a hybrid grown in the cool, semiarid northern areas are substantially different from those needed to protect a hybrid grown in warm areas. Because specific hybrids may be commercially used for only a few years, their susceptibility or resistance to some "minor" diseases may never be known. Monogenic resistance is available for the control of some corn diseases and is used to various degrees. However, the emphasis of many corn-breeding programs is on polygenetically inherited, partial resistance. As a result of decades of selection, adequate levels of polygenic resistance to diseases present in a geographical area are often available among the inbred lines adapted to and commercially used in that area. When lines with acceptable levels of resistance are crossed, the progeny selected as new inbred lines usually have similar levels of resistance. In these cases, the selection process for disease resistance is a matter mostly of discarding very susceptible materials rather than of selecting a very high level of resistance. Most corn hybrids marketed in the United States have been previously tested in a number of locations for yield and other agronomic characteristics. Even when evaluation for disease resistance is not intentional, it is very likely that, in some locations, yield is limited by disease and hybrids are discarded from consideration for commercial use because of low yield.

With the concerted efforts of plant pathologists and plant breeders in both the public and private sectors, widespread and significant damage to the U.S. corn crop from diseases is the exception rather than the rule. Nevertheless, as a result of changing cultural practices, the constantly changing genetics of corn hybrids, and evolving pathogen populations (not to mention climatic change and introduction of new pathogens), the importance of specific corn diseases is continuously changing and requires constant vigilance.

Selected References

Oerke, E. C. 2006. Crop losses to pests. J. Agric. Sci. 144:31-43.

Sprague, G. F., and Dudley, J. W., eds. 1988. Corn and Corn Improvement, 3rd ed. American Society of Agronomy, Madison, WI.

(Prepared by D. G. White, M. L. Carson, and G. P. Munkvold)

Diagnostic Key to Corn Diseases

The purpose of this diagnostic key is to assist with the recognition of many of the common diseases of corn that occur in the United States. Several minor diseases that are not often encountered also are included. Accurate disease diagnosis is important for optimal understanding of when and where diseases occur and how they are best managed. Unfortunately, diagnosis on the basis of symptoms alone is often not reliable. The best diagnoses are made on the basis of symptoms and subsequent laboratory identification of the pathogen associated with the symptoms.

Lesions on Leaves

1 Lesions long and narrow

1a Mature lesions: 5 cm or greater in length

Wavy margins .... Stewart's Bacterial Wilt
Wavy margins with dark or water-soaked specks .... Goss's Bacterial Wilt and Blight
Yellow streaks on entire length of leaf .... Immature Lesions of Stewart's Bacterial Wilt
.... Herbicides Acting as Synthetic Auxins (Growth Regulators)
Yellow or clear veinal striping .... Herbicides Inhibiting Chlorophyll Formation
Lesions between veins .... Bacterial Leaf Blight
.... Mature Gray Leaf Spot on Very Susceptible Genotype
.... Low-Temperature "Sunscald" and Cold Banding
Cigar-shaped lesions .... Northern Corn Leaf Blight
Cigar-shaped lesions with black pycnidia .... Diplodia Leaf Streak
Downy growth on both sides of leaf .... Sorghum Downy Mildew
Leaf burn necrosis on tips of leaves .... Herbicides Inhibiting Photosynthesis

1b Mature lesions: less than 5 cm in length

Rectangular in shape .... Gray Leaf Spot
Grayish tan: generally less than 2 mm wide .... Northern Corn Leaf Spot Race 3

2 Mature lesions oval in shape: generally less than 2.5 cm in length .... Gray Leaf Spot
.... Southern Corn Leaf Blight
.... Northern Corn Leaf Spot Races 1 and 2
.... Anthracnose Leaf Blight
.... Yellow Leaf Blight

3 Mature lesions round in shape
- **3a** Tan center, rings black to purple with yellow halo: lesions generally less than 0.5 cm in diameter .... Eyespot
- **3b** Spots are brown, small, and chlorotic: generally less than 0.5 cm in diameter .... Physoderma Brown Spot
- **3c** Lesions white, turning to tan and brown .... Holcus Spot
- **3d** Pustules are dark brown with sporulation on both upper and lower leaf surfaces .... Common Rust
- **3e** Pustules are light cinnamon brown to orange with sporulation mostly on lower leaf surface .... Southern Rust

Abnormal Color of Leaves

1 Mosaic coloration of leaves .... Diseases Caused by Viruses
.... Genetic Stripe, Leaf Spots, and Flecks

2 Purplish cast on leaves of seedlings .... Phosphorus Deficiency Caused by Cool Soils
.... Herbicides Inhibiting Seedling Growth

3 Mature leaves purplish to reddish .... Diseases Caused by Viruses

4 Necrotic streaks on leaves of seedlings ..... Seedling Disease Caused by Seedborne *Aspergillus* spp. and *Penicillium* spp.

5 Abnormal coloration of leaves emerging from whorl
- **5a** Leaves are white. .... Herbicides Inhibiting Amino Acid Synthesis
- **5b** Leaves are necrotic .... Bacterial Stalk Rot
  .... Herbicides Inhibiting Photosynthesis
  .... Herbicides Inhibiting Amino Acid Synthesis
  .... Herbicides Inhibiting Chlorophyll Formation
  .... Herbicide Injury Followed by Bacterial Soft Rot
  .... Stalk-Boring Insects

Galls on Leaves

White blisters on leaves, turning darker with age .... Common Smut
Raised warts on underside of leaves, green to white, following veins, sometimes with prominent trichomes .... Maize Rough Dwarf

Root Diseases

1 Stunting of roots with or without rot .... Seedling Disease Caused by *Aspergillus* spp. and *Penicillium* spp.
.... Pythium Seed Rot and Seedling Blights
.... Seed Rot and Seedling Blight Caused by *Stenocarpella maydis*
.... Imbibitional Chilling Damage
.... Fusarium Root Rot
.... Diseases Caused by Nematodes
.... Herbicides Acting as Synthetic Auxins (Growth Regulators)
.... Herbicides Inhibiting Amino Acid Synthesis
.... Herbicides Inhibiting Seedling Growth

2 Rot of mesocotyl .... Seedling Disease Caused by *Aspergillus* spp. and *Penicillium* spp.
.... Pythium Root Rot
.... Seed Rot and Seedling Blight Caused by *Stenocarpella maydis* or *Fusarium* spp.

3 Red color of senescing roots
- **3a** Dark red .... Red Root Rot
- **3b** Pink to red .... Fusarium Root Rot

Stalk or Leaf Sheath Symptoms

1 Stalks rotted
- **1a** Black specks on stalk surface
  Specks may be scraped off with thumb .... Gibberella Stalk Rot
  Specks embedded in stalk .... Diplodia Stalk Rot
- **1b** Black specks (sclerotia) inside and on surface .... Charcoal Rot
- **1c** Black shiny discoloration just under surface .... Anthracnose Stalk Rot
- **1d** No distinguishing features: some white mycelial growth on stalk surface or internally .... Fusarium Stalk Rot

2 Leaf sheaths discolored but stalks not rotted
- **2a** Purplish brown blotches .... Purple Leaf Sheath
- **2b** Reddish brown blotches also occurring on leaves .... Physoderma Brown Spot

3 Stalks soft rotted and plants killed before flowering
- **3a** Stalks twisted or distorted .... Pythium Stalk Rot
- **3b** Internodes soft or slimy: putrid odor .... Bacterial Stalk Rot

Ear Rots and Other Ear Problems

1	White mycelium on large areas of ear, typically originating at base	Diplodia Ear Rot
2	Red mycelium on and between kernels, typically originating at tip	Gibberella Ear Rot or Red Ear Rot
3	Scattered white or lavender-colored kernels	Fusarium Ear Rot or Pink Ear Rot
4	Green growth on kernels	
4a	Yellow-green	Aspergillus Ear Rot
4b	Powdery green to blue-green	Penicillium Ear Rot
4c	Dark green to dark bluish green: may cover entire ear	Trichoderma Ear Rot
5	Black fungal growth on kernels	
5a	Grayish near base with shredding of cob tissue	Nigrospora Ear or Cob Rot
5b	Dark growth starting at base of kernel	Cladosporium Rot
5c	Peppery growth on and between kernels	Rhizopus Ear Rot
5d	Dark brown to black powdery mold on damaged kernels	Ear Rot Caused by *Aspergillus* Section *Nigri*
6	Kernels with red streaks	Kernel Red Streak
7	Kernels with ruptured seed coat	
7a	Silks embedded in kernel	Silk Cut
7b	Top of kernel ruptured	Popped Kernel
8	Multiple ear shoots	Branched Ear
9	Ear distortion	
9a	White galls becoming black spore masses	Common Smut
9b	Spore mass is dark and loose, replacing ear	Head Smut
9c	Ear narrow in middle	Herbicides Inhibiting Amino Acid Synthesis
10	Tassel Malformation	Tassel Ear
		Head Smut
		Crazy Top

Stunting of Plants

1	Root damage leading to stunted plants	Diseases Caused by Nematodes
		Herbicides Inhibiting Amino Acid Synthesis
2	Seeds planted in cold wet soils leading to stunted plants	Imbibitional Chilling Damage
3	Rot of mesocotyl leading to stunted plants	Seedling Disease Caused by Seedborne *Aspergillus* spp. and *Penicillium* spp.
		Seed Rot and Seedling Blight Caused by *Stenocarpella maydis*
4	Stunting of plants with little or no evidence of root rot and usually with abnormal coloration of leaves	Some Virus or Phytoplasma Diseases
5	Stunting of plants in conjunction with excessive tillering	Crazy Top

(Prepared by D. G. White)

Part I. Infectious Diseases

Diseases Caused by Bacteria

Bacteria are composed of a group of more than 4,000 prokaryotic species, most of which lack chlorophyll and are saprophytic. Bacteria are found in air, water, and soil and also on or within all plants and animals. They are remarkably variable in their characteristics, can adapt to a wide range of conditions, and commonly persist in heterogeneous populations. In the past, classification and identification of plant-pathogenic bacteria have been difficult. Historically, bacterial pathogens of corn were most often identified by symptoms, cell and colony morphology, host specificity, growth on differential media, and a wide array of sophisticated immunological and physiological tests. Currently, the move has been to more rapid molecular-based tests, including polyacrylamide gel electrophoresis, nucleic acid hybridization, polymerase chain reaction, microarray panels, and an assortment of immunoassay-based "dip-stick" tests.

Plant-pathogenic bacteria are unicellular rods up to 3 µm long that do not form spores. They may or may not have flagella to aid in motility. In a warm, moist environment, populations of bacteria can double by binary fission within a few hours. Bacteria are disseminated by splashing or flowing water, windblown sand or soil, animals (including insects, mites, and nematodes), and humans, through a wide range of cultural practices and by transport of diseased plant material. Free moisture and moderate temperatures are general requirements for the development of many bacterial diseases. Bacteria can enter plants through weather-related wounds such as those produced by hail or blowing sand, as well as wounds caused by insects, nematodes, or cultural practices such as pruning. Bacteria may enter healthy plants through natural openings, such as stomata and hydathodes. Bacteria can multiply rapidly inside plants, where they may cause death of cells (necrosis), abnormal growth (tumors), blockage of water-conducting tissue (wilting), or breakdown of tissue structure (soft rots). Bacteria often migrate throughout the plant; that is, they become systemic. Numerous bacterial genera produce a slime layer or biofilm, consisting of extracellular polysaccharides outside the cell walls. This capsule serves to protect the bacterial cell against desiccation and is an important factor in pathogenicity and virulence. Bacteria are pathogenic primarily through the action of enzymes or toxins and the secretions of extracellular polysaccharides that result in chlorosis, water-soaking, wilting, and other symptoms. When conditions are unfavorable for growth and multiplication, they remain dormant on or within soil and living or dead plants, on tools and equipment, or within insects.

Most plant-pathogenic bacteria are vulnerable to high temperatures, dry conditions, metals such as copper, and sunlight. Plant-pathogenic bacteria survive poorly when soil debris is highly decomposed. They are subject to predation by soil-inhabiting animals and can be inhibited by antibiotics secreted by microorganisms, including actinomycetes.

Fig. 3. Bacterial ooze (*Pantoea stewartii*) from diseased leaf tissue mounted in water. (Courtesy D. G. White)

Examination of diseased tissue is the first step in the diagnosis of a bacterial disease. Bacterial streaming (ooze) from infected leaf tissue mounted in water can be seen with a microscope (Fig. 3). The presence of bacterial ooze is usually diagnostic for bacterial causal organisms.

Selected References

Bradbury, J. F. 1986. Guide to Plant Pathogenic Bacteria. CAB International, Mycological Institute, London.

Goto, M. 1992. Fundamentals of Bacterial Plant Pathology. Academic Press, San Diego, CA.

Holt, J. G., Krieg, N. R., Sneath, P. H. A., Staley, J. T., and Williams, S. T. 1994. Bergey's Manual of Determinative Bacteriology, 9th ed. Williams & Wilkins, Baltimore, MD.

Schaad, N. W., Jones, J. B., and Chun, W., eds. 2001. Laboratory Guide for Identification of Plant Pathogenic Bacteria, 3rd ed. American Phytopathological Society, St. Paul, MN.

Schumann, G. L., and D'Arcy, C. J. 2010. Essential Plant Pathology, 2nd ed. American Phytopathological Society, St. Paul, MN.

Sigee, D. C. 1993. Bacterial Plant Pathology: Cell and Molecular Aspects. Cambridge University Press, New York.

(Prepared by D. J. Jardine and L. E. Claflin)

Stewart's Bacterial Wilt

Stewart's bacterial wilt (also called Stewart's disease, bacteriosis, or bacterial wilt) was first reported in the United States on Long Island, New York, in 1897. The disease has also been reported in Argentina, Austria, Brazil, Canada, China, Costa Rica, Greece, Malaysia, Mexico, Puerto Rico, Italy, Guyana, Peru, Poland, Romania, Russia, Thailand, and Vietnam. Losses in the United States were severe during the 1930s but have since been sporadic because of the use of more resistant hybrids and seed-applied insecticides. Stewart's bacterial wilt causes major problems with the production of seed to be sold outside the areas where the disease occurs. Numerous countries continue to require phytosanitary certificates showing that the imported corn seed was grown in the absence of the disease or that the seed was free of *Pantoea stewartii* in laboratory tests. This restriction was imposed because *P. stewartii* is seed transmitted, although at extremely low frequency.

Symptoms

Two phases of Stewart's bacterial wilt occur. The wilt phase occurs when plants infected during early vegetative growth stages become systemically infected. Usually these plants rapidly wilt (Figs. 4 and 5) and may closely resemble plants with symptoms of drought, cold weather chilling injury, nutritional deficiency, or insect injury. Leaves have linear, pale green to yellow streaks with irregular wavy margins that follow veins and may extend the length of the leaf. Lesions rapidly desiccate and become necrotic (Fig. 5). Plants that are not killed may produce bleached, dead tassels. Cavities may form near the soil line in the stalk pith of severely infected plants (Fig. 6). Bacteria colonize the vascular system and may spread throughout the plant. Field or dent corn is generally not as susceptible as sweet corn to this phase of the disease, although some field corn inbred lines and hybrids are very susceptible and some sweet corn hybrids are quite resistant.

The second and more common phase of this disease occurs on the leaves at any growth stage and thus is called the leaf blight phase. It is typically associated with the local infection of leaf tissue due to vector transmission. Symptoms usually are most apparent after tasseling. Lesions on leaves are gray-green to yellow-green and develop as streaks with wavy margins along the veins (Fig. 7). The streaked areas, which usually originate from feeding scars of the corn flea beetle (Fig. 8), die and become straw colored (Fig. 9). If the disease is severe, entire leaves may die and dry up. When leaves die prematurely, plants are predisposed to fungal stalk rots. However, the leaf blight phase is not as damaging as the seedling wilt phase, which can kill plants.

Causal Organism

Stewart's bacterial wilt is caused by *Pantoea stewartii* subsp. *stewartii* (syns. *Erwinia stewartii* and *Xanthomonas stewartii*). The bacterium is a nonmotile, facultatively anaerobic, gram-negative rod (0.4–0.8 × 0.9–2.2 µm). Colonies on yeast extract–dextrose–calcium carbonate agar are yellow and convex. Colonies may be "running," or fluidal. *P. stewartii* produces extracellular polysaccharides that contribute to water-soaked lesions and occlusion of xylem vessels. Masses of bacteria may ooze as yellow, moist beads from cut ends of infected stalks or

Fig. 4. Early-season plant death resulting from systemic infection by *Pantoea stewartii*. (Courtesy G. P. Munkvold)

Fig. 5. Wilt phase of Stewart's bacterial wilt. (Courtesy D. G. White)

Fig. 6. Internal cavity resulting from systemic infection by *Pantoea stewartii*. (Courtesy G. P. Munkvold)

Fig. 7. Early symptoms of Stewart's bacterial wilt, leaf blight phase. (Courtesy G. P. Munkvold)

may "stream" from cut edges of infected leaf tissue. A saprophytic species, *Pantoea agglomerans* (syns. *Enterobacter agglomerans* and *Erwinia herbicola*), is commonly found on plant tissues and can easily be confused with *P. stewartii*. However, *P. agglomerans* is usually positive for nitrate reductase, phenylalanine deaminase, and esculin hydrolysis and is motile. Specific identification techniques, such as use of monoclonal antibodies in enzyme-linked immunosorbent assay (ELISA) testing and several polymerase chain reaction (PCR) methods, have been developed for detecting *P. stewartii* in corn tissue. Closely related, nonpathogenic *Pantoea* spp. can be found on corn and can complicate detection efforts because of occasional false-positive reactions even when using specific ELISA- or PCR-based methods. Accurate detection of this pathogen in corn seed is an essential practice for the export of seed.

Disease Cycle and Epidemiology

The corn flea beetle, *Chaetocnema pulicaria*, is the primary vector and overwintering site of the bacterium. The beetles overwinter as adults, and bacteria survive in their gut. In the spring, adult corn flea beetles emerge at the same time corn is planted. Flea beetles feed on corn and other grasses, thereby spreading the bacterium. There are two to three generations of corn flea beetles during the growing season in the North American Corn Belt. Disease prevalence is directly related to the size of flea beetle populations, which in turn are heavily dependent on winter survival and in-season weather. Flea beetle population levels tend to peak late in the season with the second or third generation. Populations are larger when the weather is dry. Other species of beetles can act as vectors, but their role is believed to be insignificant.

P. stewartii enters the plant through feeding wounds made by corn flea beetles (Figs. 10 and 11). Foliar symptoms and wilting of tissue develop as bacterial cells plug the xylem tissues. Beetles acquire the bacterium from infected leaves and spread the disease to healthy leaves and plants throughout the growing season. A minimum of 6 h of feeding time is reportedly required for acquisition. As many as 30% of beetles may be carrying the bacterium during the spring. This percentage usually increases during the growing season. High levels of nitrogen and phosphorus have been reported to enhance disease incidence and severity. High temperatures and soil moisture levels also may contribute to severity. High levels of calcium and potassium have been reported to decrease severity.

Reported hosts for *P. stewartii* include flint, sweet, and yellow dent corns, popcorn, teosinte, and eastern gamagrass. Sorghum, Sudan grass, sugarcane, yellow foxtail, German millet, and common millet are susceptible when artificially inoculated. Various poaceous weeds have been shown to act as symptomless carriers of the bacterium. Some of these hosts may also serve as overwintering sites, although the role of the overwintering plant host is not clearly known.

P. stewartii can be seed transmitted. However, levels of plant-to-seed transmission are low and are related to the susceptibility of the seed parent (female) plant. Rates of seed-to-seedling transmission of the bacterium are even lower than those of plant-to-seed transmission and are enhanced by injuries that create wounds to emerging seedlings. Intensive epidemiological studies and pest risk analysis have led to the conclusion that the risk of seed transmission in commercial seedlots is negligible. Even if *P. stewartii* were introduced into a new area, it is not likely that it would become established in the absence of a suitable insect host for overwintering and dissemination.

Fig. 8. Early spread of Stewart's bacterial wilt leaf blight symptoms from corn flea beetle feeding scars. (Courtesy G. P. Munkvold)

Fig. 9. Leaf lesions of Stewart's bacterial wilt leaf blight. (Courtesy D. G. White)

Fig. 10. Corn flea beetle and feeding scars. (Courtesy D. G. White)

Fig. 11. Corn flea beetle. (Courtesy G. P. Munkvold)

Disease Forecasting

A method to forecast the incidence of Stewart's bacterial wilt was proposed in 1934 and modified in 1949. In this forecast, the occurrence and severity of Stewart's bacterial wilt is associated with the sum of the mean temperatures (°F) for December, January, and February. During mild winters (when the sum is above 100°F), a larger proportion of the overwintering population of the beetles survive and Stewart's bacterial wilt is more likely to occur. During cold winters (when the sum is 80°F or below), winter survival is adversely affected and, consequently, incidence of Stewart's bacterial wilt is likely to be very low or absent. An alternate method to predict the risk of Stewart's bacterial wilt in Iowa seed production fields was proposed in 2002 and is based on the number of winter months with a mean temperature below 24°F. Both methods of forecasting Stewart's bacterial wilt assume susceptible hybrids or inbreds and that the bacterium is present in the overwintering population of beetles. For unknown reasons, Stewart's bacterial wilt is not prevalent in the southern United States, where these forecasts are not applicable.

Management

Resistant hybrids are recommended in areas where Stewart's bacterial wilt is endemic and where warm winters favor survival of the corn flea beetle. Resistance appears to restrict the movement of bacteria in the vascular system of plants, thus preventing plants from becoming systemically infected. Most field corn hybrids are highly to moderately resistant to Stewart's bacterial wilt, whereas sweet corn hybrids can range from resistant to highly susceptible. Resistance to Stewart's bacterial wilt is inherited relatively simply and can be easily selected in a breeding program. Both major resistance genes and minor resistance genes with additive effects have been identified. Foliar insecticide applications can be economical in seed production fields, based on population thresholds derived from trapping data. However, seed treatment insecticides (e.g., clothianidin, imidacloprid, and thiamethoxam) have largely replaced foliar applications. Seed treatment insecticides effectively kill the insect vector before the bacterium is transmitted. In fact, the widespread use of seed treatment insecticides on a substantial acreage of the commercial field corn crop in the United States has substantially reduced populations of flea beetles to low enough levels so that the disease is relatively minor in importance even when winters are mild. The use of in-furrow or foliar-applied insecticides has proven to be less reliable than seed treatment insecticides.

Selected References

Coplin, D. L., Majerczak, D. R., Zhang, Y., Kim, W.-S., Jock, S., and Geider, K. 2002. Identification of *Pantoea stewartii* subsp. *stewartii* by PCR and strain differentiation by PFGE. Plant Dis. 86:304-311.

Esker, P. D., Harri, J., Dixon, P. M., and Nutter, F. W., Jr. 2006. Comparison of models for forecasting of Stewart's disease of corn in Iowa. Plant Dis. 90:1353-1357.

Freeman, N. D., and Pataky, J. K. 2001. Levels of Stewart's wilt resistance necessary to prevent reductions in yield of sweet corn hybrids. Plant Dis. 85:1278-1284.

Menelas, B., Block, C. C., Esker, P. D., and Nutter, F. W., Jr. 2006. Quantifying the feeding periods required by corn flea beetles to acquire and transmit *Pantoea stewartii*. Plant Dis. 90:319-324.

Mergaert, J., Verdonck, L., and Kersters, K. 1993. Transfer of *Erwinia ananas* (synonym, *Erwinia uredovora*) and *Erwinia stewartii* to the genus *Pantoea* emend. as *Pantoea ananas* (Serrano 1928) comb. nov. and *Pantoea stewartii* (Smith 1898) comb. nov., respectively, and description of *Pantoea stewartii* subsp. *indologenes* subsp. nov. Int. J. Syst. Bacteriol. 43:162-173.

Michener, P. M., Pataky, J. K., and White, D. G. 2002. Rates of transmitting *Erwinia stewartii* from seed to seedlings of a sweet corn hybrid susceptible to Stewart's wilt. Plant Dis. 86:1031-1035.

Pataky, J. K. 2003. Stewart's Wilt of Corn. APS Features. Online. doi: 10.1094/APSnetFeature-2003-0703

Pataky, J. K., Michener, P. M., Freeman, N. D., Weinzierl, R. A., and Teyker, R. H. 2000. Control of Stewart's wilt in sweet corn with seed treatment insecticides. Plant Dis. 84:1104-1108.

Pataky, J. K., Bohn, M. O., Lutz, J. D., and Richter, P. M. 2008. Selection for quantitative trait loci associated with resistance to Stewart's wilt in sweet corn. Phytopathology 98:469-474.

(Prepared by D. J. Jardine and L. E. Claflin)

Goss's Bacterial Wilt and Blight

Goss's bacterial wilt and blight (initially called leaf freckles and wilt) has been observed on dent corn in Nebraska since 1969. It also is found in Colorado, Illinois, Indiana, Iowa, Kansas, Minnesota, South Dakota, Texas, and Wisconsin. The disease appeared sporadically outside Nebraska until about 2005 but recently has become more common and severe in neighboring states. The first report outside the United States was in 2009 in Manitoba, Canada. Yield losses may exceed 50% when susceptible hybrids are infected at vegetative growth stages.

Symptoms

Symptoms are somewhat similar to those of Stewart's bacterial wilt. There are various degrees of leaf blight that appear as

Fig. 12. Lesions of Goss's bacterial wilt with characteristic dark flecks. (Courtesy L. E. Claflin)

Fig. 13. Leaf blight symptoms of Goss's bacterial wilt. (Courtesy A. E. Robertson)

water-soaked or gray to light yellow stripes that generally follow the leaf veins with wavy or irregular margins (occasionally with reddish borders on certain hybrids or inbreds) (Fig. 12). Within these lesions, dark green to black, water-soaked, irregular spots (freckles) usually appear and are an excellent diagnostic symptom (Fig. 13). Freckles are translucent when backlit. Lesions are usually most pronounced as plants near silking. Early infections are uncommon; however, wilted, withered, and dead plants may result from infection during the vegetative stages (Fig. 14). Systemically infected plants usually have orange vascular bundles (Fig. 15) and may develop decayed cavities in the stalk (Fig. 16). The husk tissue and kernels may also assume an orange hue. Severe infections result in rapid death and desiccation of entire leaves or whole plants (Fig. 17). Dried bacterial exudate is common on the surfaces of lesions. The symptoms are sometimes confused with leaf scorching associated with high temperatures and high winds or northern corn leaf blight lesions.

Fig. 14. Early-season leaf blighting symptoms of Goss's bacterial wilt. (Courtesy A. E. Robertson)

Fig. 15. Vascular discoloration caused by the Goss's bacterial wilt pathogen. (Courtesy T. Jackson-Ziems)

Causal Organism

Goss's bacterial wilt and blight is caused by *Clavibacter michiganensis* subsp. *nebraskensis* (syn. *Corynebacterium michiganense* subsp. *nebraskense*). The bacterium is a nonmotile, catalase-positive, oxidase-negative, gram-positive, irregular, and often wedge- or club-shaped rod, averaging 0.5 × 2.5 µm. Diversity exists; bacteriocin and bacteriophage typing have placed this pathogen into eight groups and 20 subgroups. On nutrient broth–glucose–yeast extract agar, bacterial colonies appear apricot to orange colored and are mucoid. CNS is a semiselective medium for isolating *Clavibacter michiganensis* subsp. *nebraskensis*.

Disease Cycle and Epidemiology

Clavibacter michiganensis subsp. *nebraskensis* overwinters in debris from a previous diseased crop on or near the soil

Fig. 16. Stalk cavities caused by a combination of Goss's bacterial wilt and bacterial stalk rot. (Courtesy T. Jackson-Ziems)

Fig. 17. Severe leaf blighting and premature death caused by Goss's bacterial wilt. (Courtesy A. E. Robertson)

surface. Bacterial populations in residue decline gradually throughout the winter and spring. The population decline is more rapid during the summer. The pathogen has not been detected in soil absent of residue. Spread is by splashing water or windblown infested particles. Plant injury, e.g., from hail or sandblasting, is required for penetration, and the most severe epidemics follow leaf damage caused by hail. *Clavibacter michiganensis* subsp. *nebraskensis* is seedborne and is seed transmitted at low frequencies in very susceptible dent corn inbreds, but it is unclear whether this plays a role in epidemiology of the disease in the field.

Clavibacter michiganensis subsp. *nebraskensis* is pathogenic to dent, sweet, flint, and popcorn corn and has been isolated from natural infections of green foxtail and shattercane. Plants found to be susceptible after artificial inoculation include grain sorghum, eastern gamagrass, Sudan grass, sugarcane, and teosinte. Bacteria grow most rapidly in plants at 27°C; slow growth or even death occurs in plants at temperatures below 12°C and above 40°C.

Management

The use of partially resistant hybrids is the most practical means of disease management. However, resistance does not imply immunity; although losses will be considerably less than in susceptible hybrids. Crop rotation and burying of infested debris through tillage operations, preferably immediately after harvest, are effective in reducing disease severity. Where soil erosion concerns require no-till or reduced-tillage methods, rotation can help reduce the amount of primary inoculum but does not totally eliminate the bacterium.

Selected References

Biddle, J. A., McGee, D. C., and Braun, E. J. 1990. Seed transmission of *Clavibacter michiganense* subsp. *nebraskense* in corn. Plant Dis. 74:908-911.

Carson, M. L., and Wicks, Z. W., III. 1991. Relationship between leaf freckles and wilt severity and yield losses in closely related maize hybrids. Phytopathology 81:95-98.

Gross, D. C., and Vidaver, A. K. 1979. A selective medium for isolation of *Corynebacterium nebraskense* from soil and plant parts. Phytopathology 69:82-87.

Jackson, T. A., Harveson, R. M., and Vidaver, A. K. 2007. Reemergence of Goss's wilt and blight of corn to the central High Plains. Online. Plant Health Progress. doi: 10.1094/PHP-2007-0919-01-BR

Louws, F. J., Bell, J., Medina-Mora, C. M., Smart, C. D., Opgenorth, D., Ishimaru, C. A., Hausbeck, M. K., de Bruijn, F. J., and Fulbright, D. W. 1998. rep-PCR–mediated genomic fingerprinting: A rapid and effective method to identify *Clavibacter michiganensis*. Phytopathology 88:862-868.

Ngong-Nassah, E. N., Carson, M. L., and Wicks, Z. W., III. 1992. Inheritance of resistance to leaf freckles and wilt caused by *Clavibacter michiganense* subsp. *nebraskense* in early maturing maize inbred lines. Phytopathology 82:142-146.

Smidt, M., and Vidaver, A. K. 1986. Population dynamics of *Clavibacter michiganense* subsp. *nebraskense* in field-grown dent corn and popcorn. Plant Dis. 70:1031-1036.

Suparyono and Pataky, J. K. 1989. Influence of host resistance and growth stage at the time of inoculation on Stewart's wilt and Goss's wilt development and sweet corn hybrid yield. Plant Dis. 73:339-345.

Vidaver, A. K., Gross, D. C., Wysong, D. S., and Doupnik, B. L., Jr. 1981. Diversity of *Corynebacterium nebraskense* strains causing Goss's bacterial wilt and blight of corn. Plant Dis. 65:480-483.

(Prepared by D. J. Jardine and L. E. Claflin)

Holcus Spot

Holcus spot may be found wherever corn is grown, but the disease rarely reduces yield. Symptoms often appear suddenly a few days after a thunderstorm when plants are in a vegetative growth stage.

Symptoms

Round to elliptical spots, ranging from 2 to 10 mm in diameter, often appear toward the tips of lower leaves. Initially, lesions are dark green and water-soaked, and later they become creamy white to tan and eventually dry and turn brown, often with reddish to brown margins (Fig. 18). Larger lesions may be surrounded by yellowish halos. Symptoms may be confused with those of chemical drift, e.g., paraquat injury. Diagnosis of this disease may be difficult because bacteria may be present only in the very center of a lesion. When examinations for bacterial ooze are conducted, leaf sections must include centers of the lesions. After initial symptom appearance, lesions do not expand and the disease usually does not spread noticeably to additional leaves.

Causal Organism

Holcus spot is caused by *Pseudomonas syringae* pv. *syringae* (syn. *P. holci*). The bacterium is a short rod with rounded ends, 0.6–1.2 × 1.5–3.0 μm (average 0.73 × 2.13 μm). It is gram negative and motile with one to several polar flagella and produces a fluorescent diffusible pigment on King's medium B. *P. syringae* pv. *syringae* produces a low-molecular-weight peptide-containing toxin, syringomycin, that induces necrosis. The organism is best recovered on King's medium B agar by plating surface-sterilized leaf tissue ground in sterile water. Colonies are round, smooth, glistening, raised or pulvinate, grayish white by reflected light, and greenish fluorescent by transmitted light.

Disease Cycle and Epidemiology

P. syringae pv. *syringae* overwinters in crop residue and penetrates the host through stomata or wounds created by hail or blowing sand. Its host range is extremely broad and includes grasses such as foxtail millet, pearl millet, Sudan grass, broomcorn, Johnsongrass, wheat, and sorghum. Numerous other crops, including dicotyledonous plants, are infected by various strains and pathovars of *P. syringae*, some of which are host specific. Cool (12–25°C), rainy, and windy weather, especially early in the season, favors disease development. Ice-nucleating activity has been reported for *P. syringae* pv. *syringae* on corn. The bacteria act as sites for ice nucleation and can cause rapid freezing and plant injury at temperatures above –10°C. This action prevents supercooling, which occurs in the absence of ice-nucleating substances such as small dust particles and bacteria. Consequently, ice-nucleating activity increases plant sensitivity to subzero temperatures (–2 to –5°C), resulting in in-

Fig. 18. Holcus spot symptoms. (Courtesy G. P. Munkvold)

jured tissue. Such injuries provide the bacteria a means to enter damaged tissue. Bacteria without ice-nucleating activity allow moisture to supercool without ice crystal formation.

Management

Holcus spot is usually not severe, and most hybrids have not been characterized for their resistance or susceptibility to this disease. Control measures are rarely justified. Crop rotation and tillage of debris from a previous diseased corn crop may not result in adequate control because of the broad host range of the bacterium.

Selected References

Kendrick, J. B. 1926. Holcus bacterial spot of *Zea mays* and *Holcus* species. Iowa Agric. Exp. Stn. Bull. 100:303-334.

Lindow, S. E., Arny, D. C., and Upper, D. D. 1982. Bacterial ice nucleation: A factor in frost injury to plants. Plant Physiol. 70:1084-1089.

Morgan, M. K., and Chatterjee, A. K. 1988. Genetic organization and regulation of proteins associated with production of syringotoxin by *Pseudomonas syringae* pv. *syringae*. J. Bacteriol. 170:5689-5697.

(Prepared by D. J. Jardine and L. E. Claflin)

Bacterial Leaf Blight

Bacterial leaf blight is a minor disease that has been reported in Alabama, Arkansas, Florida, Georgia, Illinois, Kansas, Montana, Nebraska, New York, Pennsylvania, Texas, and Virginia in the United States and is found worldwide on corn and other hosts.

Symptoms

Water-soaked, linear lesions appear on the leaves as they emerge from the whorl. The lesions turn brown and sometimes then gray or white (Fig. 19). On some genotypes, lesions have red borders. Lesions seldom elongate after the leaf matures, and new lesions are rarely observed after the plant tassels. Infected leaves of susceptible plants often shred after maturity.

Causal Organism

Pseudomonas avenae subsp. *avenae* (syns. *Bacillus avenae*, *P. alboprecipitans*, and *Acidovorax avenae* subsp. *avenae*, among others) causes bacterial leaf blight. The pathogen is a gram-negative rod (0.6 × 1.6 μm), motile, and oxidase and arginine dihydrolase negative. Colonies on yeast extract–dextrose–calcium carbonate agar are cream colored with tan to brown centers, convex, and smooth. The optimum temperature for growth is 36°C. No fluorescent pigment is produced on King's medium B agar. A white precipitate may be formed in culture media. A semiselective medium, SNR, is useful for the recovery of *P. avenae* subsp. *avenae*.

Fig. 19. Bacterial leaf blight. (Courtesy D. G. White)

Disease Cycle and Epidemiology

In greenhouse tests, *P. avenae* subsp. *avenae* is pathogenic to other plants, including oats, barley, rice, wheat, rye, pearl millet, proso millet, sorghum, yellow foxtail, and tall wheatgrass, but the bacterium has not been recovered from any of these plants in nature. In Florida and Georgia, vasey grass (*Paspalum urvillei*) serves as a primary source of inoculum. Disease development has occurred over temperatures ranging from 10 to 32°C. While no specific work on dissemination of the bacterium has been reported, it is likely spread by wind and rain, similar to other foliar bacterial diseases of corn. The bacterium does not survive well in soil or plant debris.

Management

Control measures are rarely justified. The use of resistant hybrids is the most effective management practice where the disease is severe. However, the resistance or susceptibility of most inbreds and hybrids has not been determined because of the relative minor importance of this disease in most areas. Crop rotation and tillage of debris from a previous diseased corn crop may not result in adequate control because of the broad host range of the bacterium.

Selected References

Johnson, A. G., Robert, A. L., and Cash, L. 1949. Bacterial leaf blight and stalk rot of corn. J. Agric. Res. 78:719-732.

Pataky, J. K., du Toit, L. J., and Kerns, M. R. 1997. Bacterial leaf blight on *shrunken-2* sweet corn. Plant Dis. 81:1293-1298.

Sumner, D. R., and Schaad, N. W. 1977. Epidemiology and control of bacterial leaf blight of corn. Phytopathology 67:1113-1118.

White, D. G., Pataky, J. K., and Stall, R. E. 1994. Unusual occurrences of bacterial leaf blight on maize and sorghum in central Illinois. Plant Dis. 78:640.

(Prepared by D. J. Jardine and L. E. Claflin)

Bacterial Stalk Rot

Bacterial stalk rot has been reported worldwide but is only rarely of economic importance.

Symptoms

Primary symptoms generally appear in midseason when plants suddenly lodge. One to several internodes above the soil line appear tan to dark brown, water-soaked, and soft or slimy, and the stalk tissue has a macerated appearance (Fig. 20) and a foul odor. A top rot occurs when plants are infected during rapid vegetative growth, particularly in corn plants that are sprinkler irrigated with river, lake, or impounded water (Fig. 21). Tips of the uppermost leaves may wilt or die, and a slimy soft rot occurs in the base of the whorl. Leaves forming the whorl are easily pulled from the plant and the rotted tip has a foul odor. The decay spreads rapidly downward until the plant collapses. Symptoms may be similar to those of Pythium stalk rot. Top rot symptoms caused by bacterial stalk rot can be easily confused with those caused by lipid synthesis–inhibiting herbicides. The foul odor associated with soft-rotting bacteria is generally absent in herbicide-injured plants.

Causal Organism

Erwinia carotovora subsp. *carotovora* (syns. *E. chrysanthemi* pv. *zeae*, *E. carotovora* f. sp. *zeae*, and *Pectobacterium carotovorum* subsp. *carotovorum*) causes bacterial stalk rot. The bacterium is a short, gram-negative rod (0.6–0.9 × 1.0–1.7 μm), facultatively anaerobic, oxidase negative, catalase positive, arginine dihydrolase negative, motile with peritrichous flagella, and lacking a capsule. On nutrient agar, the colonies are grayish white, raised, glistening, and smooth with entire margins. After 3–6 days of growth on potato–glucose agar (pH 6.5), the colonies are characteristically umbonate with undulate to coralloid margins (i.e., they have a "fried egg" appearance). Numerous serovars have been reported, indicating the heterogeneity of the organism. The host range is very broad and includes both monocots and dicots.

Disease Cycle and Epidemiology

The bacterium overwinters only in host stalk tissue above the soil surface. Splashing water spreads the bacterium to plant surfaces. Stomata, hydathodes, or wounds in the leaves or stalks are the most common infection sites. Contaminated irrigation or flood water also can be the source of inoculum.

The disease is most prevalent and destructive in areas with high levels of rainfall, where irrigation with surface water is applied through overhead sprinklers, or on land subject to flooding. The disease is favored by high temperatures (32–35°C) and high relative humidity. This bacterium has been reported to be seedborne. Larvae of the maize borer (*Chilo partellus*) have been implicated in the transmission of the bacterium. Observations in the Pacific Northwest indicate that the disease frequently is severe in plants growing in sulfur-deficient soils. This pathogen also causes a soft rot in cabbage, carrot, onion, papaya, sorghum, sugar beet, sweetpotato, tobacco, tomato, and many other plants.

Fig. 20. Bacterial stalk rot. (Courtesy D. G. White)

Fig. 21. Top rot symptoms of bacterial stalk rot. (Courtesy D. G. White)

Management

Fall plowing to incorporate debris in the soil and good cultural management to avoid flooding are recommended. Well water rather than surface waters should be used for overhead irrigation. Chlorine in irrigation water has been reported to reduce disease incidence and severity.

Selected References

Dickey, R. S., Claflin, L. E., and Zumoff, C. H. 1987. *Erwinia chrysanthemi*: Serological comparisons of strains from *Zea mays* and other hosts. Phytopathology 77:426-430.

Hoppe, P. E., and Kelman, A. 1969. Bacterial top and stalk rot disease of corn in Wisconsin. Plant Dis. Rep. 53:66-70.

Kelman, A., Person, L. H., and Herbert, T. T. 1957. A bacterial stalk rot of irrigated corn in North Carolina. Plant Dis. Rep. 41:798-802.

Rosen, H. R. 1921. Further observations on a bacterial root and stalk rot of field corn. Phytopathology 11:74-79.

Rosen, H. R. 1922. The bacterial pathogen of corn stalk rot. Phytopathology 12:497-499.

Sah, D. N., and Arny, D. C. 1990. Susceptibility of maize germplasm to bacterial stalk rot caused by *Erwinia chrysanthemi* pv. *zeae*. Trop. Pest Manage. 36:154-156.

(Prepared by D. J. Jardine and L. E. Claflin)

Bacterial Leaf Streak

Bacterial leaf streak was first reported in the Republic of South Africa in 1949. It has not been reported elsewhere. The disease in found primarily in western and northern Transvaal and the Orange Free State. Yield losses have not been determined, although death of foliar tissue attributable to bacterial leaf streak has approached 40%, and susceptible hybrids have been withdrawn from production.

Symptoms

Bacterial leaf streak symptoms consist of yellow-brown lesions, 2–3 mm wide, with wavy and irregular yellow margins parallel to leaf veins (Fig. 22). Lesions may extend the entire length of the leaf and often coalesce to form large, necrotic areas that closely mimic symptoms of drought injury. Wilting is rarely observed. Yellow droplets of bacterial exudate may form on lesions under humid conditions. Symptom expression is enhanced when daily temperatures exceed 32°C.

Causal Organism

Bacterial leaf streak is caused by *Xanthomonas vasicola* pv. *vasculorum*. This bacterium is motile by means of a single polar flagellum and is gram negative, rod shaped (average 0.5 ×

Fig. 22. Bacterial leaf streak. (Courtesy D. C. Nowell)

2.0 μm), oxidase negative, and catalase positive. A yellow pigment, xanthomonadin, is produced. Colonies are light yellow and viscid on yeast extract–dextrose–calcium carbonate medium. The medium MXP is semiselective for isolation of the bacterium. Corn is the only established host. Barley, pearl millet, rye, sorghum, sugarcane, and wheat have been shown to be nonhosts. The optimal growth temperature is 28°C.

Disease Cycle and Epidemiology

It is presumed that *X. vasicola* pv. *vasculorum* overseasons in infested debris and is splash dispersed to leaves of healthy plants. Research has not been conducted on whether it is seedborne. Weeds adjacent to corn fields may be alternate hosts.

Wounds are not required for entry by the bacterium, and it most likely enters through stomata and then moves through intercellular spaces to xylem parenchyma cells. If xylem vessels are invaded, the vascular tissue becomes occluded by bacterial cells and wilting of tissue may result, although wilting of plants in the field is rarely observed. Irrigation during hot weather appears to increase the disease incidence and, under these conditions, death of leaf tissue attributable to bacterial leaf streak approaches 40%.

Management

Very little information is available on management. If available, the best control would be the use of resistant hybrids or cultivars in areas where the disease is a problem.

Selected References

Coutinho, T. A. 1988. Bacterial streak of maize: A study on host specificity and cultivar susceptibility. M.S. thesis. University of Natal, Pietermaritzburg, South Africa.

Coutinho, T. A., and Wallis, F. M. 1991. Bacterial streak disease of maize (*Zea mays* L.) in South Africa. J. Phytopathol. 133:112-120.

Qhobela, M., Claflin, L. E., and Nowell, D. C. 1990. Evidence that *Xanthomonas campestris* pv. *zeae* can be distinguished from other pathovars capable of infecting maize by restriction fragment length polymorphism of genomic DNA. Can. J. Plant Pathol. 12:183-186.

(Prepared by D. J. Jardine and L. E. Claflin)

Bacterial Stripe and Leaf Spot

This minor disease has only been reported in the central and eastern United States. The disease is widespread but very infrequent in occurrence and usually of limited economic importance.

Symptoms

Symptoms appear first on lower leaves and, under favorable conditions, spread upward. Typical primary symptoms are amber- to olive-colored, oil-soaked, translucent lesions that are usually circular. Lesions later become brown and necrotic. Very susceptible inbred lines show chlorotic striping or bleaching of the upper leaves.

Causal Organism

Bacterial stripe and leaf spot is caused by *Pseudomonas andropogonis* (syn. *Burkholderia andropogonis*). The bacterium is a short rod (0.5–0.8 × 1.5–2.5 μm) with rounded ends. It is gram negative and motile with a polar-sheathed flagellum. Poly-β-hydroxybutyrate is accumulated. *P. andropogonis* is negative for fluorescence on King's medium B agar, levan production, and oxidase and arginine dihydrolase reactions. A toxin, rhizobitoxine, is produced by the bacterium.

Disease Cycle and Epidemiology

Bacterial stripe and leaf spot becomes more severe during extended periods of warm, wet weather. Bacteria enter the host through stomata when leaves are water-soaked. *P. andropogonis* is pathogenic to numerous monocots and dicots, including sorghum, Sudan grass, and Johnsongrass. Numerous countries require phytosanitary certificates to show that shipments of corn feed and seed are free of *P. andropogonis*.

Management

The disease is usually not severe, and management is seldom required.

Selected References

Ulstrup, A. J. 1960. Bacterial stripe of corn. Phytopathology 50:906-910.

Vidaver, A. K., and Carlson, R. R. 1978. Leaf spot of field corn caused by *Pseudomonas andropogonis*. Plant Dis. Rep. 62:213-216.

Xiang, L., Wong, W. C., and Hayward, A. C. 1993. Production and use of monoclonal antibodies to *Pseudomonas andropogonis*. J. Phytopathol. 138:21-30.

(Prepared by D. J. Jardine and L. E. Claflin)

Chocolate Spot

Chocolate spot occurs sporadically in Minnesota and Wisconsin and only in fields with potassium-deficient soils. It has not been reported elsewhere.

Symptoms

Dark brown, elongated spots up to 3 cm long surrounded by broad, yellow halos develop on the leaves (Fig. 23). Lesions are most numerous along the leaf edges and toward the tip areas, where symptoms of potassium deficiency also first appear. Lesions may later coalesce, causing death of large areas.

Causal Organism

Pseudomonas syringae pv. *coronafaciens* (syn. *P. coronafaciens* pv. *zeae*) causes chocolate spot. The bacterium is a gram-negative rod (0.6 × 1.0–2.7 μm) that may form short chains. It produces fluorescent pigment on King's medium B agar, is motile by one to five polar flagella, and has an oxidative mode of glucose metabolism. It forms levan, is oxidase and arginine dihydrolase negative, and induces a hypersensitive reaction when infiltrated into tobacco leaves. A toxin, tabtoxin, is produced by the bacterium. *P. syringae* pv. *coronafaciens* is pathogenic to many cereals and is known primarily as the causal agent of halo blight of oat.

Epidemiology

Chocolate spot occurs only in potassium-deficient soils. Leaves wounded by wind-whipping appear to be more susceptible to infection.

Fig. 23. Chocolate spot. (Courtesy D. C. Arny)

Management

Proper levels of potassium should be maintained based on a soil test.

Selected References

Bowden, R. L., and Stromberg, E. L. 1982. Chocolate spot of corn in Minnesota. Plant Dis. 66:744.

Karlen, D. L., Arny, D. C., and Walsh, L. M. 1973. Incidence of chocolate spot (*Pseudomonas syringae*), northern corn leaf blight (*Helminthosporium turcicum*), and lodging of corn as influenced by soil fertility. Commun. Soil Sci. Plant Anal. 5:359-368.

Ribeiro, R. de L. D., Durbin, R. D., Arny, D. C., and Uchytil, T. F. 1977. Characterization of the bacterium inciting chocolate spot of corn. Phytopathology 67:1427-1431.

Schaad, N. W., and Cunfer, B. M. 1979. Synonymy of *Pseudomonas coronafaciens*, *Pseudomonas coronafaciens* pathovar *zeae*, *Pseudomonas coronafaciens* subsp. *atropurpurea*, and *Pseudomonas striafaciens*. Int. J. Syst. Bacteriol. 29:213-221.

(Prepared by D. J. Jardine and L. E. Claflin)

Diseases Caused by Mollicutes

Spiroplasmas and phytoplasmas are in the bacterial class Mollicutes. Mollicutes are small, gram-positive prokaryotes that lack cell walls and have reduced genomes. This class also includes mycoplasmas and ureoplasmas that are pathogenic to humans.

Spiroplasmas and phytoplasmas replicate in both their plant and insect vector hosts. They are generally pathogenic in plant hosts but have positive or neutral effects on vector fecundity and longevity. Phytoplasmas and spiroplasmas are obligately transmitted by leafhoppers (Cicadellidae), planthoppers (Fulgoroidea), or psyllids (Psyllidae). After acquisition and a 2- to 6-week latent period, the pathogens can be transmitted for the remainder of the vector's lifetime.

Only a few spiroplasma species infect plants. Most species are found in commensal association with insects and, to a lesser extent, with ticks and higher crustaceans. Spiroplasmas are motile, facultative anaerobes with helical filaments 100–200 nm in diameter and 3–5 µm long. They have genomes of 0.8–2.2 megabases (Mb) and can be cultured on several defined solid and liquid media. Spiroplasmas that are pathogenic to plants can cross the midgut barrier in the insect vector and multiply in the hemolymph and other organs, including the salivary glands, before being injected into a new plant host. In plants, spiroplasmas are confined to sieve elements within phloem tissue. Three species are known to cause disease in plants: *Spiroplasma kunkelii* infects maize and closely related grasses; *S. citri* infects citrus, periwinkle (*Catharanthus roseus*), and a few other dicots; and *S. phoeniceum* infects periwinkle. Spiroplasma infection is confirmed using serological (e.g., enzyme-linked immunosorbent assay [ELISA]) or molecular (polymerase chain reaction [PCR]) assays.

Phytoplasmas cause more than 700 diseases in hundreds of plant species. These diseases can be very severe and, in some cases, lethal. Phytoplasmas are small (less than 1 µm in diameter), spherical bacteria with genomes of 0.5–1.4 Mb. A complete genome sequence is currently available for three phytoplasmas. Phytoplasma genomes lack genetic code for some metabolic pathways, and it is thought that this is one reason they have not yet been cultured on artificial media. Currently, identification and classification of phytoplasmas are based on ribosomal RNA gene (rDNA) sequence analysis. Based on rDNA sequence, they are divided into 28 '*Candidatus* Phytoplasma species'.

Phytoplasma disease symptoms in plants include a loss of vigor, yellowing or reddening, low yield, virescence and phyllody of flowers, parthenogenesis, and proliferation of leafy shoots (witches'-broom). Phytoplasmas are most frequently detected and identified in plants and insects using a nested PCR assay to amplify the rDNA, followed by sequence analysis.

Selected References

Gasparich, G. E. 2010. Spiroplasmas and phytoplasmas: Microbes associated with plant hosts. Biologicals 38:193-203.

Lee, I. M., Gundersen-Rindal, D. E., Davis, R. E., and Bartoszyk, I. M. 1998. Revised classification scheme of phytoplasmas based on RFLP analyses of 16S rRNA and ribosomal protein gene sequences. Int. J. Syst. Bacteriol. 48:1153-1169.

(Prepared by M. G. Redinbaugh)

Corn Stunt

Corn stunt, formerly known as Rio Grande corn stunt, was first described from Texas in 1945. The disease is found in the southern United States, California, Mexico, and Central and South America. Significant yield losses have been reported from Mexico and Central and South America.

Symptoms

Symptoms differ among maize genotypes and are difficult to distinguish from those caused by the maize bushy stunt phytoplasma. Chlorosis of leaf margins is the first symptom observed, followed by reddening of the older leaves (Fig. 24). Chlorotic spots develop at the leaf base and then coalesce to form chlorotic stripes along the length of the leaf. Plants become stunted (Fig. 25) and may develop multiple tillers, barren ear shoots, and sterile tassels. Poor seed set and premature

Fig. 24. Corn stunt. (Courtesy C. Collar)

death usually occur. Symptom development is strongly influenced by interactions with two other pathogens also transmitted by the corn leafhopper (*Dalbulus maidis*): the maize bushy stunt phytoplasma and *Maize rayado fino virus*. Corn stunt diagnosis is based on serological assays using antisera specific to the cultured spiroplasma or by PCR assays.

Causal Organism

Corn stunt is caused by *Spiroplasma kunkelii*. Morphology is pleomorphic, varying from helical filaments 0.1–0.15 µm long to spherical cells 0.3–0.8 µm in diameter. *S. kunkelii* is a facultative anaerobe, lacks true cell walls, is motile, and has a "fried egg" appearance on semiselective media. Cholesterol is required for growth. It grows very slowly and has a doubling time of about 12 h. Optimal growth is at 30–32°C and pH 7.2–7.4. In plants, the organism is confined to sieve elements within phloem tissue. It can be cultured on several defined media.

Epidemiology

S. kunkelii is transmitted by leafhoppers, primarily *D. maidis* (Fig. 26) and the Mexican corn leafhopper (*D. elimatus*). Vectors transmit the corn stunt spiroplasma in a persistent replicative manner and, thus, are hosts of the pathogen. Maize is the primary plant host for the corn stunt spiroplasma. Other known natural hosts are the closely related *Z. perennis*, *Z. mays* subsp. *mexicana*, and *Z. diploperennis*. The disease is more prevalent where there is continuous maize cultivation, especially at middle to low altitudes. The pathogen can also overwinter in insect vectors, with vectors surviving on wild grasses or alfalfa in the absence of maize. Perennial *Zea* spp. and possibly *Tripsacum* spp. may also serve as reservoirs. Corn stunt frequently is found in combination with the maize bushy stunt phytoplasma and *Maize rayado fino virus*, causing more severe disease than in singly infected plants.

Management

Resistant hybrids are not currently available, although differential responses of maize germplasm to corn stunt infection under greenhouse conditions suggest that resistance or tolerance is present in maize. Development of resistant hybrids is hampered by the difficulty in working with the corn stunt spiroplasma and in identifying highly resistant germplasm. Although leafhopper control has been suggested for disease management, it has not been effective under field conditions. Pathogen reservoir removal and maize cultivation-free periods may provide some control of vector populations.

Selected References

de Oliveira, E., de Oliveira, C. M., Magalhaes, P. C., de Andrade, C. D. T., and Hogenhout, S. A. 2005. Spiroplasma and phytoplasma infection reduce kernel production, and nutrient and water contents of several but not all maize cultivars. Maydica 50:171-178.

Nault, L. R. 1980. Maize bushy stunt and corn stunt: A comparison of disease symptoms, pathogen host ranges, and vectors. Phytopathology 70:659-662.

(Prepared by M. G. Redinbaugh)

Fig. 25. Stunting resulting from corn stunt infection (left); healthy plants on the right. (Courtesy E. Alves, DuPont Pioneer)

Fig. 26. Adult corn leafhopper, *Dalbulus maidis*, the vector of corn stunt. (Courtesy C. Collar)

Maize Bushy Stunt

Maize bushy stunt was first reported from Texas in 1975. The disease is also known as Mesa Central stunt, Louisiana corn stunt, and Mississippi corn stunt and is found in the southern United States, Mexico, and Central and South America.

Symptoms

Symptoms of maize bushy stunt are difficult to distinguish from those caused by the corn stunt phytoplasma and vary by maize genotype. Infection usually begins with marginal chlorosis of whorl leaves and then the reddening of older leaf tips. Emerging leaves may develop marginal chlorosis, reddening, twisting, tearing, and shortening. Plants characteristically develop a bushy appearance because of tiller proliferation in leaf axils and at the bases of plants (Fig. 27). Plants are stunted with numerous small ears that are small in diameter and produce small grains. Root systems may be poorly developed. Diseased plants grown under high temperatures may lead to premature plant death. Interactions with other corn leafhopper (*Dalbulus maidis*)-transmitted pathogens, including the corn stunt spiroplasma and *Maize rayado fino virus*, may exacerbate symptoms. Because of the difficulties distinguishing symptoms of maize bushy stunt and corn stunt in different maize genotypes, definitive diagnosis relies on a nested PCR assay.

Causal Organism

Maize bushy stunt is caused by the maize bushy stunt phytoplasma ('*Candidatus* Phytoplasma asteris', 16S rDNA group I-B).

Epidemiology

Maize bushy stunt is transmitted primarily by *D. maidis* and *D. elimatus* in a propagative, persistent manner. Thus, maize and the vector are both hosts for the pathogen. The plant host range of maize bushy stunt is limited to *Zea mays* and teosinte (*Z. mays* subsp. *mexicana*). The disease is favored by continuous maize cultivation. The phytoplasma can overwinter in the vector, which migrates or feeds on wild grasses or alfalfa

Fig. 27. Maize bushy stunt. (Courtesy C. De Leon)

Fig. 28. Early symptoms of maize redness, caused by the stolbur phytoplasma. (Courtesy G. Avramov, DuPont Pioneer)

in the absence of maize. Maize bushy stunt is more frequently found on corn grown at high elevations. Maize bushy stunt frequently is found in combination with the corn stunt spiroplasma and *Maize rayado fino virus*.

Management

Maize genotypes have differential reduction in height and yield after maize bushy stunt phytoplasma infection under greenhouse conditions, suggesting that some resistance or tolerance to the pathogen is present. However, little is known about the genetics of resistance to maize bushy stunt, and resistant hybrids are not available. Leafhopper control has not generally been effective for disease management under field conditions. Removal of pathogen reservoirs and avoiding continuous maize cultivation may provide some control of the vector populations. Most phytoplasmas are sensitive to tetracycline and chloramphenicol, but antibiotic treatment is not environmentally or economically feasible for disease management.

Selected Reference

Harrison, N. A., Richardson, P. A., Tsai, J. H., Ebbert, M. A., and Kramer, J. B. 1996. PCR assay for detection of the phytoplasma associated with maize bushy stunt disease. Plant Dis. 80:263-269.

(Prepared by M. G. Redinbaugh)

Fig. 29. Maize redness, caused by the stolbur phytoplasma. (Courtesy I. Tosevski)

Maize Redness

Symptoms associated with maize redness were first reported from Serbia in the late 1950s, from Bulgaria, Romania, and Hungary in the 1960s, and more recently from Italy. The disease has also been referred to as corn reddening, corn blindness, and premature wilting of corn. Maize redness outbreaks in the south Banat region of Serbia in 2002 and 2003 caused yield losses of 40–90%.

Symptoms

Maize redness symptoms appear during the late summer. The initial and characteristic symptom is reddening of the midrib on leaves just above the ear (Fig. 28), followed by reddening of the leaves and stalks (Fig. 29). Infected plants subsequently desiccate and die. Ear development is abnormal and seed set is greatly reduced (Fig. 30). However, dwarfing and phyllody are characteristically absent. Diagnosis of maize redness is confirmed using a nested PCR assay.

Causal Organism

Maize redness is caused by the stolbur phytoplasma (‘*Candidatus* Phytoplasma solani’, 16S rDNA group XII-A).

Epidemiology

In Serbia, maize redness has been associated with warm, dry summers. The stolbur phytoplasma causing maize redness also

Fig. 30. Poor pollination and kernel filling associated with maize redness disease. (Courtesy G. Avramov, DuPont Pioneer)

infects Johnsongrass (*Sorghum halepense*) and wheat (*Triticum aestivum*). The stolbur phytoplasma also infects a number of other plant hosts, including solanaceous plants, grapes, and several broadleaf weeds. The primary vector for transmission of the maize redness phytoplasma is the cixiid planthopper, *Reptalus panzeri*. *R. quinquecostatis* may be a secondary vector. Adult insects lay eggs on soil surrounding maize roots in early September. Hatching nymphs feed on maize roots and become infected with the phytoplasma. The disease overwinters in *R. panzeri* nymphs and in Johnsongrass and wheat. Large numbers of nymphs can survive on winter wheat planted into maize fields where the vector is present and then emerge as phytoplasma-infected adults the subsequent summer.

Management

No resistant hybrids are currently available to control maize redness. Removal of overwintering hosts might provide some control but has not been investigated. Current research is exploring the possibility of controlling leafhopper populations using crop rotations, seed treatments, and insecticides.

Selected References

Duduk, B., and Bertaccini, A. 2006. Corn with symptoms of reddening: New host of stolbur phytoplasma. Plant Dis. 90:1313-1319.

Jović, J., Cvrković, T., Mitrović, M., Krnjajić, S., Redinbaugh, M. G., Pratt, R. C., Gingery, R. E., Hogenhout, S. A., and Toševski, I. 2007. Roles of stolbur phytoplasma and *Reptalus panzeri* (Cixiinae, Auchenorrhyncha) in the epidemiology of maize redness in Serbia. Eur. J. Plant Pathol. 118:85-89.

Jović, J., Cvrković, T., Mitrović, M., Krnjajić, S., Petrović, A., Redinbaugh, M. G., Pratt, R. C., Hogenhout, S. A., and Toševski, I. 2009. Stolbur phytoplasma transmission to maize by *Reptalus panzeri* and the disease cycle of maize redness in Serbia. Phytopathology 99:1053-1061.

(Prepared by M. G. Redinbaugh)

Diseases Caused by Fungi and Oomycetes

The fungi and oomycetes are a diverse group of eukaryotic organisms that share similar morphological, nutritional, and reproductive characteristics. These organisms obtain nutrients by the absorption of organic matter digested outside of their cells by the excretion of digestive enzymes, as compared with plants, which make their own food through photosynthesis, and animals, which ingest organic matter prior to digestion. These organisms usually have a morphology of filamentous hyphae, collectively called a mycelium, but some exist as single cells, as do the yeasts, or as a plasmodium (a multinucleate mass of motile protoplasm). The cells of these organisms are often bound by cell walls, and they reproduce by the formation of sexual spores, asexual spores, or both. At one time, all of these organisms were placed in the kingdom Fungi, but comparisons of nucleic acid sequences and physiological and histological characteristics have shown that some of these organisms belong to groups with very distinct genetic origins. Those falling within the current concept of the kingdom Fungi have cell walls composed of chitin, use glycogen as their main storage compound, and have mitochondria with platelike cristae. By contrast, members of the phylum Oomycota (or water molds) have cell walls composed of β-glucans, use mycolaminarins as their storage compounds, and have mitochondria with tubular cristae. Because of these differences, the oomycota are now placed in the kingdom Stramenopila (Chromista). Plasmodial organisms, such as those in the genus *Plasmodiophora*, have been grouped with the protists and placed in other taxa outside of the kingdom Fungi.

The classification of fungi is continually being revised at almost all levels as new information increases our understanding of the organisms and their genetic relationships to one another. Classification systems based primarily on morphological characteristics are being modified to accommodate findings based on genetic sequence data. The main groups (phyla) of fungi are distinguished by their forms of sexual reproduction, vegetative nuclear condition (haploid or dikaryotic), and hyphal septation (septate versus coenocytic). Additional characteristics, including life cycles, formation of fruiting bodies, asexual spores, physiological characteristics, hyphal anastomosis groupings, and many other characteristics, are used to place fungi into various taxonomic groups.

The structures produced by fungi range in size from single microscopic cells to large fruiting bodies, such as mushrooms, conks, and puffballs. Many of the specialized structures are associated with the sexual reproductive stage (the teleomorph) or the asexual reproductive stage (the anamorph). Many fungi reproduce by means of spores. There is a large variety of spore sizes, shapes, and colors among the fungi. Some spores consist of a single cell, while others are multicellular. Asexual spores are produced following mitotic divisions of nuclei, resulting in spores that are genetically identical to the organism that produced them. Asexual spores are often produced in great numbers and used for dissemination to new locations. Other asexual spores have thick walls and allow the fungus to survive under adverse conditions. Sexual spores are the result of genetic recombination during meiosis and serve to increase the genetic diversity within a population. However, sexual spores can also be used for purposes of dissemination and survival. Spores can be formed unprotected directly on hyphae or within a spore-producing structure. The form and nature of a spore-producing structure is very useful in the identification of the fungus. Nonspore structures, such as sclerotia, thickened hyphae, and rhizomorphs, can also serve as means of survival and dissemination.

Some fungi produce only sexual spores, while for others, only the asexual stage has been observed. However, many fungi have both sexual (teleomorphic) and asexual (anamorphic) stages of reproduction. In fact, some fungi produce sexual spores and several different kinds of asexual spores. Because some fungi can produce two or more distinct kinds of spores, often first described by different mycologists, there have traditionally been two or more valid scientific names for many fungi. Now that DNA sequence data can be used to determine the genetic relatedness of these fungi, regardless of the types of spores they produce, a decision has been made to abandon the multiple name system and to use a single scientific name for each fungus. This system is being implemented but there are many details that have not yet been determined.

Fungi cause the majority of contagious plant diseases. Of the more than 100,000 species of fungi that have been described, more than 8,000 are known to be plant pathogens. Fungi can function as parasites, obtaining nutrients directly from the tissues of a living host, or as saprophytes, obtaining nutrients from nonliving organic matter such as dead plant or animal tissues. Some plant-pathogenic fungi are obligate parasites, requiring a living host to complete their life cycles, but many have the ability to exist as both parasites and saprophytes, depending on environmental conditions and the available opportunities.

Fungi, as a group, can attack all parts of a plant and cause a wide variety of symptoms, from root rots and leaf spots to wilting and stunting. Some fungi cause localized lesions of restricted size, while others are able to spread systemically within the plant, causing symptoms throughout. Symptoms of chlorosis, necrosis, or both often result from fungal pathogens that produce toxins or enzymes that spread into the plant tissue in advance of the fungus, killing cells and releasing nutrients for the fungus to use. These types of pathogens are said to be necrotrophic because they obtain their nutrition by killing plant tissues. Stunting or growth deformations are often the results of infection by fungal pathogens that have a more intimate relationship with their host, extracting nutrients from the living cells without immediately killing them. These types of pathogens are said to be biotrophic because they extract their nutrition from living cells. Most obligate parasites are biotrophic. However, the terms "biotrophic" and "necrotrophic" should be viewed as opposite ends of a continuum rather than as discreet, exclusive groups, with various pathogens functioning at different locations along that continuum.

The method and site of infection are closely related to the method of pathogen dissemination. Soilborne, root-infecting fungi are usually disseminated by the movement of soil or infested plant debris. Many of the fungi that infect the aboveground parts of the plant are disseminated by air currents, splashing rain, or vectors such as insects. Some fungi are disseminated on infected seeds or other plant propagation tissues.

Fungi and oomycetes known to infect corn are also listed in the Appendix. Readers will note that in both the Appendix and in the text, alternate names of fungi are identified exclusively as "synonyms," rather than as specific life cycle stages (e.g., "anamorph" and "teleomorph"). This practice reflects the policy for naming fungi issued in 2011 by the International Botanical Congress: namely, that the use of multiple names for the same fungus be abandoned. (Readers interested in learning more about this policy are referred to the 2011 article by Hawksworth [see the Selected References].) Going forward, accommodating this new policy may mean that the most familiar names of some fungi will change. Changes to the names of fungi, as well as other updates to taxonomy and nomenclature, will be made in the Common Names of Diseases list for corn, which is available on the website of The American Phytopathological Society.

Selected References

Agrios, G. N. 2005. Plant Pathology, 5th ed. Academic Press, New York.

Alexopoulos, C. J., Mims, C. W., and Blackwell, M. 1996. Introductory Mycology. John Wiley & Sons, New York.

Barnett, H. L., and Hunter, B. B. 1998. Illustrated Genera of Imperfect Fungi, 4th ed. American Phytopathological Society, St. Paul, MN.

Hawksworth, D. L. 2011. A new dawn for the naming of fungi: Impacts of decisions made in Melbourne in July 2011 on the future publication and regulation of fungal names. IMA Fungus 2:155-162.

Kirk, P. M., Cannon, P. F., Minter, D. W., and Staplers, J. A. 2008. Dictionary of the Fungi, 10th ed. CAB International, Wallingford, United Kingdom.

Rossman, A. Y., Palm, M. E., and Spielman, L. J. 1987. A Literature Guide for the Identification of Plant Pathogenic Fungi. American Phytopathological Society, St. Paul, MN.

Ulloa, M., and Hanlin, R. T. 2012. Illustrated Dictionary of Mycology, 2nd ed. American Phytopathological Society, St. Paul, MN.

Webster, J., and Weber, R. W. S. 2007. Introduction to Fungi, 3rd ed. Cambridge University Press, Cambridge.

(Prepared by D. M. Eastburn)

Seed Rots and Seedling Blights (Damping-Off)

Seedling emergence and establishment of a vigorous stand of corn can be impeded by a wide variety of pathogens that attack seeds and seedlings. The recognized pathogens in this group are filamentous fungi or oomycetes, but these diseases are usually associated with a complex of microorganisms that includes yeasts, bacteria, and nematodes. Seed rots and seedling blights are not usually of widespread importance, but they are economically important in localized areas every year. Usually associated with planting in cool, wet soils, seed and seedling diseases cause yield losses by reducing stand density and plant vigor and have additional economic impacts when replanting is required. Imbibitional chilling injury can occur under cold conditions, impairing emergence and contributing to seedling disease susceptibility. Seedling blights and seed rots rarely occur when injury-free seed with high germination rates and fungicidal seed treatment is planted in a well-prepared seedbed with warm, moist soil.

Symptoms

The first noticeable symptom is usually poor emergence, often in areas that are poorly drained (Fig. 31). On the field scale, seed rots and seedling blights typically appear as missing, dead, or chlorotic, stunted seedlings that occur in patches or scattered individually. Uneven plant height also can be a symptom of seedling disease. Aboveground symptoms also include chlorosis, stunting, wilting, and leaf necrosis that may appear as scorching or as distinct lesions in some cases (Figs. 32 and 33).

Seed may be rotted and die prior to germination or seedlings may be killed before or shortly after emergence (pre- or postemergence damping-off). Seed rot is characterized by dark discoloration of the seed coat, softening or liquefaction of the

Fig. 31. Stand reduction resulting from saturated soil and seedling disease. (Courtesy G. P. Munkvold)

endosperm, and lack of germination. Rotted seeds that fail to germinate can be difficult to locate because of their rapid deterioration in moist soil. After germination, the radicle and shoot are vulnerable to infection and may show symptoms of discoloration and decay, sometimes with a slimy or shriveled appearance. Organisms that cause seed rots or seedling blights typically cannot be distinguished by symptom appearance, although in some cases, fungal mycelium or sporulation may be evident on the seed surface or on belowground seedling tissues (Fig. 34).

Seedlings that do emerge may have rotted seminal roots, nodal roots, or both. The mesocotyl often shows general symptoms of decay or distinct sunken lesions or shriveled areas (Fig. 35). In some cases, the decay proceeds above the first whorl of nodal roots. Affected plants are less vigorous and may also die.

Another condition that is occasionally observed, particularly on plants from damaged seed produced during a drought year, are striped leaves (Fig. 36). These stripes are narrow and white and usually appear only on young plants. Upper leaves of the plant are free from striping, particularly those above the ear shoot. This disease was first described as being caused by *Aspergillus flavus* and *Aspergillus tamarii* and may be confused with genetic virescence and with damage caused by certain herbicides. Symptoms are basically the same in that striping occurs on leaves of seedlings. However, symptoms caused by *Aspergillus* spp. are associated with pericarp damage and penetration of the embryo by *Aspergillus* spp. Plants that are not severely affected normally outgrow this condition.

Causal Organisms

Many species of fungi and bacteria may be involved in seed rot and seedling blight diseases. These organisms may be seedborne (pathogens that are in or on the seed at planting) or soilborne (pathogens that are in the soil in which the seed is planted).

Seedborne fungi are usually ear rot pathogens. The most common fungi associated with seed are *Fusarium verticillioides*, *F. proliferatum*, and *F. subglutinans*. Other fungi, such as

Fig. 32. Yellowing of a corn seedling caused by seedling disease. (Courtesy G. P. Munkvold)

Fig. 33. Seedling chlorosis, necrosis, and postemergence damping-off. (Courtesy G. P. Munkvold)

Fig. 34. Sporulation of a *Penicillium* sp. on a seedling mesocotyl and seed coat. (Courtesy G. P. Munkvold)

Fig. 35. Shriveling and decay of a mesocotyl caused by seedling disease. (Courtesy G. P. Munkvold)

Fig. 36. Foliar symptoms of Aspergillus seedling blight. (Courtesy D. G. White)

Penicillium spp., *Aspergillus* spp., *Rhizopus* spp., *Rhizoctonia* spp., *Bipolaris* spp., *Alternaria* spp., *Nigrospora* spp., *Trichoderma* spp., *F. graminearum*, several other *Fusarium* spp., and *Stenocarpella maydis*, also are associated with hybrid seed. A common symptom of Penicillium seedling blight is necrosis of the leaf tip, which may occur in streaks or large areas. Later, these areas become brittle. Another symptom is chlorosis of entire plants, and these seedlings usually die. This condition has been associated with damage to the pericarp and penetration by *Penicillium oxalicum*. Plants that survive are often stunted and yellow.

Soilborne fungi and oomycetes involved with seed rots and seedling blights include those that cause kernel rots as well as others. Different species cause seed decay and seedling root rot, depending on location, soil conditions, and crop sequence. The most common soilborne organisms that can cause seed rot and decay of seedling roots are *Pythium* spp., including *P. irregulare*, *P. debaryanum*, *P. ultimum*, *P. rostratum*, and *P. paroecandrum*; *Phytopythium vexans* (syn. *Pythium vexans*); and at least 10 other *Pythium* spp. Other common soilborne seedling pathogens include *Rhizoctonia solani* (AG-2 and AG-4), *F. oxysporum*, *F. solani*, and several other *Fusarium* spp.

Disease Cycle and Epidemiology

Seed rot and seedling blight pathogens survive in or on seeds, directly in the soil, or in crop residue in the soil. Survival may be as mycelium, conidia, or other spore types, depending on the pathogen. The seedborne fungi can be found in association with a large number of kernels that may or may not display symptoms. External contamination or symptomless infection may not affect germination, but these fungi cause emergence problems at higher levels of infection and can be transmitted from seed to seedling. The amount of damage caused by these fungi depends on the extent of rotting that has occurred before harvest. If rot is extensive, the embryo may be killed and the seed will not germinate. Seed that is not severely rotted germinates when planted in environments that are favorable for germination, but when it is planted in unfavorable environments, germination is slow or absent because the internal fungi have an opportunity to rot the developing seedling.

A number of factors interact to affect emergence under adverse conditions. These include genetics, the environment in which the seed was produced, stage of maturity at seed harvest, structural integrity, deterioration (after harvest), use of herbicides, soil compaction, insect damage, and presence of pathogens. Susceptibility to seed rots and seedling blights can be affected by seed quality; for example, these diseases can be more prevalent in seed produced under drought conditions, seed with mechanical damage, or seed damaged by an early frost before harvest. Seed rots and seedling blights are favored when seed is planted in cool (below 13°C), poorly drained, or compacted soil or wherever misplacement of fertilizer or misuse of pesticides is involved. Infection of seeds or seedling tissues takes place as they come into contact with passive soilborne pathogen propagules or when motile *Pythium* zoospores find the seeds or seedling tissues. Frequently, the greatest stress leading to emergence problems is related to cool, wet soils. Seeds imbibe water at temperatures above freezing, but seed metabolism is extremely slow below 13°C. Cell membranes in embryo tissue may rupture during imbibition, causing disruption of protein production and energy transformation in the activated cells. Under warm conditions, the membranes may be repaired and the damage to germination limited, but low temperatures slow repair. This imbibitional chilling damage can result in death of the seed, weak seedlings, and delayed emergence, even in the absence of fungi. Imbibitional chilling damage also predisposes the seed to infection by various seed and seedling pathogens. All seeds exude carbohydrates and proteins after imbibition, but seeds with damaged membranes leak more into the soil. Exudates from damaged seeds stimulate spore germination of numerous fungi, including *Pythium* spp. In many situations, it is difficult to separate imbibitional chilling damage from that caused by pathogens. Seedling blight caused by *Penicillium oxalicum* is favored by high temperatures, unlike most other seed rot and seedling blight diseases.

Seed with mechanical damage, insect feeding injury, or other types of damage is more readily attacked by fungi because of greater exudation and because damaged areas serve as sites for penetration by fungi. Cool conditions slow the growth of the plant and allow fungi more time to kill the seedling. High moisture levels result in an increase in exudates, which stimulate germination of oospores, and in low-oxygen conditions, which favor the growth of *Pythium* spp. and reduce root development. Pythium seed rot and seedling blight is most serious in cool, wet soils, particularly with pericarp-damaged seed.

Several seedling pathogens in the genus *Fusarium* are mycotoxin-producing fungi, and there is conflicting evidence for the role of mycotoxins in seedling disease. It is likely that some mycotoxins contribute to the aggressiveness of *Fusarium* spp. to corn seedlings, but more research is needed.

Management

Improved methods of hybrid seed production and handling, improved mechanical planters, and use of fungicide seed treatments have greatly reduced losses from seed rots and seedling blights. High-quality seed is the first critical step in the management of seed rot and seedling blight diseases. High-quality seed is the result of careful seed production, of careful handling and conditioning practices, and of germination and vigor testing for quality control. Severely rotted seed is usually removed from hybrid seed lots by using gravity tables, color sorters, or other seed conditioning processes. Gravity separation is an important tactic for removing symptomless infected seed as well. Seed lots containing rotted seed also are easily identified by germination and vigor tests.

Seed treatment is an essential tactic for reducing the impact of both seedborne and soilborne pathogens, and nearly all commercial corn seed is treated with a combination of fungicides. Most standard treatments are a combination of active ingredients that must include at least one (e.g., mefenoxam or metalaxyl) with activity against *Pythium* spp. Other ingredients typically are broad-spectrum fungicides with activity against *Fusarium* spp. and other fungi. Systemic ingredients can provide a longer duration of protection than contact fungicides, but many seed treatment fungicides are contact only. Nonfungicidal seed treatment components (seed enhancements) or biological seed treatments also can contribute to reducing the risk of seed and seedling disease. Physical seed treatments, such as hot water or steam, can reduce seedborne contamination, but these are rarely used with corn. Most standard corn seed treatments also include a systemic insecticide, which can contribute to seed and seedling disease management by preventing insect injuries.

Rapid seedling emergence and resistance to various seed rot and seedling blight diseases are heritable characteristics. Breeding efforts in this area are complicated by the large number of organisms and environmental factors involved. No single method can be used to evaluate a particular genotype for rapid emergence under all conditions and for resistance to seed and seedling diseases. However, potential problems in experimental hybrids are identified through testing over wide geographical areas, and these hybrids are seldom advanced to the commercial level.

Planting practices can affect seed and seedling disease occurrence. The risk of emergence problems is reduced if planting is done in soil above 13°C, when soil moisture is not excessive. No-till production can lead to greater disease levels (Fig. 37) because of reduced soil temperature and increased soil moisture, especially at earlier planting dates. Spring tillage

Fig. 37. Chlorosis and stunting caused by cool, wet soil and seedling pathogens under no-till conditions (background) compared with healthy growth under tilled conditions (foreground). (Courtesy G. P. Munkvold)

or row-cleaning attachments on the planter can improve emergence by allowing the soil in the spermosphere to warm more quickly and improving seed-to-soil contact. Crop rotation has little value for seed and seedling disease management, except that the residue of legumes or small grains tends to be lighter than corn residue and, therefore, does not affect soil temperature and moisture as much as corn residue. Other soil management practices that promote drainage and reduce compaction also can be beneficial.

Selected References

Berjak, P., and Villiers, T. A. 1972. Aging in plant embryos. II. Age-induced damage and its repair during early germination. New Phytol. 71:135-144.

Broders, K. D., Lipps, P. E., Paul, P. A., and Dorrance, A. E. 2007. Characterization of *Pythium* spp. associated with corn and soybean seed and seedling disease in Ohio. Plant Dis. 91:727-735.

Caldwell, R. W., Tuite, J., and Carlton, W. W. 1981. Pathogenicity of *Penicillia* to corn ears. Phytopathology 71:175-180.

Callan, N. W., Miller, J. B., Mathre, D. E., and Mohan, S. K. 1996. Soil moisture and temperature effects on *shrunken2* sweet corn seed decay and seedling blight caused by *Penicillium oxalicum*. J. Am. Soc. Hortic. Sci. 121:83-90.

Jing-bao, L., Zhi-yuan, F., Hui-ling, X., Yan-min, H., Zong-hua, L., Liu-jing, D., Shang-zhong, X., and Ji-hua, T. 2011. Identification of QTLs for maize seed vigor at three stages of seed maturity using a RIL population. Euphytica 178:127-135.

McDonald, M. B. 1975. A review and evaluation of seed vigor tests. Proc. Assoc. Off. Seed Anal. 65:109-139.

Munkvold, G. P., and O'Mara, J. K. 2002. Laboratory and growth chamber evaluation of fungicidal seed treatments for maize seedling blight caused by *Fusarium* species. Plant Dis. 86:143-150.

Obendorf, R. L. 1972. Factors associated with early germination in corn under cool conditions. Proc. Corn Sorghum Res. Conf. 27:132-139.

Rodriguez-Brljevich, C., Kanobe, C., Shanahan, J. F., and Robertson, A. E. 2010. Seed treatments enhance photosynthesis in maize seedlings by reducing infection with *Fusarium* spp. and consequent disease development in maize. Eur. J. Plant Pathol. 126:343-347.

Simon, E. W. 1984. Early events in germination. Pages 77-115 in: Seed Physiology, Vol. 2. R. Murray, ed. Academic Press, Sydney, Australia.

Solorzano, C. D., and Malvick, D. K. 2011. Effects of fungicide seed treatments on germination, population, and yield of maize grown from seed infected with fungal pathogens. Field Crops Res. 122:173-178.

Zhang, B. Q., Chen, W. D., and Yang, X. B. 1998. Occurrence of *Pythium* species in long-term maize and soybean monoculture and maize/soybean rotation. Mycol. Res. 102:1450-1452.

(Prepared by J. L. Dodd and D. G. White;
Revised by G. P. Munkvold)

Root Rots

Root rots are among the least studied and least understood fungal diseases of corn, even though they occur on every corn plant in every field every year. Yield loss cannot be accurately estimated because techniques are not available to compare yields of healthy plants with yields of diseased plants. Many studies of root rots have been based on the isolation of various fungi from roots during the growing season, sometimes followed by the inoculation of plants in the greenhouse to determine pathogenicity. Fungi recovered from roots include seedling blight and stalk rot pathogens as well as a number of other pathogens and secondary invaders. Pathogenicity studies are difficult to interpret since the inoculation of plants under stress or the introduction of fungi into sterilized soil where no competition from other organisms exists may result in levels of root rot severity that do not normally occur in the field.

Researchers agree that (i) root rot is a disease complex involving many different fungi and may also involve bacteria, nematodes, and root-feeding insects; (ii) the groups of fungi that develop on roots differ depending upon the growth stage of the host, environmental conditions, genotype, and previous crops; and (iii) fungi vary in their ability to cause root rot. The following four root rot diseases are considered to be distinct enough that they can be discussed individually. It is clear from the number of genera involved that root rot diseases (much like stalk rot diseases) result from the activities of numerous organisms.

(Prepared by D. G. White)

Pythium Root Rot (and Feeder Root Necrosis)

Species of *Pythium* and related genera are well-known for causing seed rot, damping-off, seedling blight, and root rot. Among these diseases, root rot presents the least conspicuous aboveground symptoms. Broadly defined, root rot also includes feeder root necrosis. Pythium root rot is considered an important cause of low corn yields in poorly drained soils and in soils with conservation tillage.

Symptoms

Severely infected plants have visible lesions and discoloration of the root system (Fig. 38) that may be accompanied by aboveground symptoms, such as yellowing and stunting. The cortex (outer part of the root) may be rotted and discolored, while the stele (inner part of the root) may remain white. When the root is placed between the thumb and forefinger and gently pulled, the

Fig. 38. Root discoloration caused by Pythium root rot. (Courtesy G. P. Munkvold)

Fig. 39. Symptoms of Pythium root rot in which the rotted cortex was pulled away from the stele. (Courtesy D. G. White)

cortex pulls off, leaving the white stele (Fig. 39). Often, infections are limited to juvenile tissue at the root tips (feeder roots) without causing aboveground symptoms. The infected feeder roots are either absent, because of *Pythium* infection, or easily dislodged during digging and washing, making it difficult to isolate the pathogen. The presence of *Pythium* spp. may be diagnosed by observing oospores in rotted tissue (Fig. 40). However, saprophytic *Pythium* spp. may also be present in rotted roots.

Causal Organisms

Pythium and related genera belong to the phylum Oomycota, commonly known as water molds. Although the members produce filamentous hyphae that superficially resemble those of true fungi, their cell walls contain cellulose rather than chitin, as found in true fungi. They are taxonomically classified in the kingdom Stramenopila (Chromista), more closely related to the brown algae than to the true fungi. More than 30 different *Pythium* spp. can be isolated from diseased corn seedlings, but pathogenicity has not been demonstrated for many of the species. At least 18 species have been reported to cause

Fig. 40. Oospores of a *Pythium* sp. in rotted root tissue. (Courtesy T. Paulitz)

seedling blight and root rot, including *Phytopythium helicoides* (syn. *Pythium helicoides*) and *Phytopythium vexans* (syn. *Pythium vexans*), and the following *Pythium* spp.: *P. acanthicum*, *P. adhaerens*, *P. angustatum*, *P. aphanidermatum*, *P. arrhenomanes*, *P. graminicola*, *P. irregulare*, *P. lutarium*, *P. paroecandrum*, *P. pulchrum*, *P. rostratum*, *P. splendens*, *P. sylvaticum*, *P. tardicrescens*, *P. torulosum*, and *P. ultimum*. Although identification of groups of species of *Pythium* and related genera is relatively easy, differentiation of certain morphologically similar species presents some challenges. Primary species causing Pythium root rot of corn vary widely among studies, but *P. arrhenomanes*, *P. irregulare*, *P. lutarium*, *P. sylvaticum*, *P. torulosum*, and *P. ultimum* are among those considered important in the midwestern United States.

Disease Cycle and Epidemiology

Pythium spp. overwinter in soil and plant debris as oospores. The oospores germinate and produce mycelium or sporangia, which produce and release zoospores. Both zoospores and mycelium can infect corn roots. *P. arrhenomanes* can be readily isolated after heavy rain or irrigation, suggesting that the pathogen is more active when moisture levels are high. *Pythium* spp. vary in growth rate. Thus, species with faster growth rates may be isolated at higher frequencies than species with slow growth rates. *Pythium* spp. are primary invaders and generally are poor competitors with other soilborne organisms. The disease is favored by high moisture levels and low temperatures, which result in low oxygen levels in the soil. This favors *Pythium* spp. by inhibiting other competitive microorganisms.

Management

The most effective ways to manage Pythium root rot are to improve soil drainage and avoid planting into cold soils. Treatment of seed with systemic fungicides that are effective against *Pythium* spp. protect seeds and seedlings. However, the fungicides lose their effect as the plant grows. *Pythium* spp. and isolates of a given species vary in their sensitivity to different fungicides. Appropriate rotation or combinations of fungicides with different modes of action should be practiced to improve efficacy and prevent development of fungicide resistance in *Pythium* spp. Very little information is available on resistance in corn hybrids. Rotation with other crops provides little or no control.

Selected References

Broders, K. D., Lipps, P. E., Paul, P. A., and Dorrance, A. E. 2007. Characterization of *Pythium* spp. associated with corn and soybean seed and seedling disease in Ohio. Plant Dis. 91:727-735.

Chen, W., and Hoy, J. W. 1993. Molecular and morphological comparisons of *Pythium arrhenomanes* and *P. graminicola*. Mycol. Res. 97:1371-1378.
Deep, I. W., and Lipps, P. E. 1996. Recovery of *Pythium arrhenomanes* and its virulence to corn. Crop Prot. 15:85-90.
Harvey, P. R., Warren, R. A., and Wakelin, S. 2008. The *Pythium–Fusarium* root disease complex—An emerging constraint to irrigated maize in southern New South Wales. Aust. J. Exp. Agric. 48:367-374.
Hendrix, F. F., Jr., and Campbell, W. A. 1983. Some pythiaceous fungi, new roles for old organisms. Pages 123-329 in: Zoosporic Plant Pathogens, A Modern Perspective. S. T. Buczacki, ed. Academic Press, London.
Rao, B., Schmitthenner, A. F., Caldwell, R., and Ellett, C. W. 1978. Prevalence and virulence of *Pythium* species associated with root rot of corn in poorly drained soil. Phytopathology 68:1557-1563.
Zhang, B. Q., and Yang, X. B. 2000. Pathogenicity of *Pythium* populations from corn–soybean rotation fields. Plant Dis. 84:94-99.

(Prepared by W. Chen)

Rhizoctonia Crown and Brace Root Rot

Rhizoctonia crown and brace root rot (Rhizoctonia root rot) of corn has been reported in France, Germany, New Zealand, and Japan and in irrigated fields in Georgia (United States) and may also be present in tropical and subtropical regions. Yield losses of grain as high as 12% have been reported in producers' fields. Recently, root rot on corn was found in inoculated field plots in Minnesota (United States) with no aboveground symptoms or yield loss, which suggests the disease may be more widespread than is apparent.

Symptoms

Initial symptoms of Rhizoctonia crown and brace root rot on seedlings and juvenile plants include tan to brown elliptical lesions on the mesocotyl and lateral seminal roots; occasionally, small, medallion-shaped lesions also form on the shoot or stalk (Fig. 41). Under severe conditions, lesions darken and expand and result in terminal decay, and the seedlings die. More commonly, older plants display large reddish brown to black lesions or cankers on adventitious crown and brace roots (may appear crook shaped) and seminal roots (Fig. 42). Mild to severe disintegration of roots takes place 2–5 cm below the soil surface, resulting in loss of the terminal root system. As the disease progresses, plants generate new lateral and tertiary roots, which often turn brown and decay. Root systems of severely infected plants cannot anchor the plant in the soil or absorb nutrients and water. Plants are stunted and chlorotic and tend to lean or lodge in scattered patches, especially after winds or heavy rains, thus revealing rotted roots, which is helpful in distinguishing diseased from healthy plants. Hyphae and sclerotia also have been observed on roots.

Fig. 41. Stalk lesions caused by *Rhizoctonia solani.* (Courtesy C. E. Windels)

Causal Organism

Rhizoctonia crown and brace root rot is caused by *Rhizoctonia solani* AG 2-2 IIIB (syn. *Thanatephorus cucumeris*). Young hyphae are robust and tend to branch near the distal septum at about a 30° angle, but as hyphae mature, branches form at a 90° angle. Other microscopic features typical of *R. solani* include multinucleate cells (Fig. 43); dolipore septa; and constriction of hyphae and formation of septa near the origin of hyphal branches. Ends of preexisting cells can form simple or branched chains of melanized, barrel-shaped, monoloid cells. These cells may form on or above the surface of a substrate and within plant tissue, where they survive as vegetative hyphae or organize into compact masses of sclerotia. Mycelial growth on potato dextrose agar initially is cream to light tan, but within 2–3 weeks, cultures are dark brown and partially to completely

Fig. 42. Rhizoctonia crown and brace root rot. (Courtesy C. E. Windels)

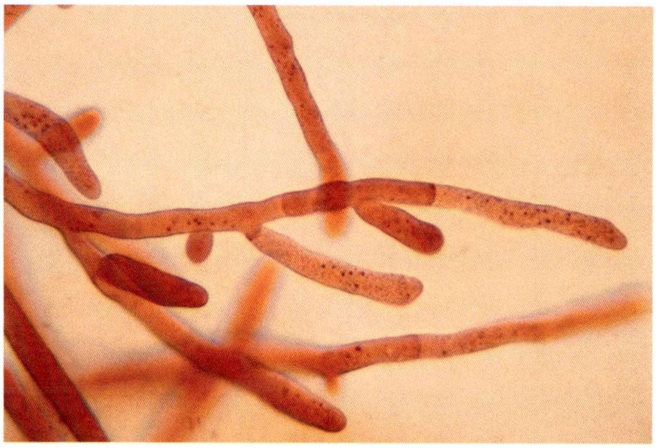

Fig. 43. Multinucleate mycelium of *Rhizoctonia solani.* (Courtesy C. E. Windels)

covered in sclerotia up to 5 mm in diameter. Surface of cultures ranges from velvetlike to crusty, with delicate to dense aerial mycelium.

Disease Cycle and Infection Habit

The fungus survives at least for several months as sclerotia, monoloid cells, or mycelium in or on debris of previously infected plants or by saprophytic colonization of crops and weeds. Reports of duration of survival vary, depending on the isolate, substrate, size of inoculum unit, and environmental conditions. In Minnesota, less than 1% of sclerotia (≥1 mm in diameter) of one of three isolates buried in field soil germinated after 2 years compared with 64% of sclerotia of the three isolates stored dry in a refrigerator; sclerotia <1 mm under the same conditions averaged 0% viability in the field and 25% in a refrigerator. In Georgia, the fungus survived more than 9 months in fallow soil and for several months in peanut debris left in the soil after harvest. The greater the size of infected debris in the soil, the longer the fungus is protected and remains viable. Propagation of the fungus also is extended by rotation of a wide range of susceptible crops, such as beans and other legumes, sugar beet, weeds, and plant species that remain symptomless. Production of basidiospores by *T. cucumeris* can occur but has not been reported as a source of inoculum on corn.

Infection of corn by *R. solani* AG 2-2 IIIB has not been described; however, it is known for other crops. Sclerotia or hyphae in plant debris are stimulated to grow by plant exudates. Hyphae grow over the root surface, aggregate in dense cushions, and produce one or more infection pegs that apply pressure and produce cellulolytic and pectolytic enzymes to penetrate the epidermis and cortex. Entrance also takes place through natural openings and wounds. Intercellular tissue is destroyed as *R. solani* grows into the root, root tissue dissolves, and symptoms develop.

Aggressiveness of *R. solani* AG 2-2 IIIB on corn varies with isolates, origin of isolates, and environmental conditions. The fungus is active across a wide range of soil temperatures and moisture levels. In Germany, the disease is most severe in sugar beet–corn rotations when corn is grown during dry seasons and temperatures are continuously high (up to 36°C). In Georgia, the disease is an important component of the pathogen complex that infects corn in monoculture or irrigated corn rotated with other susceptible crops. Isolates of *R. solani* AG 2-2 IIIB cause corn seedling rot from 8 to 28°C and infect corn roots from 8 to 34°C. The fungus also is aggressive at 20–34°C (night/day), which explains the appearance of the disease on corn later in the growing season.

Management

Germplasm is not routinely screened for resistance to Rhizoctonia crown and brace root rot; however, German greenhouse and field trials indicate that resistance is present. This may be a promising way to manage the disease in the future.

No fungicides are registered to control the disease on corn. The herbicide pendimethalin, however, has been shown to increase the level of disease when corn is sown in cold soils.

Cultural practices can be helpful in reducing the disease and include rotations of at least 2 years with nonhost crops, weed control, and cultivation and drainage to keep soil dry. Specific tillage practices have little effect on disease severity, but practices that favor decomposition of crop residues may be helpful.

Selected References

Buddemeyer, J., Pfähler, B., Petersen, J., and Märländer, B. 2004. Genetic variation in susceptibility of maize to *Rhizoctonia solani* (AG 2-2IIIB)—Symptoms and damage under field conditions in Germany. J. Plant Dis. Prot. 111:521-533.

Buhre, C., Kluth, C., Bürcky, K., Märländer, B., and Varrelmann, M. 2009. Integrated control of root and crown rot in sugar beet: Combined effects of cultivar, crop rotation, and soil tillage. Plant Dis. 93:155-161.

Fowler, M. 1980. Root rot—Maizegrowers must live with it. N. Z. J. Agric. 141:45-47.

Ithurrart, M. E., Büttner, G., and Petersen, J. 2004. Rhizoctonia root rot in sugar beet (*Beta vulgaris* ssp. *altissima*)—Epidemiological aspects in relation to maize (*Zea mays*) as a host plant. J. Plant Dis. Prot. 111:302-312.

Liu, Z. L., Sinclair, J. B., and Chen, W. 1992. Genetic diversity of *Rhizoctonia solani* anastomosis group 2. Phytopathology 82:778-787.

Perraton, B., and Lucas, P. 1983. Importance of *Rhizoctonia solani* Kühn in some lodging symptoms on maize. Agronomie 3:791-796.

Pfähler, B., and Petersen, P. 2004. Rapid greenhouse screening of maize for resistance to *Rhizoctonia solani* AG2-2IIIB. J. Plant Dis. Prot. 111:292-301.

Sneh, B., Burpee, L., and Ogoshi, A. 1991. Identification of *Rhizoctonia* species. American Phytopathological Society, St. Paul, MN.

Sumner, D. R., and Bell, D. K. 1982. Root diseases induced in corn by *Rhizoctonia solani* and *Rhizoctonia zeae*. Phytopathology 72:86-91.

Sumner, D. R., and Bell, D. K. 1986. Influence of crop rotation on severity of crown and brace root rot caused in corn by *Rhizoctonia solani*. Phytopathology 76:248-252.

Sumner, D. R., and Bell, D. K. 1994. Survival of *Rhizoctonia* spp. and root diseases in a rotation of corn, snap bean, and peanut in microplots. Phytopathology 84:113-118.

Sumner, D. R., and Dowler, C. C. 1983. Herbicide, planting date, and root disease interactions in corn. Plant Dis. 67:513-517.

Windels, C. E., and Brantner, J. R. 2006. Crop rotation effects on *Rhizoctonia solani* AG 2-2. Sugarbeet Res. Ext. Rep. 36:286-290.

(Prepared by C. E. Windels)

Fusarium Root Rot

Fusarium spp. are commonly isolated from corn roots. Some species are pathogens, while others are secondary invaders. Several of the pathogenic species can infect stalks and ears as well as roots. Many of the *Fusarium* spp. that are pathogenic on corn may also infect other plants.

Symptoms

When symptoms appear, they range from a very slight brownish discoloration to dark black discoloration to completely rotted roots (Fig. 44). These symptoms usually are indistinguishable from other root rots. Roots infected by *F. graminearum* may display a red to pink pigment, which can be confused with symptoms of red root rot (caused by *Setophoma terrestris* [syn. *Phoma terrestris*]). *Fusarium* spp. can also be isolated from symptomless roots.

Causal Organisms

Numerous *Fusarium* spp. cause Fusarium root rot. The most commonly reported are *F. oxysporum* and *F. solani*. Others include *F. acuminatum*, *F. equiseti*, and *F. graminearum*. Some species, such as *F. proliferatum*, *F. subglutinans*, and *F. verticillioides* tend to be associated with root rot in younger plants.

Disease Cycle and Epidemiology

Most root rot beyond the seedling stage is caused by *Fusarium* spp. that survive in the soil or crop residue. Survival may be as chlamydospores, mycelium, or conidia, depending on the species. Infection occurs as plant roots come into contact with fungal propagules in the soil. Root injury is not necessary, but infection is enhanced when roots are damaged by cultivation or insects. Root rot of young plants may be caused by seedborne infections. Susceptibility to root rot increases when plants are stressed, probably because of an inadequate allocation of photosynthate to the roots. Herbicide injury also increases susceptibility. As plants near maturity, roots become more susceptible and the amount of root rot increases. Root-infecting *Fusarium* spp. often move into the base of the corn plant, resulting in a crown rot (Figs. 44 and 45) or stalk rot. Fusarium root rot develops under a wide range of temperature and moisture conditions. Some *Fusarium* infection and root rot symptoms can be found in every field every year but are commonly not severe enough to cause economic losses. Losses occur when disease severity increases because of extremes in soil moisture (too high or too low) or when other stresses predispose plants to infection.

Management

It is extremely difficult to evaluate breeding materials for susceptibility or resistance to Fusarium root rots. It is likely that, through the extensive testing of corn hybrids over wide geographical areas, those that have some resistance to Fusarium root rot diseases are selected. Control of corn rootworm feeding through transgenic resistance or insecticidal seed treatment can reduce root infection by *Fusarium* spp. Seed treatment with fungicides can reduce seedling blight, but the treatments do not provide control for an extended period of time. Tillage has relatively little effect, but Fusarium root rot is less common when corn is rotated with other crops, such as legumes. Irrigation should be managed to avoid excessive soil moisture or soil moisture deficits. Soil management practices that improve drainage or reduce the occurrence of soil compaction can reduce the severity of Fusarium root rot.

Fig. 44. Fusarium root and crown rot. (Courtesy G. P. Munkvold)

Fig. 45. Crown rot in a young corn plant, caused by *Fusarium* spp. (Courtesy G. P. Munkvold)

Selected References

Futrell, M. C., and Kilgore, M. 1969. Poor stands of corn and reduction of root growth caused by *Fusarium moniliforme*. Plant Dis. Rep. 53:213-215.

Kedera, C. J., Leslie, J. F., and Claflin, L. E. 1994. Genetic diversity of *Fusarium* section *Liseola* (*Gibberella fujikuroi*) in individual maize stalks. Phytopathology 84:603-607.

Kommedahl, T., and Windels, E. C. 1981. Root-, stalk-, and ear-infecting *Fusarium* species on corn in the USA. Pages 94-103 in: Fusarium: Diseases, Biology, and Taxonomy. P. E. Nelson, T. A. Toussoun, and R. J. Cook, eds. Pennsylvania State University Press, University Park.

Kurtz, B., Karlovsky, P., and Vidal, S. 2010. Interaction between western corn rootworm (Coleoptera: Chrysomelidae) larvae and root-infecting *Fusarium verticillioides*. Environ. Entomol. 39:1532-1538.

Leslie, J. F., Pearson, C. A. S., Nelson, P. E., and Toussoun, T. A. 1990. *Fusarium* spp. from corn, sorghum, and soybean fields in the central and eastern United States. Phytopathology 80:343-350.

Palmer, L. T., and Kommedahl, T. 1969. Root-infecting *Fusarium* species in relation to rootworm infestations in corn. Phytopathology 59:1613-1617.

Soonthornpoct, P., Trevathan, L. E., and Ingram, D. 2000. The colonization of maize seedling roots and rhizosphere by *Fusarium* spp. in Mississippi in two soil types under conventional tillage and no-tillage systems. Phytoprotection 81:97-106.

Sumner, D. R., and Hook, J. E. 1985. Irrigation management and root and stalk rot of corn. Plant Dis. 69:239-243.

Warren, H. L., and Kommedahl, T. 1973. Prevalence and pathogenicity to corn of *Fusarium* species from corn roots, rhizosphere, residues, and soil. Phytopathology 63:1288-1290.

Young, T. R., and Kucharek, T. A. 1977. Succession of fungal communities in roots and stalks of hybrid field corn grown in Florida. Plant Dis. Rep. 61:76-80.

(Prepared by G. P. Munkvold and J. F. Leslie)

Red Root Rot

Red root rot has increased in incidence and severity since the late 1980s. The disease has been most prevalent in the U.S. Atlantic states, especially the Delmarva region (Delaware, Maryland, and Virginia peninsula), where yield reductions of 15–20% have occurred in localized areas.

Symptoms

Red root rot is usually not apparent until just prior to senescence. The major symptom, as implied by the name, is the reddish pink discoloration of the root system (Fig. 46A) and basal stalk tissue (Fig. 46B). There is a reduction in total root mass, which is obvious when a comparison is made with roots of a nearby healthy plant. As the rot continues, the color of the roots often progresses from reddish pink to deep carmine and is accompanied by necrosis and shriveling of the root tissue. A similarly colored rot can be observed in basal stalk tissue and generally does not extend more than three elongated internodes above the soil line. The red color may be confused with that associated with infection by *Fusarium graminearum*. However, the discoloration associated with red root rot is usually darker. Aboveground, the most obvious symptom of the disease appears during the late stages of ear fill and is characterized by premature death of the plant. Pycnidia of the fungus may form in clusters at the stalk base (Fig. 47). As with many stalk rot diseases, foliage and stalks become grayish green and leaves have an overall wilted appearance. Foliar symptoms develop quickly, usually in a 4- to 5-day interval, and death of the most susceptible plants happens within a week. This may be followed by severe lodging, depending on the weather conditions. Symptoms have been observed most frequently in the Delmarva region when corn is grown in a high-yield (i.e., high-population, high-fertility, irrigated) environment. The rate of disease development varies greatly among genotypes.

Causal Organism

Several studies have indicated that red root rot is caused by a complex of fungi that includes *Setophoma terrestris* (syns. *Phoma terrestris*, *Pyrenochaeta terrestris*) as the primary pathogen in association with *Pythium* and *Fusarium* spp. Mycelium of *S. terrestris* is hyaline, septate, glutted, and anastomosing. The fungus produces subglobose pycnidia (170–350 μm) that are ostiolate, papillate, dark brown to black, and carbonaceous and may occur singly and are often gregarious (Fig. 47). Conidia (1.8–2.3 × 4.5–5.5 μm) are continuous, hyaline, oblong to ovoid, bigluttulate, and sessile in pycnidia and escape as a gelatinous cirrus through the ostiole.

Fig. 46. A, Red root rot symptoms on brace roots. **B,** Red root rot symptoms in a rotted stalk. (Courtesy G. P. Munkvold)

Fig. 47. Pycnidia of *Setophoma terrestris* (syn. *Phoma terrestris*). (Courtesy G. P. Munkvold)

Microsclerotia are resting structures of the fungus found in diseased roots or in artificial agar. They are dark, thick walled, and multicellular and initiate from swellings of intercalary hyphal cells followed by repeated divisions. They are important in the long-term survival of the fungus in nature. *S. terrestris* produces a red pigment in infected plant roots, which is important in disease diagnosis, and also produces a pinkish color on Watson's agar with wheat straw, which is a positive test for identification. In culture on artificial media, the most vigorous colony growth takes place at 24–30°C.

Disease Cycle and Epidemiology

S. terrestris is recognized as a widespread saprophyte and weak parasite on the underground parts of a wide variety of hosts. The most studied and economically destructive disease caused by this pathogen is pink root of onion, and much remains to be learned about the disease on corn. The fungus is found in many types of soils and survives well under a wide range of temperature and pH conditions. The fungus can survive in the soil for years, presumably as microsclerotia, which may serve as the overwintering structures and the primary inoculum. There is a general lack of information about the importance of the pycnidial stage, and a sexual stage has not been found. In some research, *S. terrestris* was detected in corn roots as early as the midsilking stage but not consistently. It is generally considered a late-season pathogen that can be successfully recovered from infected corn roots beginning in early August or later. The rate of colonization of root and basal stalk tissues increases as the host undergoes senescence. Deterioration of infected roots results in loss of vigor, accelerated senescence, wilting of leaves, stalk lodging, and reduced yield. The optimal temperature for hyphal infection has been reported to be 25–28°C. In both corn and onion, *S. terrestris* propagules invade root epidermal and cortical tissues. Hyphal advance and penetration of host cells are facilitated primarily by the enzymatic breakdown of host cell wall tissue. Recent research indicates that early infection of corn roots by *Pythium* spp. (and perhaps other fungi) debilitates the roots to the extent that *S. terrestris* is able to invade earlier, more effectively, or both as an important secondary pathogen, resulting in red root rot.

Management

It is extremely difficult to evaluate breeding materials for susceptibility or resistance to *S. terrestris*. As with many other root-infecting fungi, it is likely that, through extensive testing of corn hybrids over wide geographical areas, those that have some resistance are selected. Studies on inheritance of resistance indicate that resistance is a polygenic trait with additive gene action. Crop rotation with soybeans seems to provide some control.

Selected References

Campbell, K. W. 1991. Evaluation of maize genotypes for red root rot resistance in the Delmarva region. M.S. thesis. University of Delaware, Newark.

Campbell, K. W., Carroll, R. B., Veitenheimer, E. E., Hawk, J. A., and Whittington, D. P. 1991. Red root rot, a new disease of maize in the Delmarva Peninsula. Plant Dis. 75:1186.

Craig, J., and Koehler, B. 1958. *Pyrenochaeta terrestris* and *Phaeocytosporella zeae* on corn roots. Plant Dis. Rep. 42:622-623.

Johann, H. 1943. *Phoma terrestris* in the roots of mature maize plants. Phytopathology 33:526-528.
Mao, W., Carroll, R. B., and Whittington, D. P. 1998. Association of *Phoma terrestris*, *Pythium irregulare*, and *Fusarium acuminatum* in causing red root rot of corn. Plant Dis. 82:337-342.
Miller, T. D. 1993. Development of a rapid screening method for selecting maize lines resistant to *Phoma terrestris*. M.S. thesis. University of Delaware, Newark.
Morgan-Jones, G., and White, J. F. 1983. Studies in the genus *Phoma*. III. *Paraphoma*, a new genus to accommodate *Phoma radicina*. Mycotaxon 18:57-65.
Newby, D. L. 1996. The incidence of red root rot in Delaware and on the eastern shore of Maryland and the initiation and progression of infection of *Zea mays* by *Phoma terrestris* and *Pythium ultimum*. M.S. thesis. University of Delaware, Newark.
Newby, D. L., Carroll, R. B., and Whittington, D. P. 1997. Incidence of *Phoma terrestris*, causal agent of red root rot in corn roots and soil in Delaware and on the eastern shore of Maryland. Plant Dis. 81:232.
Sumner, D. R., Gascho, G. J., Johnson, A. W., Hook, J. E., and Threadgill, E. D. 1990. Root diseases, populations of soil fungi, and yield decline in continuous double-crop corn. Plant Dis. 74:704-710.
Watson, R. D. 1961. Rapid identification of the onion pink rot fungus. Plant Dis. Rep. 45:289.

(Prepared by R. B. Carroll)

Foliar Diseases

Although there are exceptions, many generalizations can be made about fungal leaf spots and blights of corn and their management. Many causal fungi overwinter in residues or debris from a previous diseased crop; therefore, cultural practices such as burying diseased crop residues with fall tillage and rotation to nonhost crops should substantially reduce the amount of primary inoculum. While these measures may afford some control, these diseases can and often do cause significant damage. Many of these pathogens are capable of spreading rapidly over a wide geographical area and can produce large amounts of secondary inoculum that, under ideal environmental conditions, can lead to a destructive epidemic. The primary means of managing the most important foliar diseases is the use of genetically resistant hybrids; however, the use of fungicides, particularly the triazoles and strobilurins, is gaining popularity in the United States, especially in intensively managed fields with a high yield potential. These products can provide excellent control of many foliar diseases if applied in a timely manner. The decision to apply a fungicide should be based on the probability of a economic loss to foliar diseases, which depends on a number of factors, including susceptibility of the hybrid, cropping history of the field, distribution and prevalence of the disease in the area, grain price, cost of application, and occurrence of weather conditions conducive for disease development.

Several foliar diseases of corn are caused by fungi that were formerly known as *Helminthosporium* spp. The so-called Helminthosporia fungi that cause diseases on corn subsequently have been placed into two distinct genera (*Bipolaris* and *Exserohilum*) that can be distinguished by characteristics of sexual and asexual reproductive stuctures. For example, conidia of *Exserohilum* (syn. *Setosphaeria*) spp. can be distinguished from those of *Bipolaris* (syn. *Cochliobolus*) spp. by their strongly protruding hila (Fig. 48).

Selected References

Alcorn, J. L. 1988. The taxonomy of "Helminthosporium" species. Annu. Rev. Phytopathol. 26:37-56.
Drechsler, C. 1923. Some graminicolous species of *Helminthosporium*. J. Agric. Res. 24:641-740.
Ellis, M. B., and Holliday, P. 1972. *Cochliobolus carbonum* (conidial state: *Drechslera zeicola*). Descriptions of Pathogenic Fungi and Bacteria, no. 349. Commonwealth Mycological Institute, Kew, England.
Leonard, K. J., Thakur, R. P., and Leath, S. 1988. Incidence of *Bipolaris* and *Exserohilum* species in corn leaves in North Carolina. Plant Dis. 72:1034-1038.

(Prepared by M. L. Carson)

Southern Corn Leaf Blight

Southern corn leaf blight is generally distributed worldwide in corn-producing areas. It is most serious in warm temperate and tropical areas. During most of its history in the United States, southern corn leaf blight was not considered to be a major problem. However, it reached epidemic proportions in 1970 when a previously undescribed race (race T), which is virulent on corn with Texas male-sterile cytoplasm (cms-T), became prevalent in North America. At the time, more than 80%

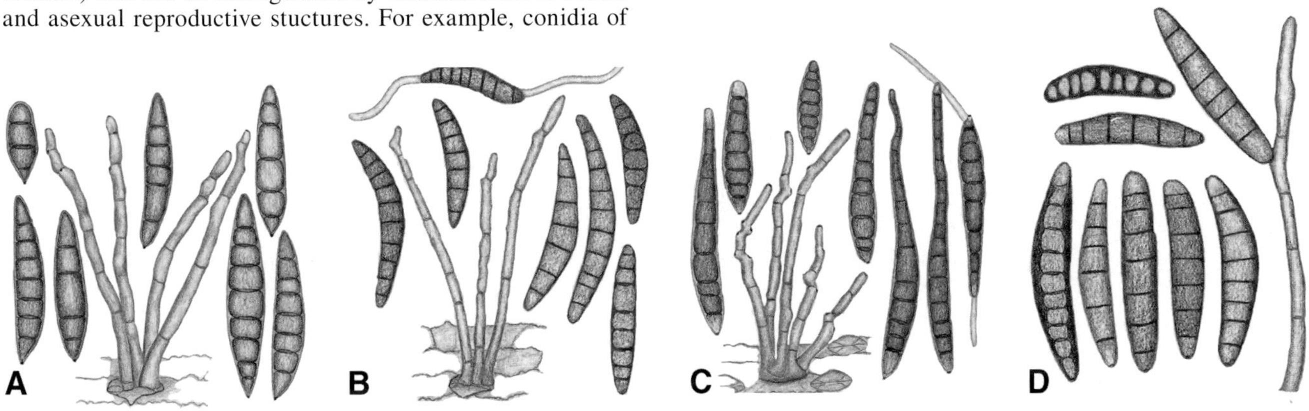

Fig. 48. Conidiophores and conidia of **A,** *Exserohilum turcicum* (syn. *Setosphaeria turcica*); **B,** *Bipolaris maydis* (syn. *Cochliobolus heterostrophus*); **C,** *Exserohilum rostratum* (syn. *Setosphaeria rostrata*); and **D,** *Bipolaris zeicola* (syn. *Cochliobolus carbonum*). (A–C, redrawn from Drechsler [1923]; and D, redrawn from Ellis and Holliday [1972])

of corn hybrids sold in the United States contained cms-T. The southern corn leaf blight epidemic of 1970 is possibly the most economically important occurrence of a disease on an annual field crop grown in the United States.

Symptoms

Southern corn leaf blight lesions vary in size and shape among inbred lines and hybrids with different genetic backgrounds. Lesions caused by different isolates of the fungus also vary (Figs. 49 and 50). Race O, which is the most common race, produces tan, elongated (2–6 × 3–22 mm) lesions between the veins and the lesions have limited parallel margins and possibly buff to brown borders. Race O usually attacks only the leaves, whereas race T can attack multiple parts of the plant. Lesions produced by race T on corn containing cms-T are tan and spindle shaped or elliptical (6–12 × 6–27 mm) and have margins that are initially water-soaked and then turn into yellow green or chlorotic halos. Race T lesions often have dark, reddish brown borders and may occur on all aboveground plant parts. Race T can also cause ear and cob rot (Fig. 51), resulting in substantial harvest and shelling losses. Either race can significantly reduce grain yields, primarily by decreasing kernel weight, when there is substantial loss of leaf tissue during the grain-filling period. Seedlings from kernels infected by race T may wilt and die within 3–4 weeks after planting. Severe blighting of the leaves caused by either race predisposes the plants to stalk rot. The two races are morphologically similar, but race T is specifically pathogenic to corn containing cms-T, while race O is not. Race C, specifically virulent on China male-sterile cytoplasm (cms-C) corn, has been reported in China. Race T produces a pathotoxin (Bmt toxin), a series of linear polyketides highly toxic only to susceptible cms-T corn. Race O does not produce this toxin.

Fig. 49. Southern corn leaf blight. (Courtesy G. P. Munkvold)

Fig. 50. Southern corn leaf blight. (Courtesy G. P. Munkvold)

Causal Organism

Southern corn leaf blight is caused by *Bipolaris maydis* (syns. *Cochliobolus heterostrophus*, *Drechslera maydis*, *Helminthosporium maydis*, and *Ophiobolus heterostrophus*). Conidia (15–20 × 70–160 µm) are typically olivaceous brown, curved, and spindle shaped, tapering to rounded ends. They contain five to eleven septa (Fig. 52) and exhibit bipolar germination. The fungus is heterothallic with two mating type alleles (MAT1-1 and MAT1-2) at a single locus. Pseudothecia are not known to occur in nature but can be readily produced in the laboratory. Four to eight ascospores are produced in cylindrical, hyaline asci. Ascospores (6–7 × 130–340 µm) are dark and have five to nine septa. Conidia can be obtained by placing infected tissue in a moist chamber for 24–48 h. Three physiological races of *B. maydis* have been reported: race O, the common race, which attacks primarily leaf tissue; race T, which is specifically virulent on corn containing cms-T and was the cause of the 1970 southern corn leaf blight epiphytotic in North America; and race C, which purportedly is specifically virulent on corn containing cms-C and thus far has been found only in China.

Disease Cycle and Epidemiology

The fungus overwinters as mycelium and conidia associated with infested corn debris from a previous crop. Conidia are disseminated by wind or splashing water to growing plants, where

Fig. 51. Ear rot caused by *Bipolaris maydis* (syn. *Cochliobolus heterostrophus*), race T. (Courtesy G. P. Munkvold)

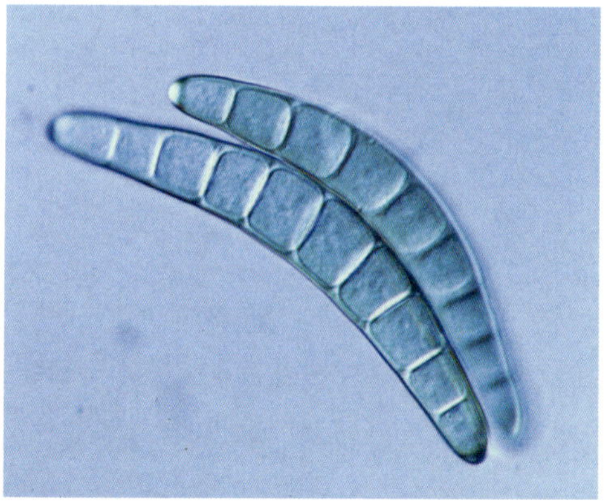

Fig. 52. *Bipolaris maydis* (syn. *Cochliobolus heterostrophus*) conidia. (Courtesy D. G. White)

primary infections occur. Conidia produced on lesions are secondary inocula. The disease cycle can be completed in as little as 60–72 h under ideal conditions.

Southern corn leaf blight occurs worldwide but is most important in regions with a warm (20–32°C), damp climate. The disease may occasionally be found in northern regions of the U.S. Corn Belt but usually is less damaging. Long, dry periods of sunny weather between rains are unfavorable for disease development.

Races O and T of *B. maydis* are difficult to identify by leaf symptoms alone on normal-cytoplasm corn. However, since the return to normal-cytoplasm corn after the 1970 epidemic caused by race T, the frequency of race T in North America has declined to undetectable levels because it is significantly less fit than race O on normal-cytoplasm hybrids.

Management

Although some benefit can be achieved by reducing the amount of primary inoculum through crop rotation and tilling under crop residues, genetic resistance has been the most effective means of reducing damage caused by southern corn leaf blight. There are two types of resistance, single gene and polygenic. A single recessive gene, *rhm*, located on the short arm of chromosome 6, was isolated from a composite population, Ibadan A from Nigeria, and has been incorporated into a few elite corn inbred lines. Resistance resulting from *rhm* is expressed on seedlings and young plants as a chlorotic fleck reaction. However, the resistance later breaks down and may not be effective on maturing plants. In the inbred line NC250, two recessive genes, neither of which is *rhm*, condition a high level of plant resistance in mature plants. High levels of polygenic or partial resistance are commonly found in much of the corn germplasm from tropical areas and the southern United States, and acceptable levels are also present in many hybrids sold in the Corn Belt of the central United States. Partial resistance is expressed on seedlings and young plants as reduced lesion size and length and as a reduction in disease development and severity on adult plants. Partial resistance in U.S. germplasm was found to be highly heritable; gene action is mostly additive, and relatively few loci are involved. Quantitative trait loci for resistance to southern corn leaf blight have been found on all 10 corn chromosomes. Both *rhm* and partial resistance are somewhat effective against race T of *B. maydis*. In addition to normal-cytoplasm corn, cms-C and cms-S types of corn are also resistant to race T.

Selected References

Balint-Kurti, P. J., and Carson, M. L. 2006. Analysis of quantitative trait loci for resistance to southern leaf blight in juvenile maize. Phytopathology 96:221-225.

Balint-Kurti, P. J., Krakowsky, M. D., Jines, M. P., Robertson, L. A., Molnár, T. L., Goodman, M. M., and Holland, J. B. 2006. Identification of quantitative trait loci for resistance to southern leaf blight and days to anthesis in a maize recombinant inbred line population. Phytopathology 96:1067-1071.

Balint-Kurti, P. J., Zwonitzer, J. C., Wisser, R. J., Carson, M. L., Oropeza-Rosas, M. A., Holland, J. B., and Szalma, S. J. 2007. Precise mapping of quantitative trait loci for resistance to southern leaf blight, caused by *Cochliobolus heterostrophus* race O, and flowering time using advanced intercross maize lines. Genetics 176:645-657.

Byrnes, K. J., Pataky, J. K., and White, D. G. 1989. Relationships between yield of three maize hybrids and severity of southern leaf blight caused by race O of *Bipolaris maydis*. Plant Dis. 73:834-840.

Carson, M. L. Stuber, C. W., and Senior, M. L. 2004. Identification and mapping of quantitative trait loci conditioning resistance to southern leaf blight of maize caused by *Cochliobolus heterostrophus* race O. Phytopathology 94:862-867.

Craig, J., and Fajemisin, J. M. 1969. Inheritance of chlorotic lesion resistance to *Helminthosporium maydis* in maize. Plant Dis. Rep. 53:742-743.

Fisher, D. E., Hooker, A. L., Lim, S. M., and Smith, D. R. 1976. Leaf infection and yield loss caused by four Helminthosporium leaf diseases of corn. Phytopathology 66:942-944.

Holley, R. N., and Goodman, M. M. 1989. New sources of resistance to southern corn leaf blight from tropical hybrid maize derivatives. Plant Dis. 73:562-564.

Hooker, A. L., Smith, D. R., Lim, S. M., and Musson, M. D. 1970. Physiological races of *Helminthosporium maydis* and disease resistance. Plant Dis. Rep. 54:1109-1110.

Klittich, C. J. R., and Bronson, C. R. 1986. Reduced fitness associated with *TOX*1 of *Cochliobolus heterostrophus*. Phytopathology 76:1294-1298.

Leonard, K. J. 1977. Races of *Bipolaris maydis* in the southeastern U.S. from 1974–1976. Plant Dis. Rep. 61:914-915.

Smith, D. R., and Hooker, A. L. 1973. Monogenic chlorotic-lesion resistance in corn to *Helminthosporium maydis*. Crop Sci. 13:330-331.

Smith, D. R., Hooker, A. L., and Lim, S. M. 1970. Physiologic races of *Helminthosporium maydis*. Plant Dis. Rep. 54:819-822.

Wei, J.-K., Liu, K.-M., Chen, J.-P., Luo, P.-C., and Lee-Stadelmann, O. Y. 1988. Pathological and physiological identification of race C of *Bipolaris maydis* in China. Phytopathology 78:550-554.

(Prepared by M. L. Carson)

Northern Corn Leaf Blight

Northern corn leaf blight is found in areas with humid climates wherever corn is grown. In the United States, the disease has demonstrated cycles of increasing and decreasing severity; it was very important during the 1940s, in 1951, and in localized areas of western North Carolina and the coastal area of Texas during the late 1980s and early 1990s. The disease can also be serious on sweet corn grown in Florida and on late-season processing sweet corn grown in the central United States. Northern corn leaf blight continues to be a significant disease in major corn-growing areas in the United States.

Symptoms

Early symptoms are pale gray-green, elliptical or cigar-shaped lesions that are initially less than 1 cm long (Fig. 53). These expand until they are 3–15 cm long (Fig. 54). As the lesions mature, they may become tan with distinct, dark zones of fungal sporulation (Fig. 55). The lesions may vary in morphology when genes for resistance, *Ht1*, *Ht2*, or *Ht3*, are present in the plant (Fig. 56). These genes condition long, chlorotic, linear streaks that on some host genotypes may coalesce and can be mistaken for lesions caused by the Stewart's bacterial wilt pathogen. The disease usually begins as a few scattered lesions on lower leaves and progresses to the upper leaves. The disease can develop very rapidly after anthesis, resulting

Fig. 53. Early symptoms of northern corn leaf blight. (Courtesy G. P. Munkvold)

in complete blighting of the leaves (Figs. 57–59). Late in the season, severely infected fields may appear to have been killed by an early frost. Often, several lesions occur in bands across leaves (Fig. 54) as a result of infection of tissues that formed the leaf whorl.

Causal Organism

Northern corn leaf blight is caused by *Exserohilum turcicum* (syns. *Bipolaris turcica*, *Drechslera turcica*, *Helminthosporium turcicum*, *Setosphaeria turcica*, and *Trichometasphaeria turcica*). Conidia (20 × 105 μm) are olive-gray, have three to eight septa, are spindle shaped with a protruding hilum (Fig. 60), and exhibit bipolar germination. Pseudothecia of the fungus have not been found in nature but can be produced in the laboratory. They are black and globose (350–500 × 360–720 μm). One to eight hyaline, fusoid ascospores with three septa are produced in each ascus. *E. turcicum* can also attack sorghum, Johnsongrass, and Sudan grass, but isolates from these hosts usually do not attack corn. At least seven physiological races of *E. turcicum* pathogenic on corn have been described. Races are named on the basis of the *Ht* genes to which they are virulent; e.g., race 0 is avirulent on all *Ht* genotypes, and race 23 is virulent on *Ht2* and *Ht3* genotypes.

Fig. 54. Northern corn leaf blight on a susceptible hybrid. (Courtesy D. G. White)

Fig. 55. Sporulation of *Exserohilum turcicum* (syn. *Setosphaeria turcica*) on leaf lesions. (Courtesy G. P. Munkvold)

Fig. 56. Chlorotic lesion resistance to northern corn leaf blight. (Courtesy D. G. White)

Disease Cycle and Epidemiology

E. turcicum overwinters as mycelia and conidia in and on leaf debris. Conidia can be carried by the wind over long distances. Secondary spread within and between fields occurs by conidia produced abundantly in leaf lesions on susceptible plants. Symptoms often progress from lower leaves to upper leaves as conidia produced in lesions on lower leaves serve as secondary inocula for the infection of upper leaves.

Northern corn leaf blight is found sporadically in most humid areas of the world where corn is grown but is most damaging when moderate temperatures (18–27°C) and prolonged periods of dew prevail during the growing season. If the disease is established before silking, losses in grain yield of up to 50% may

Fig. 57. Leaf blighting caused by severe northern corn leaf blight. (Courtesy G. P. Munkvold)

Fig. 58. Severe northern corn leaf blight on a susceptible hybrid (left), adjacent to a partially resistant hybrid (right). (Courtesy G. P. Munkvold)

occur. If infection is moderate or delayed until 6 weeks after silking, yield losses are minimal.

Management

Resistant hybrids of all maturity groups are available. There are two types of resistance to *E. turcicum* in corn. Polygenic resistance, or partial resistance, is not race specific and is quantitatively expressed as a reduction in lesion numbers, lesion size (Fig. 58), and sporulation and a lengthened latent period. Monogenic resistance is controlled by four well-characterized, single dominant genes, designated *Ht1*, *Ht2*, *Ht3*, and *HtN*. Other sources of monogenic resistance, such as *HtM1*, have been reported but are not as well characterized. The resistance conferred by *Ht1*, *Ht2*, and *Ht3* is expressed as chlorotic lesions that support only limited sporulation (Figs. 56 and 59). A single dominant suppressor gene found in inbred lines related to inbred line B14 suppresses the expression of *Ht2* and *Ht3* and has complicated the backcrossing of these genes into those inbred lines. Resistance conferred by *HtN* is expressed as a very long latent period and fewer lesions. Virulence to each of the *Ht* genes has been reported. Acceptable levels of polygenic, partial resistance to northern corn leaf blight have been found in a wide array of germplasm (Fig. 58). Quantitative trait loci for partial resistance have been found across much of the corn genome, depending on the source of resistance. Increased latent or incubation periods are components of partial resistance that seem to be the most consistently expressed on both seedlings and adult plants. Polygenic and monogenic resistance can act additively to reduce disease development. Where practical, fungicides may be applied, starting when lesions are first observed.

Fig. 59. Severe northern corn leaf blight on a susceptible hybrid (left), adjacent to a hybrid with major-gene resistance (right). (Courtesy G. P. Munkvold)

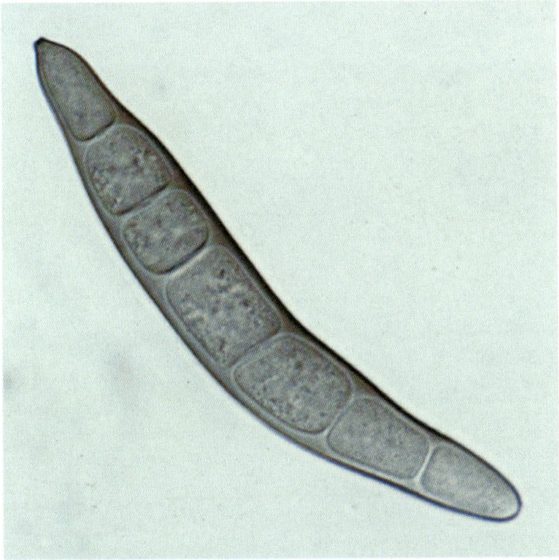

Fig. 60. *Exserohilum turcicum* (syn. *Setosphaeria turcica*) conidium with protruding hilum. (Courtesy G. P. Munkvold)

Selected References

Boosalis, M. G., Sumner, D. R., and Rao, A. S. 1967. Overwintering of conidia of *Helminthosporium turcicum* on corn residue and in soil in Nebraska. Phytopathology 57:990-996.

Carson, M. L. 1995. A new gene in maize conferring the "chlorotic halo" reaction to infection by *Exserohilum turcicum*. Plant Dis. 79:717-720.

Carson, M. L. 2006. Response of a maize synthetic to selection for components of partial resistance to *Exserohilum turcicum*. Plant Dis. 90:910-914.

Carson, M. L., and Van Dyke, C. G. 1994. Effect of light and temperature on expression of partial resistance of maize to *Exserohilum turcicum*. Plant Dis. 78:519-522.

Ceballos, H., and Gracen, V. E. 1989. A dominant inhibitor gene inhibits the expression of *Ht2* against *Exserohilum turcicum* race 2. Plant Breed. 102:35-44.

Gevers, H. O. 1975. A new major gene for resistance to *Helminthosporium turcicum* leaf blight of maize. Plant Dis. Rep. 59:296-299.

Hooker, A. L. 1963. Inheritance of chlorotic-lesion resistance to *Helminthosporium turcicum* in seedling corn. Crop Sci. 3:660-662.

Hooker, A. L. 1977. A second major gene locus in corn for chlorotic-lesion resistance to *Helminthosporium turcicum*. Crop Sci. 17:132-135.

Hooker, A. L. 1981. Resistance to *Helminthosporium turcicum* from *Tripsacum floridanum* incorporated into corn. Maize Genet. Coop. Newsl. 55:87-88.

Leonard, K. J., Levy, Y., and Smith, D. R. 1989. Proposed nomenclature for pathogenic races of *Exserohilum turcicum* on corn. Plant Dis. 73:776-777.

Ogliari, J. B., Guimaraes, M. A., Geraldi, I. O., and Camargo, L. E. A. 2005. New resistance genes in the *Zea mays–Exserohilum turcicum* pathosystem. Genet. Mol. Biol. 28:435-439.

Pataky, J. K., Perkins, J. M., and Leath, S. 1986. Effects of qualitative and quantitative resistance on the development and spread of northern leaf blight of maize caused by *Exserohilum turcicum* races 1 and 2. Phytopathology 76:1349-1352.

Pataky, J. K., Carson, M. L., and Mosely, P. R. 1991. Race 23N of *Exserohilum turcicum* in Florida. Plant Dis. 75:863.

Raymundo, A. D., and Hooker, A. L. 1982. Single and combined effects of monogenic and polygenic resistance on certain components of northern corn leaf blight development. Phytopathology 72:99-103.

Robbins, W. A., and Warren, H. L. 1993. Inheritance of resistance to *Exserohilum turcicum* in PI209135, "Mayorbella" variety of maize. Maydica 38:209-213.

Robert, A. L., and Findley, W. R., Jr. 1952. Diseased corn leaves as a source of infection in artificial and natural epidemics of *Helminthosporium turcicum*. Plant Dis. Rep. 36:9-10.

Thakur, R. P., Leonard, K. J., and Jones, R. K. 1989. Characterization of a new race of *Exserohilum turcicum* virulent on corn with resistance gene *Ht*N. Plant Dis. 73:151-155.

Welz, H. G., Wagner, R., and Geiger, H. H. 1993. Virulence variation in *Setosphaeria turcica* populations collected from maize in China, Mexico, Uganda, and Zambia. (Abstr.) Phytopathology 83:1356.

(Prepared by M. L. Carson)

Northern Corn Leaf Spot

Northern corn leaf spot is usually not a serious problem on corn hybrids but may be extremely important on inbred lines used in hybrid seed production. The disease can be found wherever corn is grown.

Symptoms

Symptoms differ among the five known races of the northern corn leaf spot pathogen. Lesions caused by race 1 are tan and oval to circular (1.2 × 2.5 cm), usually with concentric zones of sporulation (Figs. 61 and 62). Lesions can appear on leaves, leaf sheaths, and husks. Race 1 also attacks ears of susceptible plants, causing a black, felty mold over the kernels (Fig. 63). Race 2 produces oblong (up to 0.5 × 2.5 cm), chocolate-colored spots on maturing plants (Fig. 64). Ear infections caused by race 2 are indistinguishable from those caused by race 1 but are rare. Race 3 appears widely in the northern Corn Belt and portions of the southeastern United States. It produces narrow, linear (0.5–2 × 15–20 mm) lesions on the leaf blades, sheaths, and husks (Fig. 65). The lesion type varies with the genotype of the host and the isolate. The lesions are grayish tan, and each is surrounded by a light to darkly pigmented border. Symptoms caused by race 4 are similar to those produced by race 2, but race 4 lesions often have concentric zones of sporulation. Infected leaves are reported to develop a reddish brown coloration, suggesting the possible involvement of a toxin. Race 0 produces small flecks on immature leaves, is usually not found on mature plants, and is not considered to be important.

Fig. 61. Lesions caused by *Bipolaris zeicola* (syn. *Cochliobolus carbonum*), race 1. **A,** Blighted leaves, and **B,** lesions on husk and leaf sheaths. (Courtesy G. P. Munkvold)

Causal Organism

Northern corn leaf spot is caused by *Bipolaris zeicola* (syns. *Cochliobolus carbonum*, *Drechslera zeicola*, and *Helminthosporium carbonum*). Conidia (7–18 × 25–100 μm) are olive to dark brown and slightly curved with rounded ends, contain two to twelve septa (Fig. 66), and exhibit bipolar germination. The perithecial stage is not known to occur in nature but is readily produced in culture when the appropriate mating types

Fig. 63. Ear rot caused by *Bipolaris zeicola* (syn. *Cochliobolus carbonum*). (Courtesy K. Korus)

Fig. 62. Lesions caused by *Bipolaris zeicola* (syn. *Cochliobolus carbonum*), race 1. (Courtesy G. P. Munkvold)

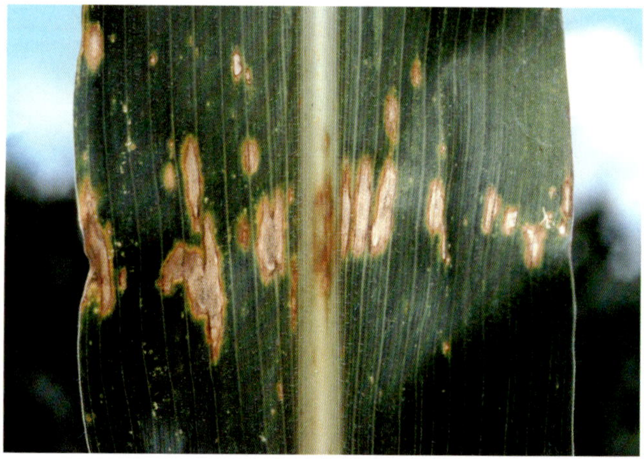

Fig. 64. Lesions caused by *Bipolaris zeicola* (syn. *Cochliobolus carbonum*), race 2. (Courtesy D. G. White)

are paired. Perithecia (320–430 × 355–550 µm) are black and ellipsoidal to globose with a well-defined ostiolar beak. Asci contain one to eight ascospores (6–10 × 180–300 µm), which are hyaline and filiform and contain five to nine septa (Fig. 66). Race 1 produces a cyclic tetrapeptide toxin that confers specific virulence to corn containing the *hm1* or *hm2* genes. Susceptibility to race 1 is relatively rare in most corn germplasm. Race 2 of *B. zeicola* is probably the most common race in North America and much of the world. Race 3 is distinguished by its ability to produce long, linear lesions on leaves of susceptible corn. Even though it produces larger lesions than race 2, it sporulates less on corn leaf tissue. Race 4 was described as responsible for damaging epidemics in seed production fields in the upper Midwest. It is considered a new race because of its apparent virulence on inbred lines previously considered resistant to race 2 and being less virulent on W64A, a race 2 susceptible inbred line. A fifth race, race 0, was reported from North Carolina and is distinguished by its low level of pathogenicity on corn, producing only small flecks or minute spots on leaves. Race 0 is also distinct from the other races in culture because it does not sporulate or grow well on potato dextrose agar. Because of its low pathogenicity on corn, race 0 is thought to have another grass species (as yet unidentified) as its primary host. Isozyme and genetic marker analyses of races of *B. zeicola* indicate that each race is distinct from the others, with the exception of races 2 and 4, which cannot be consistently separated.

Fig. 65. Lesions caused by *Bipolaris zeicola* (syn. *Cochliobolus carbonum*), race 3, on inbred line B73. (Courtesy D. G. White)

Disease Cycle and Epidemiology

B. zeicola overwinters as a saprophyte on corn residues and kernels. Even though *B. zeicola* is considered less parasitically fit than *B. maydis*, it apparently is a better saprophyte and is better able to survive on crop residues. It is commonly isolated from stalks as well as leaves. Typically, small lesions are found early in the growing season, but the pathogen does not sporulate abundantly until leaf tissue begins to mature. Conidia produced in lesions serve as secondary inoculum and are disseminated by the wind. Race 3 appears to be adapted to the cool, humid conditions of the northern Corn Belt and the Appalachian highlands of the United States, while races 1 and 2 are more broadly adapted and may be found in the warm, humid climates in the Corn Belt and lower elevations of the southeastern United States.

Moderate temperatures and high relative humidity favor disease development. Spores are produced abundantly on older lesions on mature leaves during damp weather. Northern corn leaf spot is generally considered to be of minor importance to corn grain production in the United States. Because susceptibility to race 1 is in only a few inbred lines that are no longer in commercial use, the frequency of race 1 has declined to very low levels. Damage from races 2, 3, and 4 to highly susceptible inbred lines in hybrid seed production fields can be substantial when weather is highly favorable for disease development. Significant outbreaks of the disease were found in the northern Corn Belt of the United States in 1973 (race 2) and in 1989 (race 4) when there was unusually high rainfall late in the summer.

Management

Because hybrid corn is seldom damaged by northern corn leaf spot, control is usually of concern only on inbred lines used in hybrid seed production. Susceptibility to race 1 is conditioned by two recessive genes on different chromosomes. Susceptibility to race 1 is found in the inbred lines Pr, K61, K44, Mo21A, N31, NC37, GA203, and GA209 and has also been found to be segregating in the Iowa Stiff Stalk Synthetic population, a source of many elite inbred lines in the United States. Resistance to race 2 is widely available; only a few inbred lines, such as W64A, are highly susceptible. Resistance to race 3 is apparently under polygenic control, and the gene action involved is mostly additive. Inbred lines related to the popular public inbred line B73 appear to be most susceptible to race 4. Race 0 is thought to have an unknown species of grass as its primary host and is unimportant on corn.

Fig. 66. *Bipolaris zeicola* (syn. *Cochliobolus carbonum*). Top row: Perithecia, asci, and ascospores. Bottom row: Conidiophores and conidia. (Courtesy P. Crous)

Selected References

Dodd, J. L., and Hooker, A. L. 1990. Previously undescribed pathotype of *Bipolaris zeicola* on corn. Plant Dis. 74:530.

Ellis, M. B., and Holliday, P. 1972. *Cochliobolus carbonum* (conidial state: *Drechslera zeicola*). Descriptions of Pathogenic Fungi and Bacteria, no. 349. Commonwealth Mycological Institute, Kew, England.

Hamid, A. H., Ayers, J. E., and Hill, R. R., Jr. 1982. The inheritance of resistance in corn to *Cochliobolus carbonum* race 3. Phytopathology 72:1173-1177.

Hooker, A. L., Mesterhazy, A., Smith, D. R., and Lim, S. M. 1973. A new Helminthosporium leaf blight of corn in the northern corn belt. Plant Dis. Rep. 57:195-198.

Jones, M. J., and Dunkle, L. D. 1993. Analysis of *Cochliobolus carbonum* races by PCR amplification with arbitrary and gene-specific primers. Phytopathology 83:366-370.

Nelson, R. R., Blanco, M., Dalmacio, S., and Moore, B. S. 1973. A new race of *Helminthosporium carbonum* on corn. Plant Dis. Rep. 57:822-823.

Pringle, R. B., and Scheffer, R. P. 1967. Isolation of the host-specific toxin and a related substance with nonspecific toxicity from *Helminthosporium carbonum*. Phytopathology 57:1169-1172.

Smith, D. R., Kinsey, J. G., and Kaufmann, K. D. 1985. Segregation of *Helminthosporium carbonum* race 1 susceptibility in the corn synthetic BS13. Plant Dis. 69:1101.

Welz, H. G., and Leonard, K. J. 1993. Phenotypic variation and parasitic fitness in races of *Cochliobolus carbonum* on corn in North Carolina. Phytopathology 83:593-601.

(Prepared by M. L. Carson)

Rostratum Leaf Spot

Rostratum leaf spot is occasionally found on mature corn leaves, particularly in the southeastern United States, but it is not considered to be a serious problem of corn.

Symptoms

Lesions appear on mature plants as small, pale yellow, elongated spots (1–2 × 2–5 mm) that gradually extend to form longitudinal bands or stripes between the veins (Fig. 67). In severe cases, the lesions coalesce and extend across the veins. Mature lesions (2–3 × 2–40 mm) are straw colored with light brown margins.

Causal Organism

Rostratum leaf spot is caused by *Exserohilum rostratum* (syns. *Bipolaris rostrata*, *Drechslera rostrata*, *Exserohilum halodes*, *Helminthosporium rostratum*, and *Setosphaeria rostrata*). Conidia (14–22 × 32–184 μm) are dark, olivaceous, and straight or rarely curved and taper toward both ends. They have a prominent, dark, and protruding hilum and three to fifteen septa. The proximal and distal septa are often thicker and more prominent than the others.

Fig. 67. Rostratum leaf spot, caused by *Exserohilum rostratum* (syn. *Setosphaeria rostrata*). (Courtesy D. G. White)

Disease Cycle and Epidemiology

The disease cycle is similar to those of southern corn leaf blight and northern corn leaf blight (see these disease sections above). Infection is most severe at 30°C in a water-saturated atmosphere. Older tissue is more susceptible than young. *E. rostratum* also attacks leaves of love grasses, sorghum, Sudan grass, pearl millet, and several other grass hosts.

Management

Differences in the resistance of commercial corn hybrids has been reported, but no information on the nature or inheritance of this resistance is available.

Selected References

Leonard, K. J. 1976. Synonymy of *Exserohilum halodes* with *E. rostratum*, and induction of the ascigerous state, *Setosphaeria rostrata*. Mycologia 67:402-411.

Whitehead, M. D., and Calvert, O. H. 1959. *Helminthosporium rostratum* inciting ear rot of corn and leaf-spot of thirteen grass hosts. Phytopathology 49:817-820.

(Prepared by M. L. Carson)

Physoderma Brown Spot

Physoderma brown spot or Physoderma disease is generally of minor importance in the United States, although serious localized outbreaks have occasionally been reported. Severe infection occurs in Central America. The disease is most common in areas of abundant rainfall and high mean temperatures in conjunction with reduced tillage practices where corn is planted into crop residues of a previous diseased corn crop.

Symptoms

Lesions first appear as very small, oblong to round, yellowish spots on the leaf blade (Figs. 68 and 69), leaf sheath, stalk, and sometimes on the outer ear husks and tassels. Lesions may occur in bands across the leaf blade. Infected tissues turn chocolate brown to reddish brown and coalesce to form large, irregular, angular blotches. Cells of infected tissues disintegrate and expose pustules containing large numbers of golden brown to dark brown sporangia. Stalks infected at the nodes beneath the leaf sheaths frequently break (Fig. 70).

Causal Organism

Physoderma brown spot is caused by *Physoderma maydis* (syn. *P. zeae-maydis*). *P. maydis* is a primitive true fungus belonging to the phylum Blastocladiomycota. Sporangia (18–24 × 20–30 μm) produced in pustules in infected tissue are smooth, brown, and flattened on one side and have a circular operculum (Fig. 71). Upon germination, which occurs only in light, a sporangium releases 20–50 zoospores (3–4 × 5–7 μm), each of which has a single polar flagellum that is three to four times the length of the spore. The zoospores germinate by producing fine, fibrous hyphae. The mycelium is coenocytic.

Disease Cycle and Epidemiology

Thick-walled sporangia overwinter in infested tissue or in soil and germinate in the presence of moisture and light to produce numerous zoospores. Zoospores swim in water on leaf surfaces, attack young corn tissues (usually within the leaf

Fig. 68. Physoderma brown spot. (Courtesy A. E. Robertson)

Fig. 69. Physoderma brown spot. (Courtesy A. E. Robertson)

whorl), and form infectious hyphae. The resulting mycelium enters mesophyll or parenchyma cells, forming larger vegetative structures from which sporangia develop 16–20 days after infection. Spread by secondary zoospores produced in thin-walled sporangia may be evident only during very wet weather.

Infection is generally in meristematic tissues in the leaf whorl, where water may be present for extended periods of time. Plants are most susceptible 50–60 days after germination and become increasingly resistant with age. Zoospores of *P. maydis* can infect host tissue only on a diurnal cycle with the proper combination of free water in the leaf whorl, light, and high temperatures (23–30°C), often resulting in alternating lateral bands of infected and noninfected leaf tissues. Sporangia require water on leaf surfaces and relatively high temperatures (23–30°C) for germination. Teosinte is also susceptible to *P. maydis*.

Fig. 70. Stalk breakage caused by *Physoderma maydis*. (Courtesy D. Mueller)

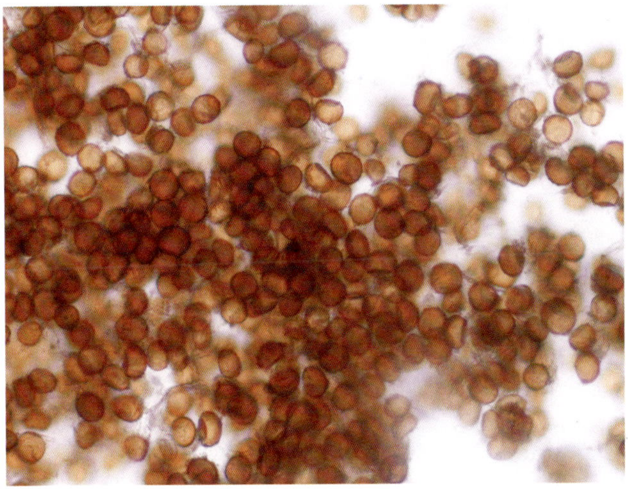

Fig. 71. *Physoderma maydis* sporangia in diseased leaf tissue. (Courtesy E. Saalau-Rojas)

Management

Resistance is available in germplasm adapted to areas where the disease is a serious problem. Resistance is inherited as a polygenic trait with mostly additive gene action.

Fall plowing of diseased crop debris buries most of the inoculum, making it unavailable for dissemination to the leaf whorl.

Selected References

Broyles, J. W. 1962. Penetration of meristematic tissues of corn by *Physoderma maydis*. Phytopathology 52:1013-1016.

Burns, E. E., and Shurtleff, M. C. 1973. Observations of *Physoderma maydis* in Illinois: Effects of tillage practices in field corn. Plant Dis. Rep. 57:630-633.

Harvey, P. H., Thompson, D. L., and Hebert, T. T. 1955. Reaction of inbred lines of corn to brown spot. Plant Dis. Rep. 39:973-976.

Hebert, T. T., and Kelman, A. 1958. Factors influencing the germination of resting sporangia of *Physoderma maydis*. Phytopathology 48:102-106.

Sparrow, F. K. 1974. Observations on chytridiaceous parasites of phanerogams. XXI. The endobiotic stage of *Physoderma maydis*. Mycologia 66:693-697.

Thompson, D. L. 1969. Quantitative genetic estimates for brown spot resistance in corn. Crop Sci. 9:246-247.

(Prepared by M. L. Carson)

Fig. 72. Eyespot lesions with tan (**A**) or reddish brown (**B**) borders. (Courtesy G. P. Munkvold)

Fig. 73. Eyespot lesions with backlighting. (Courtesy G. P. Munkvold)

Eyespot

Reported as "brown-spot" in Japan in 1959, eyespot has since been found in several north central and northeastern states in the United States and in Canada, Argentina, Brazil, northern and eastern Europe, and New Zealand. Yield losses caused by eyespot can be quite severe (50%) when susceptible hybrids are artificially inoculated. Average losses resulting from natural infection are estimated to be as great as 9% on susceptible hybrids. In addition to direct losses, severe eyespot predisposes corn to stalk rots.

Symptoms

Early symptoms consist of small, translucent, circular to oval lesions (1–4 mm in diameter) with yellowish halos (Fig. 72A). Initially, a lesion is water-soaked and later develops a tan to cream center surrounded by a brown or purple ring with a narrow, yellow halo, giving the appearance of an eyespot (Fig. 72B). The halo is easily observed when diseased leaves are lit from behind (Fig. 73). The spots usually develop in zones or patches, early or late in the season. Lesions may fuse to form large necrotic areas (Fig. 74). Lesions may develop on the leaf sheaths and outer husks but are usually found in much larger numbers on the leaves of maturing plants. The foliar symptoms may be confused with those of Curvularia leaf spot or physiological or genetic leaf spots.

Fig. 74. Leaf blighting caused by eyespot. (Courtesy G. P. Munkvold)

Causal Organism

Eyespot is caused by *Kabatiella zeae* (syn. *Aureobasidium zeae*). Conidia (3–4 × 18–33 μm, average 3.6 × 27 μm) are long, curved to falcate with pointed ends, hyaline, and nonseptate. They are produced successively on short, simple conidiophores. Sporulation can be induced by placing diseased tissue in a humidity chamber for 24–48 h. On potato dextrose agar at 24°C, growth is slow (2–3 mm per day). Some colonies are yellow or pink and turn dark blue to black with age, and the growth is tough and leathery.

Disease Cycle and Epidemiology

K. zeae overwinters in corn debris. Conidia produced in the debris are disseminated by splashing rain to nearby seedlings, where they may germinate and infect. Lesions appear 4–10 days after infection. Secondary spread is primarily by splashing of the conidia. Only corn has been reported as a host for *K. zeae*. Disease development is favored by cool, humid weather and is usually most severe when corn is grown continuously under conservation tillage conditions.

Management

Resistant hybrids are available in the United States where the disease has been a problem. The inbred lines W64A, W117, and Wf9 are susceptible, while Oh43, B37, and C123 have historically been used as sources of resistance. Resistance appears to be quantitatively inherited with mostly additive to partially dominant gene action involved and is highly heritable. As few as two genes may be involved. Quantitative trait loci have been found on all corn chromosomes except chromosome 9. There is a significant negative correlation between maturity (days to flowering) and resistance to eyespot. Tillage and crop rotation both reduce the level of early infection.

Selected References

Arny, D. C., Smalley, E. B., Ullstrup, A. J., Worf, G. L., and Ahrens, R. W. 1971. Eyespot of maize, a disease new to North America. Phytopathology 61:54-57.

Chiang, M. S., Hudon, M., and Deraux, A. 1990. Inheritance of resistance to Kabatiella eyespot of maize. Phytoprotection 71:107-112.

Chinchilla, C. M. 1987. Survival of *Kabatiella zeae* Narita and Hiratsuka in maize residues. Turrialba 37:311-316.

dos Santos, I., da Silva, A., and Malaga, G. 2007. Occurrence of maize eyespot caused by *Kabatiella zeae* in Parana and Santa Catarina state, Brazil. Fitopathol. Bras. 32:359.
Melani, M. D. 2006. Quantitative trait loci for resistance to eyespot in maize. Ph.D. thesis. North Dakota State University, Fargo.
Pronczuk, M., Bojanowski, J., and Warzecha, R. 2004. Effect of leaf infection by *Kabatiella zeae* on stalk rot prevalence and grain yield of maize hybrids. J. Phytopathol. 152:410-415.
Reifschneider, F. J. B., and Arny, D. C. 1980. Host range of *Kabatiella zeae*, causal agent of eyespot of maize. Phytopathology 70:485-487.
Reifschneider, F. J. B., and Arny, D. C. 1983. Inheritance of resistance of maize to *Kabatiella zeae*. Crop Sci. 23:615-616.
Reifschneider, F. J. B., and Arny, D. C. 1983. Yield loss of maize caused by *Kabatiella zeae*. Phytopathology 73:607-609.

(Prepared by M. L. Carson)

Anthracnose Leaf Blight

Anthracnose leaf blight has been considered important in the United States since the early 1970s, although the disease was described in the 1930s. Yield losses caused by anthracnose leaf blight of up to 19 and 37% have been observed on hybrids and inbred lines, respectively. The disease is also considered important in Europe, Southeast Asia, and South America.

Symptoms

Symptoms vary greatly, depending on the genotype, age of the leaf, and environment. Small, oval to elongated, water-soaked spots appear on the leaves at any stage of growth. The lesions are semitransparent at first, and then chlorotic. The spots may enlarge to up to 15 mm long and become tan at the center with red, reddish brown, or chocolate brown borders, often surrounded by a yellow halo (Figs. 75 and 76). Lesions may originate on any part of the leaf blade, including the leaf margins (Fig. 77). Enlarging lesions may coalesce, blighting the entire leaf. Leaf symptoms are most common early in the season on the lower leaves and late in the season on the upper leaves. Severely diseased leaves on very susceptible inbred lines or hybrids wither and die within a few days late in the season. Fruiting bodies (acervuli) develop in abundance on dead tissues (Fig. 78) and can be a distinguishing sign, when they are present. Under conditions of high humidity and moderate temperatures, they produce pinkish to orange masses of conidia. Dark, hairlike structures (setae) up to 175 µm long rise above the surface of the lesion and are easily observed with a hand lens (Fig. 79), allowing a preliminary field diagnosis of the disease. On resistant plants, the lesions are usually smaller and are chlorotic to necrotic. The fungus may also cause severe top dieback and stalk rot, which are considered to be more important causes of yield loss.

Fig. 75. Lesions of anthracnose leaf blight showing the typical chlorotic halo and acervuli. (Courtesy B. M. Anderson)

Causal Organism

Anthracnose leaf blight is caused by *Colletotrichum graminicola* (syn. *Glomerella graminicola*). *C. graminicola* on corn is

Fig. 76. Anthracnose leaf blight. (Courtesy G. P. Munkvold)

Fig. 77. Anthracnose leaf blight lesions along the leaf margins. (Courtesy G. P. Munkvold)

Fig. 78. Acervuli and conidial masses in lesions of anthracnose leaf blight. (Courtesy G. P. Munkvold)

Fig. 79. Setae of *Colletotrichum graminicola* acervuli. (Courtesy G. P. Munkvold)

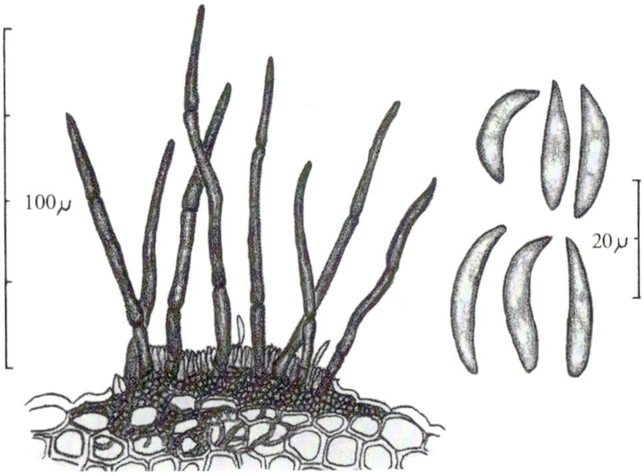

Fig. 80. Transverse section of an acervulus (left) and conidia (right) of *Colletotrichum graminicola*.

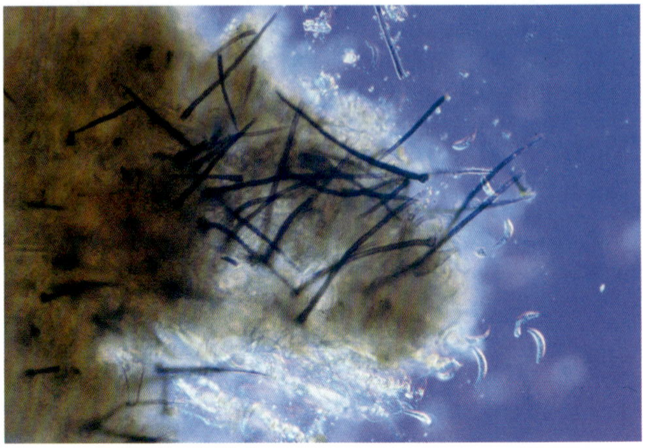

Fig. 81. Setae and conidia of *Colletotrichum graminicola*. (Courtesy G. P. Munkvold)

Fig. 82. Conidia of *Colletotrichum graminicola*. (Courtesy G. P. Munkvold)

a distinct species from *Colletotrichum* spp. on other grasses. Acervuli produced on agar and in dead host tissues are dark, circular to oval, and thick. Under moist conditions, they appear as small specks scattered in infected tissues (Fig. 78). Short conidiophores (1.6–3.3 × 4.9–13.3 μm) produced in the acervuli are erect, hyaline, nonseptate, and unbranched. Long, dark brown to black, septate setae are produced among the conidiophores (Figs. 80 and 81). Conidia (4.9–5.2 × 26.1–30.8 μm) are hyaline, nonseptate, and cylindrical, becoming sickle shaped with age (Fig. 82), and are borne singly at the tips of the conidiophores. Although isolates from corn may differ in their relative aggressiveness, there is no evidence of physiological races. The causal organism for sorghum anthracnose (*C. sublineolum*) has also been known as *C. graminicola*, but these organisms are now considered distinct species with restricted host ranges.

Disease Cycle and Epidemiology

The fungus overwinters as a saprophyte on corn residue. Spores produced in acervuli associated with debris are carried by wind or splashing rain to the leaves, where penetration takes place directly through the epidermis or through stomata. The fungus can persist in buried residues, although this is not an important source of inoculum for the leaf blight phase. Very rarely, seedborne inoculum causes seedling blight.

Disease development is generally favored by high temperatures and extended periods of cloudy, wet weather. Seedlings and mature plants are most susceptible to anthracnose leaf blight; plants undergoing rapid vegetative growth are less susceptible. Free water is necessary for spore dissemination and germination. Conidia are produced in a mucilaginous matrix, which dries and forms a protective coat around them, allowing survival and preventing germination under unfavorable conditions.

Management

Resistant hybrids are available in all areas where the disease may be a problem. Resistance to anthracnose leaf blight is common in many adapted inbred lines, is highly heritable, and may be conferred by a single dominant gene in some inbred lines. Resistance to anthracnose leaf blight is not very highly correlated with resistance to anthracnose stalk rot. Therefore, it is necessary to be aware of which phase of the disease (leaf blight or stalk rot) is being referred to when a hybrid is described as anthracnose resistant.

Crop rotation and tillage of crop residues may reduce the level of early-season infections but may have little value in reducing the severity of late-season anthracnose leaf blight and stalk rot.

Selected References

Badu-Apraku, B., Gracen, V. E., and Bergstrom, G. C. 1987. A major gene for resistance to anthracnose leaf blight in maize. Plant Breed. 98:194-199.

Bergstrom, G. C., and Nicholson, R. L. 1999. The biology of corn anthracnose: Knowledge to exploit for improved management. Plant Dis. 83:596-608.

Carson, M. L., and Hooker, A. L. 1981. Inheritance of resistance to anthracnose leaf blight in five inbred lines of corn. Phytopathology 71:488-491.

Crouch, J. A., Clarke, B. B., and Hillman, B. J. 2006. Unraveling evolutionary relationships among the divergent lineages of *Colle-*

totrichum causing anthracnose disease in turfgrass and corn. Phytopathology 96:46-60.

Lipps, P. E. 1985. Influence of inoculum from buried and surface corn residues on the incidence of corn anthracnose. Phytopathology 75:1212-1216.

Nicholson, R. L., and Moraes, W. B. C. 1980. Survival of *Colletotrichum graminicola*: Importance of the spore matrix. Phytopathology 70:255-261.

Nicholson, R. L., and Warren, H. L. 1981. The issue of races of *Colletotrichum graminicola* pathogenic to corn. Plant Dis. 65:143-145.

Politis, D. J., and Wheeler, H. 1972. The perfect stage of *Colletotrichum graminicola*. Plant Dis. Rep. 56:1026-1027.

Smith, D. R. 1976. Yield reduction in dent corn caused by *Colletotrichum graminicola*. Plant Dis. Rep. 60:967-970.

White, D. G., Yanney, J., and Anderson, B. 1987. Variation in pathogenicity, virulence, and aggressiveness of *Colletotrichum graminicola* on corn. Phytopathology 77:999-1001.

(Prepared by M. L. Carson)

Gray Leaf Spot

Gray leaf spot was initially described in Illinois in 1925, but until the 1980s, it was considered a problem only in the eastern United States, particularly in the mountains. The disease has since become well established and, in many cases, has caused damage as far west as Nebraska in the U.S. Corn Belt and has become the subject of increased interest and concern. Gray leaf spot also is found in the temperate to warm, humid areas of Mexico, Central America, northern South America, Europe, Africa, Southeast Asia, India, China, Australia, and the Philippines. Although the disease is often not a problem until some time after anthesis, it can develop rapidly, completely blighting leaves and causing the premature death of plants. Prolonged periods of dew, fog, and cloudy weather are conducive to severe gray leaf spot epidemics.

Symptoms

Gray leaf spot causes rectangular lesions (approximately 2–4 mm × 1–6 cm) on leaves of susceptible plants (Fig. 83). Early lesions (Figs. 84 and 85) are small, necrotic spots with chlorotic halos that gradually expand to full-sized lesions, which can blight entire leaves (Fig. 86). Expanding lesions are tan and turn gray as the pathogen, *Cercospora zeae-maydis* or *C. zeina*, begins to sporulate on the underside of the leaf. Symptoms

Fig. 83. Rectangular lesions of gray leaf spot, typical for susceptible genotypes such as inbred line B73. (Courtesy G. P. Munkvold)

Fig. 84. Early lesions of gray leaf spot on susceptible hybrid B73 × Mo17. (Courtesy D. G. White)

Fig. 85. Early lesions of gray leaf spot. (Courtesy G. P. Munkvold)

Fig. 86. Severe leaf blighting caused by gray leaf spot. (Courtesy G. P. Munkvold)

can vary among host genotypes; in some cases, lesions have a dark border or yellow halo (Figs. 87 and 88). The diagnostic features of gray leaf spot lesions are the sharp, parallel edges and the opacity of mature lesions. The sharp edges of the lesions result from the inability of the fungus to penetrate sclerenchyma tissue in the major veins of the leaf, which limits lateral expansion of lesions. The opacity of lesions is caused by stromatic tissue, i.e., dark, hardened mycelium produced by the fungus, in the substomatal cavities. It is possible to diagnose gray leaf spot when host tissue is dead by the opacity of the lesions (Fig. 89).

Fig. 87. Gray leaf spot lesions showing a dark border on some hybrids. (Courtesy G. P. Munkvold)

Fig. 88. Gray leaf spot lesions showing a yellow halo on some hybrids. (Courtesy G. P. Munkvold)

Fig. 89. Gray leaf spot lesions visible in dead leaf tissue. (Courtesy G. P. Munkvold)

Causal Organisms

At least two sibling species of the genus *Cercospora*, *C. zeae-maydis* and *C. zeina*, are most commonly associated with gray leaf spot. *C. sorghi* var. *maydis* has also been reported to be associated with gray leaf spot, but its pathogenicity on corn has not been demonstrated. Conidia of *C. zeae-maydis* are straight to slightly curved, hyaline, and slender (4–9 × 30–100 µm) and contain one to ten septa (Figs. 90 and 91). Conidiophores emerge in fascicles (3–14) from stomata and are 4–8 × 40–180 µm. The perfect stage has been found in overwintered corn tissue, but it is thought to be quite rare and not an important source of in-

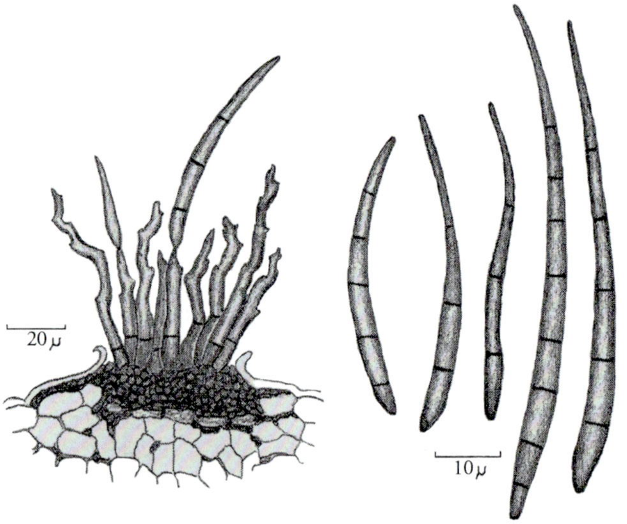

Fig. 90. Transverse section of conidiophores (left) and conidia (right) of *Cercospora zeae-maydis*.

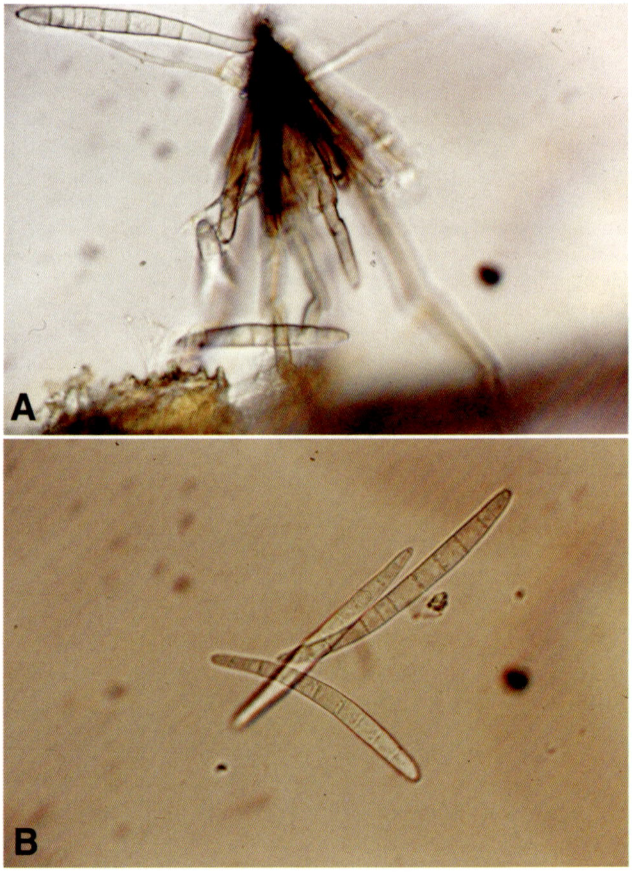

Fig. 91. *Cercospora zeae-maydis.* **A**, Conidiophores, and **B**, conidia. (Courtesy D. G. White)

Fig. 92. Gray leaf spot lesions on a resistant hybrid. (Courtesy G. P. Munkvold)

Fig. 93. Gray leaf spot lesions on a susceptible hybrid. (Courtesy G. P. Munkvold)

oculum. Conidia of *C. zeina* have dimensions similar to those of *C. zeae-maydis*, but they can be distinguished by shorter conidiophores (40–100 µm) and more broadly fusiform conidia. Also, colonies of *C. zeina* grow slower in culture than those of *C. zeae-maydis* and do not produce the red pigment associated with cercosporin production. In the United States., *C. zeae-maydis* is widely distributed while *C. zeina* is confined to the eastern states (Ohio and eastward). Corn is the only known host. Isolates of both fungi vary in aggressiveness but no physiological specialization by the two species or as races within the species has been observed.

Disease Cycle and Epidemiology

Severe gray leaf spot usually is found where corn follows corn and residue of a previous diseased crop remains on the soil surface. *C. zeae-maydis* and *C. zeina* overwinter as stromatic tissue in crop debris. In the spring, conidiophores and conidia are produced from stroma. In North Carolina (United States), conidia have been trapped over crop debris as early as May, but abundant spore production usually begins in June. Conidia are disseminated by wind or in splashing water. After landing on the leaf surface, conidia may germinate, producing germ tubes that grow toward stomata by detecting humidity gradients. Germ tubes produce appressoria over stomata, and penetration pegs penetrate the leaf through the stomata. Direct penetration of the leaves has not been reported. Penetration of host tissue may take up to a week depending upon environmental conditions. Germinated conidia of *C. zeae-maydis* can survive at 60% relative humidity on the leaf surface but require relative humidity of 95% or higher for germ tube growth and appressorium formation. Because of the prolonged incubation and latent period, secondary inoculum is produced late in the growing season and may not be as important in disease epidemiology as primary inoculum coming from crop debris. The increased incidence and severity of this disease in the midwestern Corn Belt and the eastern United States in the 1980s and 1990s has been attributed to the continuous culture of corn in conjunction with minimum tillage practices and the use of susceptible hybrids.

Management

Hybrids differ greatly in their susceptibility (Figs. 92 and 93), but high levels of resistance are not yet available for all hybrid maturities. Resistant hybrids are available and widely grown in the southern Appalachian region and the eastern and central Corn Belt, where the disease can be most damaging. Resistance to gray leaf spot appears to be polygenically controlled with mostly additive gene action. Quantitative trait loci for resistance have been found on all 10 chromosomes of corn. Public inbred lines reported to be resistant include B68, NC250, NC258, Pa875, Va14, Va17, Va85, T222, and Mo18W. Inbred lines derived from the Iowa Stiff Stalk Synthetic population are particularly susceptible, but variability for resistance has been found within this important germplasm. Very high levels of resistance have been found in corn inbred lines derived from southern and tropical germplasm. Conventional tillage with the fall burial of crop residues and crop rotation reduces the amount of primary inoculum but may not offer adequate control in areas where the pathogen is well established and conditions are favorable for disease development. Foliar fungicides provide good control and are economically justified for some high-yielding susceptible hybrids.

Selected References

Bergquist, R. R., Payne, G. A., Thompson, D. L., and Bowman, D. T. 1985. Inheritance estimates of resistance in maize (*Zea mays*) to gray leaf spot (*Cercospora zeae-maydis*). (Abstr.) Phytopathology 75:1310.

Bubeck, D. M., Goodman, M. M., Beavis, W. D., and Grand, D. 1993. Quantitative trait loci controlling resistance to gray leaf spot in maize. Crop Sci. 33:838-847.

Carson, M. L., and Goodman, M. M. 2006. Pathogenicity, aggressiveness, and virulence of three species of *Cercospora* associated with gray leaf spot of maize. Maydica 51:89-92.

Carson, M. L., Goodman, M. M., and Williamson, S. M. 2002. Variation in aggressiveness among isolates of *Cercospora* from maize as a potential cause of genotype-environment interaction in gray leaf spot trials. Plant Dis. 86:1089-1093.

Clements, M. J., Dudley, J. W., and White, D. G. 2000. Quantitative trait loci associated with resistance to gray leaf spot of corn. Phytopathology 90:1018-1025.

Coates, S. T., and White, D. G. 1998. Inheritance of resistance to gray leaf spot in crosses involving selected resistant inbred lines of corn. Phytopathology 88:972-982.

Crous, P., Groenewald, J. Z., Groenewald, M., Caldwell, P., Braun, U., and Harrington, T. C. 2006. Species of *Cercospora* associated with grey leaf spot of maize. Stud. Mycol. 55:189-197.

Donahue, P. J., Stromberg, E. L., and Myers, S. L. 1991. Inheritance of reaction to gray leaf spot in a diallel cross of 14 maize inbreds. Crop Sci. 31:926-931.

Elwanger, G. F., Johnson, M. W., Hill, R. R., and Ayers, J. E. 1990. Inheritance of resistance to gray leaf spot of corn. Crop Sci. 30:350-358.

Gevers, H. O., Lake, J. K., and Hohls, T. 1994. Diallel cross analysis of resistance to gray leaf spot in maize. Plant Dis. 78:379-383.

Graham, M. J., Hawk, J. A., Carroll, R. B., Ayers, J. E., Lamkey, K. R., and Hallauer, A. R. 1993. Evaluation of Iowa Stiff Stalk Synthetic for resistance to gray leaf spot. Plant Dis. 77:382-385.

Payne, G. A., and Waldron, J. K. 1983. Overwintering and spore release of *Cercospora zeae-maydis* in corn debris in North Carolina. Plant Dis. 67:87-89.
Rupe, J. C., Siegel, M. R., and Hartman, J. R. 1982. Influence of environment and plant maturity on gray leaf spot of corn caused by *Cercospora zeae-maydis*. Phytopathology 72:1587-1591.
Saghai Maroof, M. A., Yue, Y. G., Xiang, Z. X., Stromberg, E. L., and Rufener, G. K. 1996. Identification of quantitative trait loci controlling resistance to gray leaf spot disease in maize. Theor. Appl. Genet. 93:539-546.
Thompson, D. L., Bergquist, R. R., Payne, G. A., Bowman, D. T., and Goodman, M. M. 1987. Inheritance of resistance to gray leaf spot in maize. Crop Sci. 27:243-246.
Ulrich, J. F., Hawk, J. A., and Carroll, R. B. 1990. Diallel analysis of maize inbreds for resistance to gray leaf spot. Crop Sci. 30:1198-1200.
Wang, J., Levy, M., and Dunkle, L. D. 1998. Sibling species of *Cercospora* associated with gray leaf spot of maize. Phytopathology 88:1269-1275.
Ward, J. M. J., Stromberg, E. L., Nowell, D. C., and Nutter, F. W., Jr. 1999. Gray leaf spot, a disease of global importance in maize production. Plant Dis. 83:884-895.
Wisser, R. J., Balint-Kurti, P. J., and Nelson, R. J. 2006. The genetic architecture of disease resistance in maize: A synthesis of published studies. Phytopathology 96:120-129.

(Prepared by M. L. Carson)

Rhizoctonia Banded Leaf and Sheath Blight

Rhizoctonia banded leaf and sheath blight has been reported throughout tropical and subtropical areas of the world. Although yield losses in the United States are considered minimal to Rhizoctonia banded leaf and sheath blight, yield losses greater than 40% have been reported in India and southern China. Increased incidence and severity has been observed in the mid-south states of the United States, where corn is being produced in rotation with rice and soybeans under flood and furrow irrigation practices. High-nitrogen and low-potassium fertility along with increased plant populations has also been suspect for influencing the severity of the disease.

Symptoms

The disease develops first on the stalk at the soil line and eventually spreads to leaves, sheaths, and ears. Characteristic leaf and sheath symptoms are large gray to brown lesions with concentric dark-colored bands or rings (Figs. 94 and 95). Severe infestation of the ear can lead to ear rot and loss in yield. The fungus can form sclerotia on infected tissue (Fig. 96).

Causal Organism

Rhizoctonia banded leaf and sheath blight is caused by *Rhizoctonia solani* AG1-1A (syn. *Thanatephorus cucumeris*).

Fig. 95. Rhizoctonia banded leaf blight, caused by *Rhizoctonia solani*. (Courtesy X. Li)

Fig. 94. Rhizoctonia banded leaf blight, caused by *Rhizoctonia solani*. (Courtesy X. Li)

Fig. 96. Rhizoctonia banded sheath blight and sclerotia of *Rhizoctonia solani*. (Courtesy B. L. Renfro)

Rhizoctonia solani is non-spore forming and is identified by mycelial and sclerotial characteristics. Mycelium is light brown and cottony in appearance and is hyaline, septate, and branched at 90° angles. Sclerotia are round to spherical, dark brown to black, and typically 1–5 mm in diameter.

Management

Management of Rhizoctonia banded leaf and sheath blight with fungicides is neither always practical nor economical. This is largely because of the difficulty of getting the fungicide down through the corn canopy to the pathogen to achieve adequate control. However, in some cases, the use of fungicides such as azoxystrobin has been recommended to slow the further development and spread of the disease within the canopy. Maintaining adequate soil fertility may also reduce the impact of Rhizoctonia banded leaf and sheath blight in corn.

Selected References

Ahuja, S. C., and Payak, M. M. 1982. Symptoms and signs of banded leaf and sheath blight of maize. Phytoparasitica 10:41-49.

Baruah, P., and Lal, S. 1981. Host range of *Rhizoctonia solani* f. sp. *sasakii*, the incitant of banded sclerotial disease in maize. Indian Phytopathol. 36:494-496.

Jones, R. K., and Belmar, S. B. 1989. Characterization and pathogenicity of *Rhizoctonia* spp. isolated from rice, soybean, and other crops grown in rotation with rice in Texas. Plant Dis. 73:1004-1010.

Lal, S., Baruah, P., and Butchaiah, K. 1980. Assessment of yield losses in maize cultivars due to banded sclerotial disease. Indian Phytopathol. 33:440-443.

Slaton, N. A., Cartwright, R. D., Meng, J., Gbur, E. E., and Norman, R. J. 2003. Sheath blight severity and rice yield as affected by nitrogen fertilizer rate, application method, and fungicide. Agron. J. 95:1489-1496.

(Prepared by S. Monfort)

Diplodia (Stenocarpella) Leaf Streak

Diplodia leaf streak can be found worldwide but, in the United States, it generally considered to be a minor disease. Disease occurrences may be on the increase in the United States as soil conservation practices that leave more corn residue on the soil surface are becoming more of a standard practice. Serious yield reductions caused by Diplodia leaf streak have been reported from countries in Latin America and Africa.

Symptoms

Symptoms are somewhat similar to northern corn leaf blight lesions on very susceptible hybrids or to necrotic lesions associated with the leaf blight phase of Stewart's bacterial wilt or Goss's bacterial wilt and blight. Lesions first appear as small, roughly elliptical cream to brown spots on the leaf blade. Lesions can expand into long and narrow "streaks" (approximately 8–40 mm wide) (Fig. 97) that can nearly span the length of the leaf. Lesions generally are tapered at the ends. Scattered, dark pycnidia may form within the lesions (Fig. 98).

Causal Organism

Diplodia leaf streak is caused by *Stenocarpella macrospora* (syns. *Diplodia macrospora* and *Stenocarpella zeae*). The first observation of this fungus was made from corn stalks in Alabama (United States) in 1896. Pycnidia (150–450 µm in diameter) are dark and immersed. Dark, clavate to cylindrical-clavate, uniseptate conidia (6–13 × 43–95 µm) (Fig. 99) are extruded from pycnidia in dark cirrhi. In culture media, *S. macrospora* requires the availability of biotin or a biotin-like substance to produce pycnidia and conidia. Observations of scolecospores have been reported in culture.

Disease Cycle and Epidemiology

S. macrospora overwinters as pycnidia and mycelia on corn residue in the soil or on seed. Under wet and warm conditions, conidia are extruded from pycnidia and disseminated onto leaves by wind and splashing rain. Infection and disease development can occur in leaves of all ages. Pycnidia are produced within lesions and give rise to conidia that serve as secondary inoculum. Diplodia leaf streak severity was reported to increase in plants also affected by *Maize dwarf mosaic virus*. *S. macrospora* also has been reported to infect ear, kernel, and stalk tissues.

Management

Crop rotation and tillage practices that bury the corn residues reduce the amount of primary inoculum. Differences among

Fig. 97. Diplodia leaf streak. (Courtesy C. Bradley)

Fig. 98. Diplodia leaf streak showing pycnidia of *Stenocarpella macrospora*. (Courtesy C. Bradley)

Fig. 99. Conidia of *Stenocarpella macrospora*. (Courtesy D. Pedersen)

corn lines have been observed, but no information is available on the nature of resistance to Diplodia leaf streak.

Selected References

Latterell, F. M., and Rossi, A. E. 1983. *Stenocarpella macrospora* (=*Diplodia macrospora*) and *S. maydis* (=*D. maydis*) compared as pathogens of corn. Plant Dis. 67:725-729.

Meyer, M. D., and Pataky, J. K. 2010. Increased severity of foliar diseases of sweet corn infected with maize dwarf mosaic and sugarcane mosaic viruses. Plant Dis. 94:1093-1099.

Mora, L. E., and Moreno, R. A. 1984. Cropping pattern and soil management influences on plant diseases: 1. *Diplodia macrospora* leaf spot of maize. Turrialba 34:35-40.

Morant, M. A., Warren, H. L., and von Qualen, S. K. 1993. A synthetic medium for mass production of pycnidiospores of *Stenocarpella* species. Plant Dis. 77:424-426.

Olatinwo, R. O., Cardwell, K. F., Deadman, M. L., and Julian, A. M. 1999. Epidemiology of *Stenocarpella macrospora* (Earle) Sutton on maize in the mid-altitude zone of Nigeria. J. Phytopathol. 147:347-352.

Sutton, B. C., and Waterston, J. M. 1966. *Diplodia macrospora*. Descriptions of Pathogenic Fungi and Bacteria, Set 9, no. 83. Commonwealth Mycological Institute, Kew, England.

(Prepared by C. A. Bradley)

TABLE 1. Foliar Diseases That Are Currently of Minor Importance or Do Not Occur in the United States

Disease Causal Organism	Fruiting Structures	Symptoms	Geographic Distribution
Yellow leaf blight *Didymella maydis*	Pycnidia (60–150 μm in diameter) are immersed, reddish brown, subglobose, and ostiolate. Conidia (3–7.5 × 8–20 μm) are single celled, hyaline, oblong to ellipsoid to subcylindrical, and typically biguttulate.	Lesions are rectangular to oval (average 0.3 × 1.3 cm); yellow, cream colored, or tan; sometimes surrounded by chlorotic tissue; and may coalesce, causing blighting. Leaf sheaths and husks are also susceptible. Pycnidia appear as minute, black specks. Texas male-sterile cytoplasm corn is most susceptible (Figs. 100 and 101).	Primarily the northern and northeastern states in the United States; also Argentina and Canada; similar species have been reported from Europe, Africa, Brazil, and Taiwan
Curvularia leaf spot *Curvularia* spp. (most prevalent: *Curvularia aeria*)	Conidia (8–16 × 18–32 μm) are pale brown and straight or slightly curved, have distinct protuberant hila, are tapered at both ends, and usually have three septa (the two middle cells are larger than the end cells). Conidiophores are brown, septate, straight or flexuous, often bent, and simple or branched and bear conidia sympodially.	Lesions are small (1–2 mm), straw colored, and circular to oval; have reddish brown or dark brown margins; may coalesce to form necrotic areas up to 1 cm long; and may be surrounded by straw-colored or chlorotic halos (Figs. 102 and 103).	Prevalent in hot, humid corn-growing areas; can cause significant crop damage in the tropics
Phaeosphaeria leaf spot *Phaeosphaeria maydis*	Perithecia are immersed and spherical to subglobose. Asci (7.5–8.5 × 44.5–70 μm) are hyaline, clavate or cylindrical, straight or curved, and eight spored. Ascospores (3.5–5.0 × 14.5–17.5 μm) are hyaline, are straight or slightly curved, have three septa, and are slightly constricted at the septa, and have a distinctly swollen penultimate cell. Pycnidia (*Phoma* sp.) are spherical or globose and dark brown to black and have a round ostiole. Conidia (2.4–3.2 × 3.2–9 μm) are hyaline and ellipsoid, elongate, or round.	Lesions (0.3–2.0 cm) are scattered over the leaf and are initially small; pale green or chlorotic, becoming bleached or dried with dark reddish brown margins; and round or elongate to oblong. Spots may coalesce and become irregularly shaped. Perithecia and, less frequently, pycnidia develop in the lesions (Figs. 104 and 105).	Southern Florida and Hawaii in the United States; foothills of the eastern and central Himalayas in India; Costa Rica; Colombia; Brazil; Cameroon; South Africa; Zimbabwe; and Kenya
Ascochyta leaf spot *Ascochyta zeae*, *Ascochyta maydis*, *Ascochyta zeina*	Pycnidia are 55–160 μm in diameter. Conidia have one septum and are hyaline, ellipsoidal to irregular, and rarely constricted at the septa. They vary in size by species: *A. zeae*, 3–4.5 × 8.5–13.5 μm; *A. maydis*, 3–4.5 × 11–18 μm; and *A. zeina*, 7.5 × 18 μm.	Lesions are brown to reddish purple and ellipsoidal, becoming elongated and somewhat irregular with brown margins. Pycnidia are moderately abundant and subepidermal.	Reported only from the United States
Zonate leaf spot *Microdochium sorghi*	Conidia (1.4–3.2 × 20–195 μm) are hyaline, needle shaped, straight or slightly curved, of variable length, and borne in a pinkish salmon-colored slimy matrix in a sporodochium.	Reddish brown, water-soaked lesions (up to 2.5–5 cm) occur mainly on older leaves and form a zonate, targetlike pattern. Black sclerotia (0.1–0.2 mm in diameter) may develop within necrotic tissue.	United States, Central and South America, and Africa
Septoria leaf blotch *Septoria* spp.	Hyaline, filiform, septate conidia are produced in black, partially immersed pycnidia.	Small, light green to yellow to brown lesions may merge to produce severe blotching and necrosis. Black pycnidia form in diseased tissue.	Cool, humid environments in the tropics and subtropics

(continued on next page)

[a] Most corn grown in the United States is immune to leaf spot caused by *Stenocarpella maydis*, even though it is susceptible to ear rot and stalk rot caused by the same fungus.

Diseases of Minor Importance or Limited Occurrence

Table 1 contains information on foliar diseases of corn that are rare, cause little economic damage, or do not occur in the United States (Figs. 100–108). However, this classification does not mean that these diseases may not become important in the future, nor does it mean that these diseases may not be important elsewhere in the world. For example, in North America, Phaeosphaeria leaf spot is considered a curiosity only seen in south Florida, but it is an important disease in Brazil and parts of Africa. The history of corn pathology includes many examples of diseases that were considered of little importance but that suddenly caused widespread and significant damage to the crop because of changing cultural

Fig. 100. Yellow leaf blight. (Courtesy D. G. White)

Fig. 101. Yellow leaf blight. (Courtesy G. L. Worf)

TABLE 1. *(continued)*

Disease Causal Organism	Fruiting Structures	Symptoms	Geographic Distribution
Tar spot complex *Phyllachora maydis, Monographella maydis, Coniothyrium phyllachorae*	Perithecia of the primary pathogen, *P. maydis*, an obligate parasite, are immersed and nearly spherical and bear cylindrical asci (8–10 × 80–100 µm) with short pedicels. Ascospores (5.5–7 × 9–12 µm) are one celled, fuliginous, and broadly ellipsoidal.	Black, sunken, oval to circular glossy spots (0.5–2.0 mm in diameter) may merge to form stripes up to 10 mm long (Fig. 106).	Cool, humid areas at high elevations in Latin America; also reported from Indiana and Illinois in the United States
Hyalothyridium leaf spot *Hyalothyridium maydis*	Pycnidia (150–180 µm in diameter) are immersed, globose, thin walled, and subhyaline to pale brown with dark ostioles arising from stomata. Conidia (16–28 × 23–40 µm) are hyaline, muriform, and broadly oblong with 10–15 cells that often separate at maturity.	Small, tan, elliptical spots with brown borders to large, almost circular, light brown blotches up to 1.5 cm in diameter sometimes cover most of the leaf surface (Fig. 107). Other fungi, e.g., *Leptosphaerulina australis* and *Astromella* spp., may be involved and responsible for variable symptoms.	Humid areas of Mexico, Costa Rica, and Colombia
Fusarium leaf scorch *Fusarium verticillioides*	See the Fusarium Stalk Rot and the Fusarium Ear Rot or Pink Ear Rot disease sections in the text.	Water-soaked spots occur mainly on whorl leaves and turn white and papery with brown borders. The whorl does not open, and leaf tips turn downward. Occasionally, necrotic spots develop on higher portions of the stalks.	Warm, dry regions of the Caribbean
Didymella leaf spot *Didymella exitialis*	Pseudothecia (90–130 µm in diameter) are dark brown to black and submersed. Asci (8–12 × 40–60 µm) are eight spored, cylindrical, and bitunicate. Ascospores (4–5.5 × 12–14 µm) are uniseptate, hyaline, and ellipsoidal.	Small, elongate or elliptical leaf spots with brownish margins occur. Lesions enlarge and coalesce, forming white to cream-colored streaks or patches in which perithecia are sometimes formed.	India
Selenophoma leaf spot *Selenophoma* spp.	Pycnidia are hyaline with membranous walls. Alpha conidia (2.8–4.7 × 24–50 µm) are one celled and lunate. Beta conidia are also formed.	Eyespot lesions with concentric circles are surrounded by chlorotic halos.	Colombia, above 2,700 m
Leptosphaeria leaf spot *Paraphaeosphaeria michotii, Phaeosphaeria variiseptata*	Perithecia (50–200 µm in diameter) are globose. Ascospores (3.5–6 × 14–23 µm) are cylindrical and light brown and have two septa. A pycnidial stage (*Phoma* sp.) is also produced.	Small, tan or cream-colored lesions become large and concentric or streaked and cover large areas.	Himalayas and Illinois (United States)
Alternaria leaf spot *Alternaria alternata*	Conidia (9–18 × 20–63 µm) are obclavate, obpyriform, ovoid, or ellipsoidal; pale to midgolden brown; and smooth or verruculose and often have a short beak. They have up to eight transverse septa and commonly have several longitudinal or oblique septa. *A. alternata* is cosmopolitan.	Chlorotic streaks become necrotic on leaves of all ages. Tissue damage, e.g., from insect feeding or embedded sand grains, followed by a prolonged dew period are apparently essential for infection.	Central U.S. Corn Belt
Diplodia leaf spot *Stenocarpella maydis*	See the Diplodia (Stenocarpella) Stalk Rot disease section in the text.	Oval leaf spots occur with pycnidia scattered in the necrotic areas (Fig. 108).	Caribbean islands[a]

Fig. 102. Curvularia leaf spot. (Courtesy C. De Leon)

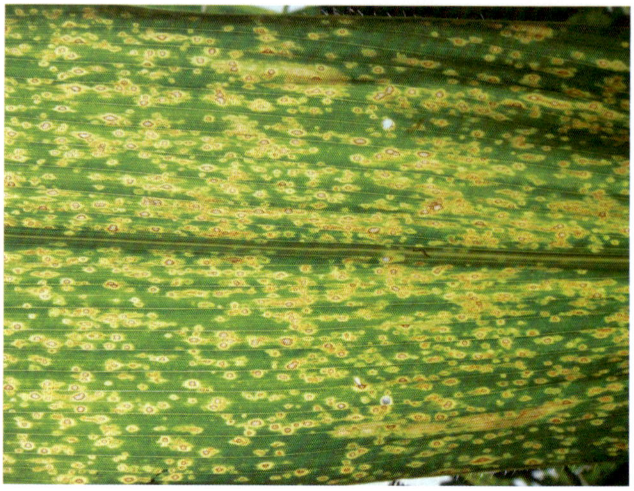

Fig. 103. Curvularia leaf spot. (Courtesy X. Xu)

Fig. 104. Phaeosphaeria leaf spot. (Courtesy L. E. Claflin)

Fig. 105. Phaeosphaeria leaf spot with perithecia of *Phaeosphaeria maydis* in the lesions. (Courtesy B. M. Anderson)

Fig. 106. Tar spot. (Courtesy B. M. Anderson)

Fig. 107. Hyalothyridium leaf spot. (Courtesy F. M. Latterell)

Fig. 108. Diplodia leaf spot and pycnidia of *Stenocarpella maydis* in the lesions. (Courtesy D. G. White)

practices, hybrid susceptibility, and evolving pathogen populations. Anthracnose, eyespot, gray leaf spot, and Goss's bacterial wilt and blight are all examples of corn diseases that were considered minor for years but became recognized as potential major problems in large areas of the corn-producing regions of the United States.

Selected References

Abadeje, S. A. 1969. Curvularia leaf spot of maize. Trans. Br. Mycol. Soc. 52:267-271.

Ahuja, S. C., and Payak, M. M. 1982. Symptoms and signs of banded leaf and sheath blight of maize. Phytoparasitica 10:41-49.

Baruah, P., and Lal, S. 1981. Host range of *Rhizoctonia solani* f. sp. *sasakii*, the incitant of banded sclerotial disease of maize. Indian Phytopathol. 36:494-496.

Carson, M. L. 2005. Yield loss potential of Phaeosphaeria leaf spot of maize caused by *Phaeosphaeria maydis* in the United States. Plant Dis. 89:986-988.

Carson, M. L., Goodman, M. M., and Glawe, D. A. 1991. Phaeosphaeria leaf spot of maize in Florida. Plant Dis. 75:968.

Carson, M. L., Stuber, C. W., and Senior, M. L. 2005. Quantitative trait loci conditioning resistance to Phaeosphaeria leaf spot of maize caused by *Phaeosphaeria maydis*. Plant Dis. 89:571-574.

Ceballos, H., and Deutsch, J. A. 1992. Inheritance of resistance to tar spot complex in maize. Phytopathology 82:505-512.

De Leon, C. 1984. Maize Diseases: A Guide for Field Identification. International Maize and Wheat Improvement Center (CIMMYT), Mexico City.

Eddins, A. H. 1930. Dry rot of corn caused by *Diplodia macrospora* Earle. Phytopathology 20:439-448.

Jones, R. K., and Belmar, S. B. 1989. Characterization and pathogenicity of *Rhizoctonia* spp. isolated from rice, soybean, and other crops grown in rotation with rice in Texas. Plant Dis. 73:1004-1010.

Kaur, R., Payak, M. M., and Nath, R. 1973. Curvularia leaf disease of maize. Kasetsart J. 7:76-80.

Lal, S., Baruah, P., and Butchaiah, K. 1980. Assessment for yield losses in maize cultivars to banded sclerotial disease. Indian Phytopathol. 33:440-443.

Latterell, F. M., and Rossi, A. E. 1979. Hyalothyridium leaf spots of maize. (Abstr.) Phytopathology 69:536.

(Prepared by M. L. Carson)

Management of Foliar Diseases of Corn with Fungicides

Fungicide applications for foliar disease management are more common with sweet corn and inbred corn lines used in hybrid seed production than with hybrid dent corn since inbred lines and sweet corn are higher in value and often more susceptible to foliar diseases (Fig. 109). In the mid-2000s, however, the values of hybrid dent corn grain increased two- to threefold and foliar fungicide applications have become more prevalent on the crop. Approximately 10% of the hybrid dent corn crop in the United States may be sprayed annually for disease management and also for direct physiological effects of fungicides on corn plants.

Fungicide active ingredients approved for use on corn in the United States primarily belong to three chemical groups: demethylation inhibitors (DMIs), quinone outside inhibitors (QoIs), and succinate dehydrogenase inhibitors (SDHIs). Commercial fungicide products that are available contain either a QoI alone, a DMI alone, or a premix of two of the three groups.

Fungicides in the QoI group are often referred to as strobilurins. QoI fungicides inhibit fungal mitochondrial respiration that stops energy production in the fungus and consequently results in its death. Since they prevent spore germination and early mycelium growth, these fungicides are usually applied preventively (preinfection) or as early as possible in the disease cycle. They are less effective on mycelium that has colonized leaf tissue. Depending on the fungicide, QoIs may move translaminarly (e.g., trifloxystrobin, kresoxim-methyl) or translaminarly and systemically to a limited extent (e.g., azoxystrobin) within leaf tissue. They have approximately 7–21 days of residual activity.

There is evidence that QoI fungicides can influence the physiology of nondiseased plants, e.g., improve water and nitrogen use efficiency, increase antioxidant activity, prevent frost damage, and delay senescence, thus contributing to greater yields. Consequently, this group of fungicides also has been marketed for plant health benefits, and many corn fields may be sprayed in the absence of disease. However, the magnitude of yield benefits from these physiological effects is variable and unpredictable.

The SDHI group of fungicides, sometimes referred to as carboxamides, are similar to QoI fungicides in that they inhibit fungal mitochondrial respiration; however, they target a different enzyme in the respiration chain. These fungicides are locally systemic, move translaminarly and upward, and have approximately 14–21 days of residual activity.

Fungicides in the DMI group, some of which may be referred to as triazoles, inhibit a specific enzyme that plays a role in sterol production. Since sterols are essential for the development of functional cell walls, application of a DMI results in abnormal fungal growth and death. DMIs have no affect on spore germination since spores contain enough sterol for the formation of germ tubes. They should be applied preventively or early in the fungal infection process. DMIs are locally systemic and have approximately 14 days of residual activity.

In inbred corn lines, foliar fungicide applications often start when disease first becomes apparent and severity is low. Applications continue at regular intervals throughout the production season. In hybrid corn, up to two applications of a fungicide

Fig. 109. Aerial fungicide application in a seed production field in Iowa. (Courtesy G. P. Munkvold)

may be applied in a growing season; one at crop developmental stage V5 to V6 (five- to six-leaf stage), and one within a 2- to 3-week period after tasseling (crop developmental stages VT to R3). Most farmers who apply fungicides to hybrid dent corn apply the product at this latter stage of crop development; however, some apply a fungicide both at V5 to V6 and at VT.

It is recommended that applications of a fungicide to hybrid dent corn should be based on integrated pest management (IPM) practices since fungicides are management tools that are useful only under conditions in which disease would be expected to significantly reduce yield potential. Scouting for foliar disease is usually done prior to, or at, tasseling. A systematic design, such as a "W" or "Z" pattern, can be followed to traverse a field when scouting fields. At each point, the ear leaf and canopy below the ear leaf should be examined on several plants for disease, and incidence and severity should be recorded overall.

The following factors should be considered before making an application of a foliar fungicide:

- hybrid susceptibility and yield potential,
- level of disease symptoms at tasseling,
- weather conditions at tasseling and projected conditions during grain fill,
- cropping and disease history,
- amount of crop residue present in the field, and
- planting date since late-planted crops may be exposed to higher inoculum levels at earlier crop developmental stages.

Most research on using fungicides with hybrid dent corn has been for gray leaf spot management. In the United States, a fungicide application is not recommended on hybrids that are resistant or moderately resistant to foliar disease. On susceptible or moderately susceptible hybrids, a fungicide application may be warranted if disease is present in the lower third of the canopy at flowering.

Fungicides may be applied to corn using a ground applicator or an aircraft. The effectiveness of a fungicide at reducing disease severity is highly dependent on spray coverage on each leaf and within the canopy. The desired droplet-size range for fungicide application is 200–300 μm, and nozzle types and operating pressure need to be carefully selected to ensure this small droplet range. Adequate carrier volume is required to ensure spray coverage within the canopy. In the United States, typical carrier volumes are 2 gallons per acre (gpa) (22.5 liters per hectare [L/ha]) and 10–15 gpa (112–174 L/ha) for aerial and ground applications, respectively. Adjuvants also may be added to fungicides to improve coverage and, consequently, efficacy.

Fungicide Resistance Management

The Fungicide Resistance Action Committee (FRAC) is an international organization of scientists from agrochemical companies that establishes fungicide management guidelines to prolong the effectiveness of fungicides that are most at risk for resistance to develop. Fungicides are classified into FRAC code groups based on their mode of action.

Resistance to various fungicide products has been reported in several fungal pathogens of numerous crops. The risk of resistance in DMI (FRAC code 3), QoI (FRAC code 11), and SDHI (FRAC code 7) fungicides is medium, high, and medium, respectively, since each group has a very specific site of action. Guidelines for fungicide resistance management include:

- The use of disease-resistant hybrids and cultural practices that reduce disease severity.
- If multiple applications are needed in the same field in the same season or in the same field in different seasons, tank mix or alternate high-risk fungicides with fungicides that belong to different FRAC codes that are effective against the disease(s) of concern and have similar lengths of residual activity.
- Do not apply reduced rates of fungicides.
- Apply fungicides preventively or early in the disease cycle.

Selected References

Mueller, D. S., and Bradley, C. A. 2008. Field Crop Fungicides for the North Central United States. Iowa State University, Ames, and University of Illinois, Urbana-Champaign. North central Integrated Pest Management Center.

Munkvold, G. P., Martinson, C. A., Shriver, J. M., and Dixon, P. M. 2001. Probabilities for profitable fungicide use against gray leaf spot in hybrid maize. Phytopathology 91:477-484.

Paul, P. A., Madden, L. V., Bradley, C. A., Robertson, A. E., Munkvold, G. P., Shaner, G., Wise, K. A., Malvick, D. K., Allen, T. W., Grybauskas, A., Vincelli, P., and Esker, P. 2011. Meta-analysis of yield response of hybrid field corn to foliar fungicides in the U.S. Corn Belt. Phytopathology 101:1122-1132.

Vincelli, P. 2002. QoI (strobilurin) fungicides: Benefits and risks. The Plant Health Instructor. doi: 10.1094/PHI-I-2002-0809-02

Ward, J. M. J., Laing, M. D., and Rijkenberg, F. H. J. 1997. Frequency and timing of fungicide applications for the control of gray leaf spot in maize. Plant Dis. 81:41-48.

Wegulo, S. N., Rivera-C, J. M., Martinson, C. A., and Nutter, F. W., Jr. 1998. Efficacy of fungicide treatments for control of common rust and northern leaf spot in hybrid corn seed production. Plant Dis. 82:547-554.

(Prepared by A. E. Robertson)

Downy Mildews

Ten species of three genera of the phylum Oomycota are reported to cause downy mildew of corn. Six species occur fairly often and four infrequently. Individually, they fill specific regional or global niches. At one time and in certain regions, some downy mildews were the major factors limiting corn production. However, since the late 1970s, the development of resistant cultivars, the use of fungicides, and improvements in cropping and cultural practices have greatly diminished the losses. Research has shown that downy mildew resistance in corn to one species of the genus *Peronosclerospora* generally is effective toward the other *Peronosclerospora* pathogens.

Most of the downy mildew pathogens of corn did not originate in the Americas, to which corn is native, but they either adapted or possessed the ability to infect corn when it was introduced into other regions of the world. As this occurred or when a downy mildew pathogen was introduced into warm, humid climates, one or more downy mildew diseases often became major problems.

Downy mildew pathogens are obligate parasites. The leaf is penetrated directly by an infection peg from an appressorium formed over or near a stoma, and this is followed by penetration and vesicle formation in the substomatal chamber. The coenocytic mycelia of the fungi grow intercellularly, and host cell membranes are invaginated by haustoria.

Delineation of species is sometimes uncertain. The principal characteristics used to distinguish species are the shape and dimensions of conidia or sporangia; the size and branching patterns of conidiophores or sporangiophores; the length of sterigmata; and the shape, color, and morphology of the oospores and the oogonial envelope. Host range and symptoms are also often used in species identification (Table 2).

The use of isozyme analysis, molecular (DNA) markers, and other chemical-based techniques may provide better means of differentiating species in the future. However, since definitive direction for combining or separating species within these genera is lacking, the diseases are discussed here as they have been previously described.

Symptoms of downy mildews are similar. Both systemic and local lesions develop. Symptoms are affected by the age of the plant at the time of infection, the species of the pathogen, and the environment. Systemically infected plants exhibit mottling, chlorotic streaking, malformation of reproductive organs, excessive tillering, or any combination of symptoms. Plant hormone levels often change after infection, resulting in these bizarre symptoms. When systemic infection occurs, plants are often barren.

Selected References

Bonde, M. R. 1982. Epidemiology of downy mildew diseases of maize, sorghum and pearl millet. Trop. Pest Manage. 28:49-60.

Bonde, M. R., Peterson, G. L., Dowler, W. M., and May, B. 1984. Isozyme analysis to differentiate species of *Peronosclerospora* causing downy mildews of maize. Phytopathology 74:1278-1283.

TABLE 2. Downy Mildews of Corn

Pathogen (Disease Name)	Sporangiophores or Conidiophores[a]	Sporangia or Conidia	Oospores	Geographical Distribution	Hosts Other Than Corn	Seedborne/ Seed-Transmitted
Peronosclerospora sorghi (sorghum downy mildew)	Hyaline, 180–300 μm long, bloated, usually dichotomously branched two to three times, septate near base, ephemeral	Conidia hyaline, oval to almost spherical, 14.4–27.3 × 15–28.9 μm; germinate by germ tube	Usually brown to subhyaline, spherical, 25–42.9 μm in diameter; germinate by germ tube	North America, Central America, Asia, Africa, South America, Australia, Europe	*Euchlaena* spp., *Panicum* spp., *Pennisetum* spp., *Sorghum* spp., *Andropogon* spp.	Yes, in pericarp and scutellum Yes, only on freshly harvested seeds
Peronosclerospora maydis (Java downy mildew)	Hyaline, long (150–550 μm), bloated, dichotomously branched two to four times, ephemeral	Conidia hyaline, spherical to subspherical, 17–23 × 27–39 μm; germinate by germ tube	Unknown	Australia, Indonesia	*Zea mays* subsp. *mexicana* × *Zea mays* hybrids, *Euchlaena* spp., *Pennisetum* spp., *Tripsacum* spp.	Yes Only before seeds are dried
Peronosclerospora philippinensis (Philippine downy mildew)	Hyaline, long (150–400 μm), bloated (widening abruptly), dichotomously branched two to four times, ephemeral	Conidia hyaline, elongate ovoid to round cylindrical, 17–21 × 27–39 μm, slightly rounded at apex; germinate by germ tube	Rare or nonexistent; spherical, smooth walled, 22.6 μm in diameter; germinate by side germ tube	Philippines, Indonesia, India, Nepal, China, Thailand, Pakistan	*Avena sativa*, *Euchlaena* spp., *Zea mays* subsp. *mexicana* × *Zea mays* hybrids, *Saccharum* spp., *Sorghum* spp.	Yes, in embryo and endosperm Yes, at 36% moisture, not at 14% moisture
Peronosclerospora sacchari (sugarcane downy mildew)	Hyaline, 160–170 μm, bloated (widening gradually), dichotomously branched two to three times, ephemeral	Conidia hyaline, elliptical, oblong to conical, 15–23 × 25–41 μm with round apex; germinate by germ tube	Yellow to yellow brown, globular to slightly angular, 40–50 μm in diameter; germinate by germ tube	Australia, Fiji Islands, Japan, Nepal, New Guinea, India, Philippines, Taiwan, Thailand	*Euchlaena* spp., *Saccharum* spp., *Sorghum* spp., *Andropogon* spp., *Tripsacum* spp.	Yes, in embryo Yes, only at moisture levels above 20%
Sclerospora graminicola (Graminicola downy mildew, green ear of pearl millet)	Hyaline, long (average 268 μm), bloated, nonseptate, irregularly dichotomously branched, ephemeral	Sporangia hyaline, broadly elliptical, 12–21 × 14–31 μm with a papillate operculum; germinate by zoospores	Pale brown, spherical, usually smooth walled, 22.5–35 μm in diameter; said to germinate by germ tube or by sporangia	Worldwide, but rare on maize and found only in Israel and the United States	*Euchlaena* spp., *Panicum* spp., *Pennisetum* spp., *Saccharum* spp., *Setaria* spp., *Sorghum* spp., *Agrostis* spp., *Chaetochloa* spp., *Echinochloa* spp., *Holcus* spp.	No No
Sclerophthora macrospora (crazy top)	Hyaline, short (average 13.8 μm), simple, hyphoid, determinate	Sporangia hyaline, lemon shaped, operculate, 30–65 × 60–100 μm; germinate by numerous zoospores	Hyaline to pale yellow, develop mainly in vascular bundles, 45–75 μm in diameter; germinate by sporangia	North America, Mexico, South America, Europe, Africa, Asia	*Avena sativa*, *Digitaria* spp., *Panicum* spp., *Setaria* spp., *Sorghum* spp., *Triticum* spp., numerous others	No No
Sclerophthora rayssiae var. *zeae* (brown stripe downy mildew)	Hyaline, short, hyphoid, determinate	Sporangia hyaline, ovate to almost cylindrical, operculate, 18.5–26 × 29–66.5 μm; germinate by zoospores	Brown, spherical, 29.5–37 μm in diameter; germinate by sporangia	India, Nepal, Pakistan, Thailand	*Digitaria* spp.	Yes, in embryo Yes

[a] Length and branching of conidiophores and sporangiophores in *Sclerospora* and *Peronosclerospora* spp. vary, depending somewhat on ambient air conditions during their formation.

Frederiksen, R. A., and Renfro, B. L. 1977. Global status of maize downy mildew. Annu. Rev. Phytopathol. 15:249-275.
Kenneth, R. G. 1981. Downy mildews of gramineous crops. Pages 367-394 in: The Downy Mildews. D. M. Spencer, ed. Academic Press, New York.
Lal, S., and Singh, I. S. 1984. Breeding for resistance to downy mildews and stalk rots in maize. Theor. Appl. Genet. 69:111-119.
Michelmore, R. W. 1988. The downy mildews. Pages 53-79 in: Advances in Plant Pathology, Vol. 6. D. S. Ingram and P. H. Williams, eds. Academic Press, New York.
Schmitt, C. G., Woods, J. M., Shaw, C. G., and Stansbury, E. 1979. Comparison of some morphological characters of several corn downy mildew incitants. Plant Dis. Rep. 63:621-625.
Shaw, C. G. 1978. *Peronospora* spp. and other downy mildews of the Gramineae. Mycologia 70:594-604.
Williams, R. J. 1984. Downy mildews of tropical cereals. Pages 1-103 in: Advances in Plant Pathology, Vol. 2. D. S. Ingram and P. H. Williams, eds. Academic Press, New York.

(Prepared by D. R. Smith and B. L. Renfro;
Revised by M. R. Bonde)

Peronosclerospora spp. and *Sclerospora* spp.

Mildews caused by *Peronosclerospora* and *Sclerospora* spp. are often referred to as the "true" downy mildews of corn. *Peronosclerospora* spp. produce conidia that germinate by germ tubes. *Sclerospora* spp. produce sporangia that germinate by producing zoospores, although direct germination occasionally occurs. Sporangia are operculate, and conidia have no apical modification. Oospores are not sufficiently distinctive to distinguish species. The oogonial wall is 1–2 µm thick and is typically fused with the thicker (3–6 µm) oospore wall. Oospores have been reported to survive in nature for 6–10 years. The asexual spores are thin walled, fragile, and short lived. Apparently, oospores need to undergo about 1 month of exposure in soil and are stimulated by root leachates to germinate. Most oospores germinate by a germ tube and can penetrate the plant directly. They can remain viable after passing through the digestive tracts of animals and can be carried on seed surfaces and in host debris. When oospores cause infection, the first seedling leaf does not become chlorotic, as occurs when the infection results from internal seedborne mycelia.

Conidia and sporangia are produced at night. Between approximately 11:00 p.m. and 12:00 a.m., one to four conidiophores or sporangiophores emerge from stomata; and between 12:00 a.m. and 1:00 a.m., spores begin to form. The majority of conidia and sporangia mature between 2:00 a.m. and 3:00 a.m. and are shed between 2:00 a.m. and 7:00 a.m., most between 2:00 a.m. and 3:00 a.m. They are disseminated primarily by wind. Because of the fragile nature of these spores, they cannot be carried more than a few hundred meters and remain viable.

Germination of conidia and sporangia usually occurs in less than 1 h, and free water is required. Leaf penetration occurs approximately 1–2 h after germination and always through stomata. Invasion of mesophyll cells follows. The first symptoms usually appear in about 3 days as chlorotic, circular spots, flecks, streaks, or any combination of the three, which elongate in both directions parallel to the veins. The outward development ceases or slows after a few days. The downward movement of mycelia continues to the shoot apex, and systemic infection results from its colonization. In resistant plants, the infection remains localized. Systemic symptoms develop after 5–14 days and do not appear in plants after 3.5–4 weeks of growth. Conidia are produced in large numbers (up to 100,000 per cm^2) in chlorotic tissue for a period of time until senescence. When there is profuse oospore production, leaves shred and asexual reproduction is sparse or absent.

Seed transmission of downy mildew species within the genus *Peronosclerospora* has received attention because of its potential hindrance to international trade of corn. However, it has been shown that oospores of all species do not survive the routine drying of corn kernels following harvest and thus present little risk.

Selected References

Bonde, M. R. 1982. Epidemiology of downy mildew diseases of maize, sorghum and pearl millet. Trop. Pest Manage. 28:49-60.
Kenneth, R. G. 1981. Downy mildews of gramineous crops. Pages 367-394 in: The Downy Mildews. D. M. Spencer, ed. Academic Press, New York.
McGee, D. C. 1988. Maize Diseases: A Reference Source for Seed Technologists. American Phytopathological Society, St. Paul, MN.
Shaw, C. G. 1978. *Peronospora* spp. and other downy mildews of the Gramineae. Mycologia 70:594-604.
Waterhouse, G. M. 1964. The genus *Sclerospora*. Diagnosis (or descriptions) from the original papers and a key. Misc. Publ. 17. Commonwealth Mycological Institute, Kew, England.

(Prepared by D. R. Smith and B. L. Renfro;
Revised by M. R. Bonde)

Sorghum Downy Mildew

Sorghum downy mildew occurs in Africa, Asia, Australia, Central America, Europe, North America, and South America. The pathogen also infects sorghum, on which it can be differentiated into pathotypes, races, or both. The disease on corn can be a significant problem in areas of the world where resistant corn genotypes are not readily available, such as in parts of Africa.

Symptoms

Systemically infected plants are chlorotic and sometimes stunted and occasionally have white-striped leaves and abnormal seed set (Figs. 110 and 111). The chlorotic area of a leaf always includes the base of the blade, and the transverse margin is usually sharply defined between the diseased and healthy tissues ("half-leaf" symptom), appearing further up the blade in each successively formed leaf until the whole blade is chlorotic. Leaves of infected plants tend to be narrower and more erect than those of healthy plants ("strap-leaf" symptoms), and shredding may occur. A white, downy growth may appear on both surfaces of infected leaves. In some regions, the tassels of diseased plants may exhibit phyllody. Tolerant plants may show symptoms of systemic infection but have normal seed production. Other plants may be barren. Long, narrow, chlorotic, local

Fig. 110. Sorghum downy mildew. (Courtesy B. M. Anderson)

lesions develop on some corn lines. Ear shanks may be elongated. Oospores, which are not common on corn, if present, tend to appear in the ear bracts and leaves within the striped areas.

Causal Organism

Sorghum downy mildew is caused by *Peronosclerospora sorghi* (syns. *Sclerospora sorghi* and *S. graminicola* var. *andropogonis-sorghi*). The conidiophores (180–300 µm long) are erect, spreading, fragile, hyaline, and usually dichotomously branched and emerge singly or in groups from stomata on the lower and sometimes the upper sides of leaves. The conidia (14.4–27.3 × 15–28.9 µm) are hyaline, oval to subspherical, and borne on elongated, tapered sterigmata about 13 µm long. They germinate by germ tubes in free water and penetrate corn leaves through stomata. The oogonia (40–55 µm in diameter) are spherical and embedded among mesophyll cells between fibrovascular bundles. The oospores (25–42.9 µm in diameter; average 36 µm) are hyaline and spherical with light yellow outer walls.

Disease Cycle and Epidemiology

The oospores of *P. sorghi* can survive several seasons in the soil. They germinate by germ tubes and attack underground parts of susceptible seedlings, which become systemically infected. No infection by oospores occurs if seedlings emerge in cool soil. Conidia on the leaves are disseminated by the wind and provide secondary inoculum, which can induce systemic infection in plants up to 4 weeks old. Oospores of *P. sorghi* are produced less frequently and less abundantly on corn than on sorghum and appear in both hosts only in systemically infected plants. The fungus has been detected within the pericarp and scutellum of seed. Oospores are also commonly found in glume tissue of systemically infected sorghum and are a major source of primary inoculum for infection of sorghum. Corn has glumes that are much less developed and are not usually present in commercial seed. Thus, there is little if any seed transmission. Transmission to corn seedlings has occurred from seed grown in sterile soil but only from freshly harvested or immature seed.

Conidia are produced at temperatures of 17–29°C (the optimum is often 24–26°C, depending on the geographical origin of the isolate). Germination of conidia requires a saturated atmosphere or free water, and moderate temperatures of 21–25°C are optimal for most isolates. High levels of systemic infection can occur at 11–32°C with a wet period of 4 h or longer. The conidia are delicate and lose their viability after 3–4 h.

P. sorghi also infects cultivated and wild sorghum, Johnsongrass, and teosinte. In the midwestern United States, shattercane is an important wild host. Oospores are disseminated over long distances on seed surfaces (especially on sorghum seed) and in debris mixed with seed. In India, the world's fifth largest producer of corn, 83% of corn production is in the state of Tamil Nadu, where the continuous cropping of corn, in addition to a highly favorable environment for downy mildew development, has led to a major problem of sorghum downy mildew of corn.

Fig. 111. Sorghum downy mildew. (Courtesy L. E. Claflin)

Management

Resistant hybrids offer the best and most economical means of disease management. Resistance varies in the germplasm studied in various corn-growing areas. In the United States, three major genes have been identified. Two exhibit dominance, and one is recessive. In Thailand, polygenic resistance has been found and is attributable mostly to dominance and additive genetic variance. Results from Venezuela closely resemble those from Thailand. In both Indian and Venezuelan studies, significant maternal effects have also been shown to condition resistance. Genotypes that emerge rapidly often escape penetration by oospores in the soil.

Planting corn when the soil temperature is below 20°C is highly recommended if high populations of oospores may be present in soil. This may allow the seeds to germinate and escape infection by oospores, which usually will not grow at temperatures below 20°C. Rotation of corn with sorghum should be avoided where the disease is found, and corn should not be planted adjacent to sorghum because it can serve as a source of inoculum. Shattercane should be vigorously controlled in areas where the disease is found. Systemic fungicides with activity against oomycota (e.g., metalaxyl or mefenoxam), applied as seed treatments, foliar sprays, or both, may be valuable in some situations.

Selected References

Bock, C. H., and Jeger, M. J. 2002. The distribution and spread of sorghum downy mildew in sorghum and maize fields in Nigeria and Zimbabwe. Eur. J. Plant Pathol. 108:745-753.

Bonde, M. R., Peterson, G. L., Dowler, W. M., and May, B. 1984. Isozyme analysis to differentiate species of *Peronosclerospora* causing downy mildews of maize. Phytopathology 74:1278-1283.

Borges, F., and Orangel, L. 1987. Diallel analysis of maize resistance to sorghum downy mildew. Crop Sci. 27:178-180.

Cohen, Y., and Sherman, Y. 1977. The role of airborne conidia in epiphytotics of *Sclerospora sorghi* on sweet corn. Phytopathology 67:515-521.

Craig, J. 1982. Inheritance of resistance to sorghum downy mildew in corn. (Abstr.) Phytopathology 72:943.

Frederiksen, R. A., and Odvody, G. 1979. Chemical control of sorghum downy mildew. Sorghum Newsl. 22:129.

Partridge, J. E., and Doupnik, B. L. 1979. Occurrence of sorghum downy mildew on shattercane and sorghum in Nebraska. Plant Dis. Rep. 63:154-155.

Perumal, R., Nimmakayala, P., Erattaimuthu, S. R., No, E.-G., Reddy, U. K., Prom, L. K., Odvody, G. N., Luster, D. G., and Magill, C. W. 2008. Simple sequence repeat markers useful for sorghum downy mildew (*Peronosclerospora sorghi*) and related species. BMC Genet. 9:77. Published online 2008 November 29. doi: 10.1186/1471-2156-9-77

Warren, H. L., Scott, D. H., and Nicholson, R. L. 1974. Occurrence of sorghum downy mildew in Indiana. Plant Dis. Rep. 58:430-432.

White, D. G., Jacobsen, B. J., and Hooker, A. L. 1978. Occurrence of sorghum downy mildew in Illinois. Plant Dis. Rep. 62:720.

Williams, A. S., and Herron, J. W. 1974. Occurrence of downy mildew of sorghum in Kentucky. Plant Dis. Rep. 58:90-91.

(Prepared by D. R. Smith and B. L. Renfro;
Revised by M. R. Bonde)

Java Downy Mildew

Java downy mildew is of great importance in Indonesia, where yield may be reduced by 40%. It also is found in Australia. The most severe damage occurs when corn is planted late

Fig. 112. Java downy mildew. (Courtesy C. De Leon)

or the rainy season begins early, especially if the crop has been overfertilized with nitrogen or planted after corn or sugarcane.

Symptoms

White to yellow streaks (Fig. 112), which become necrotic and brown, are characteristic leaf symptoms. The fungus may become systemic, causing severe chlorosis in the upper leaves. Infected plants may be stunted and sterile and often lodge. Downy growth on the chlorotic streaks is common. Plants more than 4 weeks old are highly resistant to infection.

Causal Organism

Java downy mildew is caused by *Peronosclerospora maydis* (syn. *Sclerospora maydis*). The pathogen produces two kinds of hyphae: straight and sparsely branched and lobed and irregularly branched. Clustered conidiophores arise from stomata and are dichotomously branched two to four times. The branches are robust and 150–550 µm long with basal cells 60–180 µm long. Conidia (17–23 × 27–39 µm) are hyaline and spherical to subspherical. Oospores have not been reported.

Disease Cycle and Epidemiology

Infected corn plants grown during the dry season are the primary source of inoculum in Indonesia. The fungus may survive as mycelium in kernels, but this is a minor source of inoculum. Infection by conidia takes place through stomata of young plants, and lesions elongate toward the meristem, inducing systemic infection. Conidia and conidiophores are produced superficially on wet leaves in the dark. If infection arises from seed, the cotyledonary leaf is always infected. Seed transmission occurs when freshly harvested seeds from diseased plants are used. No seed transmission has been detected from seeds dried prior to planting.

Formation of conidia requires free moisture, darkness, and temperatures below 24°C. Conidia germinate under these conditions to produce germ tubes. Other hosts include teosinte and *Tripsacum*, *Euchlaena*, and *Pennisetum* spp.

Management

Resistant hybrids and varieties are recommended in areas where the disease is found. Resistance in corn has been shown to be polygenic with additive gene action.

Selected References

Pupipat, U. 1974. Host range, geographic distribution and physiologic races of the maize downy mildews. Trop. Agric. Res. Ser. 8:63-80.

Schmitt, C. G., and Freytag, R. E. 1977. Response of selected resistant male genotypes to three species of *Sclerospora*. Plant Dis. Rep. 61:478-481.

Williams, R. J. 1984. Downy mildews of tropical cereals. Pages 1-103 in: Advances in Plant Pathology, Vol. 2. D. S. Ingram and P. H. Williams, eds. Academic Press, New York.

(Prepared by D. R. Smith and B. L. Renfro;
Revised by M. R. Bonde)

Philippine Downy Mildew

Philippine downy mildew is the most serious downy mildew disease in the Philippines, where yield losses are 15–40% or more. It also is found in China, India, Indonesia, Nepal, Pakistan, and Thailand. The causal organism is one of three corn downy mildew pathogens on the Select Agents and Toxins List in the United States created in response to the Congressional Public Health Security and Bioterrorism Preparedness and Response Act of 2002 (see Agricultural Bioterrorism Protection Act of 2002—Biennial Review and Republication of the Select Agent and Toxin List, 7 CFR, Part 331, Federal Register Vol. 73, No. 201, Oct. 16, 2008).

Symptoms

Systemic symptoms may appear in the first true leaf as complete chlorosis or chlorotic stripes 9 days after planting. Local symptoms may appear when plants are in the two- to three-leaf stage until tassels and silks are formed. In general, infected leaves have long, chlorotic streaks with a downy growth of conidia and conidiophores (Fig. 113). Tassels may be malformed and produce less pollen. Ears may be aborted, resulting in partial or complete sterility. If infection occurs early, plants are stunted and may die.

Causal Organism

Philippine downy mildew is caused by *Peronosclerospora philippinensis* (syn. *Sclerospora philippinensis*). The mycelium is branched, slender (8 µm in diameter), irregularly constricted, and inflated. The small haustoria (2 × 8 µm) are simple and vesiculiform to subdigitate. Erect conidiophores (15–26 × 150–400 µm) grow out of stomata and are dichotomously branched two to four times. Branches are robust. Sterigmata are ovoid to subulate, slightly curved, and 10 µm long. Conidia (17–21 × 27–39 µm) are elongate ovoid to round cylindrical, hyaline, and slightly rounded at the apex. *P. philippinensis* is indistinguishable from *P. sacchari* by isozyme analysis.

Disease Cycle and Epidemiology

Conidia of *P. philippinensis* germinate and form germ tubes, which penetrate leaves through stomata. Penetration is followed by intercellular invasion of the mesophyll cells. The fungus then grows through the leaf sheath into the stem and becomes established in the shoot apex, where it persists. Invasion of the shoot apex produces chlorotic streaks on the lower leaves and later on the younger leaves. The pathogen does not invade the roots. The fungus is seedborne and may be transmitted by planting seed that has not been fully dried. Conidia and conidiophores are produced on chlorotic areas on the leaf blades and sheaths. Wind and splashing water are the main means of conidial dispersal. Four-week-old plants are highly resistant. Oospores are not produced in corn tissue.

Night temperatures of 21–26°C and free moisture are essential for conidial production, germination, and infection. The optimal temperature for spore germination is 19–20°C. Physiologic races of the fungus exist. The host range of *P. philippinensis* includes oats (by inoculation), teosinte, cultivated and wild sugarcane, and cultivated and wild spe-

Fig. 113. Philippine downy mildew. (Courtesy C. De Leon)

cies of sorghum. Some of these hosts may be reservoirs for inoculum.

Management

The use of resistant hybrids and cultivars is the best management tactic. Initial studies on the inheritance of resistance showed that a few genes exhibiting partial dominance in combination with seedling vigor were involved in resistance. Subsequent studies identified polygenically inherited levels of resistance in which expression was dependent upon inoculum load. At light disease pressure, resistance was completely dominant; at moderate disease pressure, resistance was partially dominant but mostly additive; and at high disease pressure, resistance appeared recessive. Seed treatments or foliar fungicide applications, with products that have activity against oomycota (e.g., metalaxyl or mefenoxam), can be used if the disease is severe.

Selected References

Animal and Plant Health Inspection Service, U.S. Department of Agriculture. 2008. Agricultural Bioterrorism Protection Act of 2002—Biennial Review and Republication of the Select Agent and Toxin List, 7 CFR, Part 331, Federal Register Vol. 73, No. 201, Oct. 16, 2008.

Bonde, M. R., and Peterson, G. L. 1983. Comparison of host ranges of *Peronosclerospora philippinensis* and *P. sacchari*. Phytopathology 73:875-878.

Bonde, M. R., Peterson, G. L., Dowler, W. M., and May, B. 1984. Isozyme analysis to differentiate species of *Peronosclerospora* causing downy mildews of maize. Phytopathology 74:1278-1283.

De Leon, C., Granados, G., Wedderburn, R. N., and Pandey, S. 1993. Simultaneous improvement of downy mildew resistance and agronomic traits in tropical maize. Crop Sci. 33:100-102.

Exconde, O. R., and Molina, A. B., Jr. 1978. Note: Ridomil (Ciba-Geigy), a seed-dressing fungicide for the control of Philippine corn downy mildew. Philipp. J. Crop Sci. 3:60-64.

Frederiksen, R. A. 1989. Downy mildews. Pages 59-75 in: Plant Protection and Quarantine, Vol. II: Selected Pests and Pathogens of Quarantine Significance. R. P. Kahn, ed. CRC Press, Boca Raton, FL.

Kaneko, K., and Aday, B. A. 1980. Inheritance of resistance to downy mildew of maize, *Peronosclerospora philippinensis*. Crop Sci. 20:590-594.

Renfro, B. L. 1985. Breeding for disease resistance in tropical maize and its genetic control. Pages 341-365 in: Breeding Strategies for Maize Improvement in the Tropics. A. Brandolini and F. Salamini, eds. FAO, Firenze, Italy.

Weston, W. H., Jr. 1920. Philippine downy mildew of maize. J. Agric. Res. 19:97-122.

(Prepared by D. R. Smith and B. L. Renfro;
Revised by M. R. Bonde)

Sugarcane Downy Mildew

Severe outbreaks of sugarcane downy mildew on corn have limited the corn production in Australia, Fiji Islands, and Taiwan. Yield losses of 30–60% have been reported. The disease appears to be confined to tropical and subtropical regions and has also been reported from Japan, Nepal, New Guinea, India, the Philippines, and Thailand. It is a potential threat to corn production, particularly in Asia. The causal organism of sugarcane downy mildew of corn is one of four downy mildew pathogens that also cause downy mildew on sugarcane. It is one of three corn downy mildew pathogens on the Select Agents and Toxins List in the United States created in response to the Congressional Public Health Security and Bioterrorism Preparedness and Response Act of 2002 (see Agricultural Bioterrorism Protection Act of 2002—Biennial Review and Republication of the Select Agent and Toxin List, 7 CFR, Part 331, Federal Register Vol. 73, No. 201, Oct. 16, 2008).

Symptoms

Sugarcane downy mildew is characterized by local lesions that lead to systemic infection. Initial lesions are small, round, chlorotic spots on the leaves that appear 2–4 days after infection. Symptoms of systemic infection are pale yellow to white stripes or streaks at the bases of the third to sixth oldest leaves. Several streaks may form in each leaf and may extend the length of the leaf. On the leaves of some cultivars or on older leaves, the streaks may be narrow and discontinuous. On late-infected or mildly infected plants, streaks may disappear as plants approach maturity. White, downy, or powdery masses of conidia and conidiophores appear on both leaf surfaces and on the leaf sheaths and husks. This downy growth is usually produced at night when temperatures are moderate (25°C), especially when dew is present. Plants may be distorted and have small, numerous, poorly filled ears and improperly formed tassels. Ear shanks may be elongated.

Causal Organism

Sugarcane downy mildew is caused by *Peronosclerospora sacchari* (syn. *Sclerospora sacchari*). It is one of four sugarcane downy mildew pathogens of corn that also infect sugarcane. Conidia and conidiophores are superficial. The conidiophores are erect, hyaline, and 160–170 μm long, arising singly or in pairs from stomata. The conidia (15–23 × 25–41 μm) are hyaline, elliptical, and oblong or conical with a round apex and thin, smooth walls. The oospores (40–50 μm in diameter) are globular or slightly angular and yellow with walls 3.8–5 μm thick. The conidia and oospores germinate by hyaline, slender, nonseptate germ tubes that are 3.8 μm wide. *P. sacchari* is indistinguishable from *P. philippinensis* by isozyme analysis. These two species have the same, or nearly the same, host range.

Disease Cycle and Epidemiology

The fungus overseasons as mycelium in sugarcane. Conidia produced on diseased sugarcane are disseminated by wind or in splashing water to corn leaves. Under favorable conditions, germ tubes from conidia penetrate corn leaves through stomata, and tiny, round lesions covered with ectophytic mycelia form at infection sites. The mycelium spreads intercellularly by inserting knoblike haustoria into host cells. After the endophytic mycelium invades tissues at the growing point, the first streaks appear on young leaves. Once established in corn seedlings, the fungus produces large numbers of conidia that serve as secondary inocula for corn or sugarcane seedlings. The role of oospores in infection under natural conditions has not been established. The fungus has been detected in embryos, and while freshly harvested seeds from infected plants have been shown to transmit the disease, no transmission occurred when the seeds were dried to less than 20% moisture.

Moisture, temperature, and the age of the plants are the important factors in disease development. The optimum temperature for conidial formation, germination, and infection is 20–25°C in the presence of free water. Sporulation is nocturnal and decreases with increasing light intensity. Plants less than 1 month old are very susceptible but become resistant with age. Conidia are windborne but are viable for only a few hours, even under optimal conditions. Other hosts of *P. sacchari* include sugarcane, teosinte, gama grass, broom corn, and grain sorghum. Long-range dissemination is by mycelia in sugarcane sets.

Management

Use of resistant hybrids and cultivars is the most practical management practice. A single dominant allele designated *Dmr* from one of four sources in combination with a recessive gene from another source and polygenic resistance from other sources have been demonstrated to condition high levels of resistance to sugarcane downy mildew. Because of the development of resistant corn hybrids and resistant sugarcane, this disease is no longer of consequence in Australia and Thailand. Cultural practices such as the use of pathogen-free sugarcane sets and the isolation of corn in disease-free areas where sugarcane is not grown can provide some control. If practical, roguing and destruction of diseased plants and use of fungicides (systemic seed treatments and sprays) are recommended in some areas.

Selected References

Animal and Plant Health Inspection Service, U.S. Department of Agriculture. 2008. Agricultural Bioterrorism Protection Act of 2002—Biennial Review and Republication of the Select Agent and Toxin List, 7 CFR, Part 331, Federal Register Vol. 73, No. 201, Oct. 16, 2008.

Bonde, M. R., and Peterson, G. L. 1983. Comparison of host ranges of *Peronosclerospora philippinensis* and *P. sacchari*. Phytopathology 73:875-878.

Bonde, M. R., Peterson, G. L., Dowler, W. M., and May, B. 1984. Isozyme analysis to differentiate species of *Peronosclerospora* causing downy mildews of maize. Phytopathology 74:1278-1283.

Frederiksen, R. A. 1989. Downy mildews. Pages 59-75 in: Plant Protection and Quarantine, Vol. II: Selected Pests and Pathogens of Quarantine Significance. R. P. Kahn, ed. CRC Press, Boca Raton, FL.

Lal, S., Bhargava, S. K., and Upadhyay, R. N. 1979. Control of sugarcane downy mildew of maize with metalaxyl. Plant Dis. Rep. 63:986-989.

(Prepared by D. R. Smith and B. L. Renfro;
Revised by M. R. Bonde)

Rajasthan Downy Mildew

Rajasthan downy mildew has been reported as a significant problem in the Rajasthan region of India.

Causal Organism

Rajasthan downy mildew is caused by *Peronosclerospora heteropogoni*. The conidiophores are 14.3–25.0 × 81.6–142.8 µm. Conidia (14.3–20.4 × 14.3–22.4 µm) are globose and hyaline. Oospores are tuberculate and 24.5–36.7 µm in diameter. This fungus was previously identified as *P. sorghi*. However, *P. heteropogoni* infects *Heteropogon contortus* and *P. sorghi* does not. Conidia produced on *H. contortus* appear to be the primary inoculum source to infect corn, since oospores are produced abundantly on *H. contortus* but are not produced on corn. Planting corn early allows it to escape infection by developing resistance before large numbers of conidia are produced on *H. contortus*. Corn planted late can be severely damaged.

Selected Reference

Payak, M. M. 1975. Downy mildews of maize in India. Trop. Agric. Res. 8:13-18.

(Prepared by D. R. Smith and B. L. Renfro;
Revised by M. R. Bonde)

Spontaneum Downy Mildew

In about 1920, Spontaneum downy mildew was reported to be serious on corn in the Visayan Islands of the Philippines. During recent years, it has been rare on corn in the Philippines and has not been reported elsewhere on corn, except in Thailand. Other hosts include wild and cultivated sugarcane, teosinte, eulalia, *Euchlaena luxurians*, and *Sorghum* spp.

Causal Organism

Spontaneum downy mildew is caused by *Peronosclerospora spontanea* (syn. *Sclerospora spontanea*). Conidiophores are erect, hyaline, and 350–550 µm long and emerge singly or in groups from stomata on the lower leaf surfaces. Each is composed of a basal cell, a main axis, and a more or less complex dichotomously branched system. The main axis is usually expanded above the septum to 22.7–28.6 µm and constricted below the branches to about 21 µm. The conidia are hyaline, elongate, and ellipsoidal or cylindrical with rounded apices, lack a papilla, and are borne on slender, straight sterigmata about 13 µm long. Most conidia are 11–21 × 25–65 µm, compared with those of *P. sorghi*, which are 14.4–27.3 × 15–28.9 µm. The oogonia (55.1–82.1 µm) are amber brown and usually irregularly developed with several folds in the wall; the wall thickness is variable. The spherical oospores are mostly 39.2–53.3 µm (mean 45.4–48.8 µm) in diameter. The spore wall is hyaline to pale amber and 1.8–6.1 µm thick. Oospores have not been reported in corn tissue.

Selected Reference

Titatarn, S., and Syamananda, R. 1978. The occurrence of *Sclerospora spontanea* on *Saccharum spontaneum* in Thailand. Plant Dis. Rep. 62:29-31.

(Prepared by D. R. Smith and B. L. Renfro;
Revised by M. R. Bonde)

Leaf Splitting Downy Mildew

Corn is susceptible to *Peronosclerospora miscanthi* (syn. *Sclerospora miscanthi*) when inoculated, but infection has not been found to occur naturally. The fungus attacks eulalia, sorghum, and sugarcane in Taiwan and the Philippines.

Selected Reference

Shaw, C. G. 1978. *Peronosclerospora* species and other downy mildews of Gramineae. Mycologia 70(3):594-604.

(Prepared by D. R. Smith and B. L. Renfro;
Revised by M. R. Bonde)

Graminicola Downy Mildew

Graminicola downy mildew, also called green ear, is found on various grasses throughout the world. It has been found in corn only in Israel and the United States and is rare in both countries. In the United States, it passes from *Setaria* spp., and in Israel, from pearl millet. It occurs very infrequently and has been of no economic importance for the past 65–70 years.

Symptoms

Common symptoms include grayish blotching and mottling of leaves with a white, arachnoid to downy growth that develops on discolored areas, chlorotic streaking or striping, and stunting. Leaves may be thick, corrugated, and fragile. Symptoms usually appear 10 days after the plumule emerges. Affected plants occasionally outgrow the disease.

Causal Organism

Graminicola downy mildew is caused by *Sclerospora graminicola*. Sporangiophores are long (average 268 μm), nonseptate, and stalklike with a slightly bulbous foot. Sporangia (12–21 × 14–31 μm) are borne on short sterigmata, two to six on a branch, and have a prominent papillate operculum at the apex. They are broadly elliptical. Each sporangium germinates by releasing four to eight or more irregularly kidney-shaped, biciliate zoospores 9–12 μm in diameter. Sporangia are reported to germinate occasionally by germ tubes, but these reports probably result from germ tube formation from encysted zoospores trapped within sporangia. The oospores are pale brown and 22.5–35 μm in diameter and germinate by germ tubes. The oogonia are 30–60 μm in diameter with reddish brown or amber brown, heavy, uneven outer walls 4–12 μm thick.

Disease Cycle and Epidemiology

Primary infection occurs from oospores that overwinter in the soil, where they can survive for at least 3 years. Some remain viable after passage through the alimentary canals of cattle. The mode of germination is still uncertain. Corn seedlings are systemically infected. Disease development and spread are favored by light rains or heavy dews and moderate night temperatures, conditions optimal for the production of sporangia. Secondary spread to other hosts is by windborne sporangia produced superficially, but only where free moisture on plants is sustained. The fungus is not reported to be seedborne.

Temperatures of 24–32°C for 2 days after planting are conducive to heavy infection. The optimal temperature for sporangial formation and germination is 17°C; the maximum is 34°C. Species of the genera *Agrostis*, *Chaetochloa*, *Echinochloa*, *Euchlaena*, *Holcus*, *Panicum*, *Pennisetum*, *Saccharum*, *Setaria*, and *Sorghum* are susceptible to natural or artificial inoculation. Severe losses are encountered in pearl and foxtail millets. Strong physiologic specialization exists; most isolates from pearl millet do not attack *Setaria* spp. and vice versa.

Management

No control is needed for corn.

Selected References

Kenneth, R. G. 1981. Downy mildews of gramineous crops. Pages 367-394 in: The Downy Mildews. D. M. Spencer, ed. Academic Press, New York.

Shaw, C. G. 1978. *Peronospora* spp. and other downy mildews of the Gramineae. Mycologia 70:594-604.

(Prepared by D. R. Smith and B. L. Renfro;
Revised by M. R. Bonde)

Crazy Top

Crazy top is found in most areas with temperate or warm climates. It appears rarely in tropical areas but is widespread in the United States and has been reported in Canada, Mexico, South America, eastern and southern Europe, Africa, and Asia. Crazy top causes substantial losses only in very localized areas.

Symptoms

Symptoms vary greatly with time of infection and degree of host colonization by the fungus. Generally, excessive tillering (six to ten tillers per plant), rolling, and twisting of the upper leaves appear first (Fig. 114). Plants may be severely stunted (Fig. 115) or achieve full height. The most characteristic symptom is the partial or complete proliferation of the tassel, which continues until the tassel resembles a mass of leafy structures (Fig. 116). These modified, leaflike inflorescences are described as "crazy top." Phyllody of ears also may occur (Fig. 117). Leaves of severely infected plants are often narrow, strap-

Fig. 114. Crazy top and tillering. (Courtesy D. G. White)

Fig. 115. Stunting caused by crazy top. (Courtesy G. P. Munkvold)

Fig. 116. Tassel symptoms of crazy top. (Courtesy G. P. Munkvold)

Fig. 117. Phyllody symptom in a corn ear caused by crazy top. (Courtesy G. P. Munkvold)

like, and leathery. Stunting and chlorotic striping of leaves may also occur.

Causal Organism

Crazy top is caused by *Sclerophthora macrospora* (syns. *Sclerospora macrospora* and *Phytophthora macrospora*). Sporangia occur rarely on corn grown in the field but can be obtained by incubating excised young, diseased leaves. Their production decreases markedly with increased plant age and does not occur on intact leaves. When the sporangia are induced in complete darkness at an optimal temperature of 24–28°C, they are hyaline, lemon shaped, operculate, 30–65 × 60–100 µm, and attached to short, simple, sporangiophores emerging from stomata. The sporangia germinate by the release of numerous biciliate, subspherical to reniform, hyaline zoospores or occasionally by germ tubes. After a period of motility, the zoospores encyst and produce germ tubes. The oospores are hyaline to yellowish, multinucleate, somewhat globose, and 45–75 µm in diameter with granular contents. They germinate by a thin-walled tube bearing a sporangium, which in turn produces zoospores. The mycelium is coenocytic, thin walled, intercellular, and abundant in meristematic tissues.

Histopathology

The mycelium of *Sclerophthora macrospora* can be detected in brace roots, stalk tissue, leaves, and floral structures of affected plants by staining with zinc chloriodide. The mycelium stains a deep purplish blue, while the host tissue remains nearly colorless. The oogonia and oospores are situated in "nests" of several spores in vascular bundles or in their parenchymatous sheath cells within straplike leaves and leaf sheaths of stunted plants.

Disease Cycle and Epidemiology

The oospores germinate in saturated soil to produce sporangia from which zoospores escape and penetrate host tissues. After a short period of motility, they encyst, and each produces a germ tube. The mycelium develops systemically and is most abundant in meristematic tissues. The sporangia are produced at night on sporangiophores projecting from stomata in the leaves, but these are produced sparingly on corn. Oogonia are produced abundantly in infected tissues.

Crazy top develops when soils have been flooded shortly after planting or before plants are in the four- to five-leaf stage. Pooling of soil and water in the whorl of small plants can also lead to infection. Saturation for 24–48 h is sufficient for infection to occur, probably by initiating germination of soilborne oospores and providing water for the movement of zoospores to infection sites on corn. Diseased grasses may provide sporangial inoculum. Infection occurs at a wide range of soil temperatures. The optimal temperature for sporulation on excised leaves is 24–28°C; for sporangial germination, it is 12–16°C (22–25°C for finger millet in India). Seed transmission of the pathogen has been demonstrated from freshly harvested seeds but is generally considered unimportant in the dissemination of the fungus.

Sclerophthora macrospora has been reported on more than 140 species of the family Gramineae. Hosts include oats, wheat, sorghum, rice, crabgrass, witch grass, green foxtail, St. Augustine's grass, and barnyard grass. It causes considerable losses on finger millet in India.

Management

Adequate soil drainage should be provided, and planting in low, wet spots should be avoided.

Selected References

Kenneth, R. G. 1981. Downy mildews of gramineous crops. Pages 367-394 in: The Downy Mildews. D. M. Spencer, ed. Academic Press, New York.

Payak, M. M., Renfro, B. L., and Lal, S. 1970. Downy mildew diseases incited by *Sclerophthora*. Indian Phytopathol. 23:183-193.

Shaw, C. G. 1978. *Peronospora* spp. and other downy mildews of the Gramineae. Mycologia 70:594-604.

Sun, M. H., and Ullstrup, A. J. 1970. Production and germination of sporangia of *Sclerophthora macrospora* from corn. (Abstr.) Phytopathology 60:1316-1317.

Ullstrup, A. J. 1970. Crazy top of maize. Indian Phytopathol. 23:250-261.

(Prepared by D. R. Smith and B. L. Renfro;
Revised by M. R. Bonde)

Brown Stripe Downy Mildew

Brown stripe downy mildew, a disease that causes local lesions, can be very destructive in India and has been reported from most corn-growing areas of that country. A 63% loss in grain yield has been recorded in the Tarai region, Uttar Pradesh, India. Brown stripe downy mildew has also been reported in Nepal, Pakistan, and Thailand. The disease is most severe in areas that receive 100–200 cm of rain annually. The causal organism is one of three corn downy mildew pathogens on the Select Agents and Toxins List in the United States created in response to the Congressional Public Health Security and Bioterrorism Preparedness and Response Act of 2002 (see Agricultural Bioterrorism Protection Act of 2002—Biennial Review and Republication of the Select Agent and Toxin List, 7 CFR, Part 331, Federal Register Vol. 73, No. 201, Oct. 16, 2008).

Symptoms

Initially, lesions develop on the leaves as narrow, chlorotic or yellowish stripes, 3–7 mm wide; they have well-defined margins and are delimited by the veins (Fig. 118). The stripes later become reddish to purple. Lateral development of the lesions causes severe striping and blotching. Seed development may be suppressed, and the plant may die prematurely if blotching takes place prior to flowering. Sporangia on the leaves appear as a downy or woolly growth on both surfaces of the lesions. Floral or vegetative parts are not malformed, and the leaves do not shred.

Fig. 118. Brown stripe downy mildew. (Courtesy C. De Leon)

Causal Organism

Brown stripe downy mildew is caused by *Sclerophthora rayssiae* var. *zeae*. The oogonia (3.3–4.5 µm in diameter) are subglobose, thin walled, and hyaline to light straw colored. Oospores (29.5–37 µm in diameter) are spherical to subspherical with smooth, glistening walls that are 4 µm thick and confluent with the oogonia. They have hyaline contents, including a prominent oil globule. Oospores and oogonia are numerous and scattered in the leaf mesophyll or under stomata. Sporangiophores are short and determinate and arise from hyphae in the substomatal spaces. Sporangia (18.5–26 × 29–66.5 µm) are produced sympodially in groups of two to six and are hyaline, ovate to cylindrical with a prominent peduncle, and smooth walled and individually produce four to eight zoospores upon germination. The encysted zoospores are hyaline, spherical, and 7.5–11 µm in diameter.

Disease Cycle and Epidemiology

S. rayssiae var. *zeae* survives as oospores in infected debris in the soil. Upon germination, the oospore produces a sporangiophore bearing a sporangium that liberates four to eight zoospores. In the presence of enough moisture or at high temperatures, the sporangium may produce a germ tube that can also infect corn leaves. The pathogen can also overseason in crabgrass as oospores or as mycelium from which sporangia are produced. Secondary spread is by sporangia. While the fungus can be detected within the embryo of corn seeds and has been shown to be seed transmitted when seeds from infected plants are planted in sterile soil immediately after harvest, seeds dried to 14% moisture or less and stored for 4 weeks or more do not transmit this or other downy mildew diseases.

Sporangia are produced and germinate at optimal temperatures of 22–25°C and 20–22°C, respectively, while oospores are formed at higher temperatures. Soil temperatures of 28–32°C favor disease development. Sporangial production, germination, and infection require a film of moisture for 12 h or longer. Zoospores germinate at 15–30°C (optimum 22–25°C). Zinc deficiency predisposes plants to infection.

Management

Use of resistant hybrids and cultivars is the most practical management practice. In studies conducted in India, it was determined that resistance is polygenically inherited with additive gene action predominating. Partial dominance occurs in the nonadditive component, and epistasis has been detected in some crosses. Additionally, cultural practices such as planting before the rainy season begins may be of value. Fungicide seed treatments followed by systemic foliar sprays, using products with activity against oomycota (e.g., metalaxyl or mefenoxam), provide excellent control.

Selected References

Animal and Plant Health Inspection Service, U.S. Department of Agriculture. 2008. Agricultural Bioterrorism Protection Act of 2002—Biennial Review and Republication of the Select Agent and Toxin List, 7 CFR, Part 331, Federal Register Vol. 73, No. 201, Oct. 16, 2008.

Asnani, V. L., and Bhusan, B. 1970. Inheritance study on the brown stripe downy mildew of maize. Indian Phytopathol. 23:220-223.

Frederiksen, R. A. 1989. Downy mildews. Pages 59-75 in: Plant Protection and Quarantine, Vol. II: Selected Pests and Pathogens of Quarantine Significance. R. P. Kahn, ed. CRC Press, Boca Raton, FL.

Handoo, M. I., Renfro, B. L., and Payak, M. M. 1970. On the inheritance of resistance to *Sclerophthora rayssiae* var. *zeae* in maize. Indian Phytopathol. 23:231-249.

Lal, S., Saxena, S. C., and Upadhyay, R. N. 1980. Control of brown stripe downy mildew of maize by metalaxyl. Plant Dis. 64:874-876.

Payak, M. M., Renfro, B. L., and Lal, S. 1970. Downy mildew diseases incited by *Sclerophthora*. Indian Phytopathol. 23:183-193.

Putman, M. L. 2007. Brown stripe downy mildew (*Sclerophthora rayssiae* var. *zeae*) of maize. Online. Plant Health Progress. doi: 10.1094/PHP-2007-1108-01-DG

Singh, I. S., and Asnani, V. L. 1975. Combining ability for resistance to brown stripe downy mildew in maize. Indian J. Genet. Plant Breed. 35:128-130.

(Prepared by D. R. Smith and B. L. Renfro;
Revised by M. R. Bonde)

Ergot

Ergot, also known as horse's tooth (*diente de caballo*), is found in certain humid valleys of central Mexico. The disease is not common, but high levels of infection can occur in individual fields. Yield loss occurs through replacement of kernels by sclerotia and grain from infected ears is of poor quality. Seeds from infected ears may have severely reduced germination. The fungus produces toxic alkaloids that may contaminate grain.

Symptoms

Sclerotia replace the kernels, and each ear may contain one to several sclerotia (Fig. 119). Kernels adjacent to sclerotia may become shriveled and discolored. Sclerotia (2–5 × 5–8 cm) are first white to cream colored, soft, sticky, and hollow. Sclerotia are soft and pale at first; mature sclerotia are comma shaped (resembling a horse's tooth) and white to grayish brown and have pink to lavender centers. Those beneath the husks are lighter in color. Honeydew from the sclerotia is filled with small, hyaline, ovoid spores.

Causal Organism

Ergot is caused by *Claviceps gigantea*. Sclerotia germinate to form stromatic heads that contain flask-shaped perithecia arranged peripherally (Fig. 120). The asci in the perithecia are cylindrical and hyaline. Ascospores (1.5 × 176–186 µm) are fili-

Fig. 119. Ergot. (Courtesy B. M. Anderson)

Fig. 120. Sclerotium of *Claviceps gigantea* germinating to produce perithecial stromata. (Courtesy C. De Leon)

form, hyaline, and nonseptate. Macroconidia (4.2–5.8 × 8.3–27 μm) are elliptical, and microconidia (2.5–3.3 × 4.2–6.7 μm) are ovoid. Both spore types are hyaline and nonseptate. *C. gigantea* can be isolated by surface sterilizing mature sclerotia and plating fragments of the horny interior on glucose–yeast extract agar at 24°C. Colonies attain their maximum diameter (about 1.5 cm) in about 4 weeks.

Disease Cycle and Epidemiology

The fungus overwinters as sclerotia on the ground or mixed with seed. The sclerotia germinate in the spring to produce stalked stromata with heads containing many embedded perithecia with asci (Fig. 120). The ascospores serve as primary inoculum. These spores are forcibly ejected and carried by the wind to floral parts of susceptible corn plants. The developing hyphae produce macroconidia and microconidia in a sticky matrix (honeydew), which can cause secondary infections. Insects may be important in conidial transmission.

Ergot is endemic in certain high, cool, humid valleys of central Mexico. Disease development is favored by mean temperatures of 13–15°C and high levels of rainfall (1,000 mm, or 40 in., per year).

Management

No management tactics are reported.

Selected References

Fucikovsky, L., and Moreno, M. 1971. Distribution of *Claviceps gigantea* and its percent attack on two lines of corn in the state of Mexico, Mexico. Plant Dis. Rep. 55:231-233.

Fuentes, S. F., de Lourdes de la Isla, M., Ullstrup, A. J., and Rodriguez, A. E. 1964. *Claviceps gigantea*, a new pathogen of maize in Mexico. Phytopathology 54:379-381.

(Prepared by D. G. White)

Smuts

Smut diseases are caused by fungi in the order Ustilaginales of the phylum Basidiomycota. They are characterized by the formation of galls, which deform the plant organs and consist of a mixture of fungal and plant cells, eventually developing into masses of dark teliospores. Smut infections result in the destruction of plant cells, which are then completely replaced by fungal teliospores, or in the stimulation of abnormal cell proliferation, followed by the replacement of plant cells by teliospores. Some smut fungi infect seedlings before they emerge and infect the plants systemically, eventually reaching the inflorescences and causing symptoms there; others cause only local infections of the aboveground tissues and may cause symptoms in any part of the plant. The galls, when appearing in corn ears, can completely replace the ear tissues. Therefore the yield impact of smut diseases is readily evident. They are found throughout the world and, historically, were among the most economically important diseases in cereal crops. During the twentieth century, improved control methods (including systemic seed treatments) became available for many smut diseases, and their economic impact has been reduced. Although there are more than 1,000 species of smut fungi, in corn, there are just two species of smut fungi causing two distinctly different diseases. Common smut is caused by the local infection of any plant part, whereas head smut results from systemic infection, although symptoms usually appear only in the ear or tassel. A third disease covered in this section (false smut) causes smutlike symptoms but is not a true smut fungus.

Smut fungi are obligate parasites that usually have a narrow host range. The common smut pathogen, *Ustilago maydis*, has been studied extensively as a model organism for functional genomics and cell biology. Its entire genome has been sequenced, and these studies have provided valuable insight into fungal genetics and host–pathogen interactions.

Selected References

Fischer, G. W. 1953. Manual of the North American Smut Fungi. Ronald Press, New York.

Fischer, G. W., and Holton, C. S. 1957. Biology and Control of the Smut Fungi. Ronald Press, New York.
Griffiths, M. A. 1928. Smut susceptibility of naturally resistant corn when artificially inoculated. J. Agric. Res. 36:77-84.
Vanky, K. 2012. Smut Fungi of the World. American Phytopathological Society, St. Paul, MN.

(Prepared by G. P. Munkvold)

Common Smut

Common smut, also called boil smut and blister smut, is found worldwide wherever corn is grown. The disease is most severe after young, actively growing plant tissues are wounded.

Fig. 121. Common smut. (Courtesy D. G. White)

Losses from this disease vary greatly. Annual losses in field corn seldom exceed 2% when resistant cultivars or hybrids are grown. Common smut can be of considerable economic importance in sweet corn, particularly when smut galls replace kernels on ears. In certain parts of Mexico, galls of common smut on ears of corn are considered an edible delicacy known as *huitlacoche* or *cuitlacoche*. Although there are a few, relatively obscure references to adverse effects from the consumption of smut galls, there is no well-documented, replicated evidence of deleterious reactions because of the mycotoxins or other compounds produced by *U. maydis*. However, in some cases, smut-affected ears are more susceptible to infection by mycotoxin-producing fungi, such as *Fusarium* spp. or *Aspergillus flavus*.

Symptoms

All aboveground parts of the plant can be infected, particularly young, actively growing meristematic tissues. Unlike many other smuts of grasses, common smut results from a local rather than a systemic infection; and galls on different plant parts are the result of independent infections of each of those tissues. Symptoms are easily recognized. Galls form over a 2- to 3-week period. As galls are first forming on ears or stalks, they have a spongy consistency and are covered with glistening, greenish white to silvery white periderm tissue (Fig. 121). As galls begin to mature in about 14–21 days, the interior gall darkens and turns into masses of powdery, dark olive brown to black teliospores, for which the disease is named (Fig. 122). Ears affected by smut are more susceptible to subsequent infection by ear rot fungi, such as *Fusarium* spp. Galls on leaves usually remain small, 0.6–1.2 cm in diameter, and become hard and dry without rupturing (Fig. 123). Galls on other plant tissues may reach 15 cm or more in diameter before becoming powdery masses of teliospores. Early infection of the growing point may kill or stunt young plants (Fig. 124). Galls on stalks may cause plants to die back or snap, or they may be barren or produce small ears.

Causal Organism

Common smut is caused by *Ustilago maydis* (syn. *U. zeae*) Teliospores (sometimes called chlamydospores) are olive brown to black, spherical to ellipsoidal, bluntly echinulate, and 6–11 μm in diameter. Although teliospores of the common

Fig. 122. Common smut galls on ears (left) and a stalk (right). (Courtesy G. P. Munkvold)

Fig. 123. Common smut galls on corn leaves. (Courtesy G. P. Munkvold)

Fig. 124. Stunting of a young plant with common smut infection of leaves in the whorl. (Courtesy G. P. Munkvold)

smut and head smut pathogens have overlapping morphology, they can often be distinguished based on the larger size, thicker cell wall, finer echinulation, and darker pigmentation of *Sporisorium reilianum* teliospores compared with those of *U. maydis* (Fig. 125). The dikaryotic teliospores usually germinate by formation of a septate promycelium with four or more hyaline, fusiform basidiospores, which also are called sporidia (Fig. 126). Sporidia usually are uninucleate and haploid, although solopathogenic, dikaryotic sporidia have been found in nature or constructed in the laboratory. Infection may take place from hyphae arising directly from germinating teliospores, from the fusion of uninucleate sporidia of compatible mating types, or from dikaryotic sporidia. Mating and infection are regulated by two loci, designated *a* and *b* and historically called the mating locus and the pathogenicity locus, respectively. Alleles at *a* are thought to encode pheromones or pheromone receptors, while those at *b* encode DNA-binding proteins, presumably with regulatory functions.

Disease Cycle and Epidemiology

Teliospores overwinter in crop debris or soil, where they can remain viable for several years (Fig. 126). They are disseminated directly, or under favorable conditions, they germinate to produce sporidia, which are disseminated to young, developing corn tissues. Only binucleate mycelia can infect host tissues. Usually the plant is penetrated through wounds; however, direct penetration has been reported. Host cells surrounding the fungal hyphae are stimulated to enlarge and divide, and galls begin to form. The fungus may also grow down stigmas (silks) and into developing kernels, resulting in galls replacing kernels on ears. Galls enlarge on ears for about 2–3 weeks after kernels become infected. Infection of kernels and subsequent gall formation on ears are inhibited by pollination because of an abscission zone formed at the base of the stigma soon after fertilization. Hence, kernel galls are formed from a combination of pericarp and fungal tissues as the embryo and endosperm have not yet formed because of the lack of fertilization. The decreased ability of *U. maydis* to infect kernels after pollination may explain why smut often occurs at the tip end of an ear, particularly under environmental conditions that result in incomplete pollination of ears.

There is no consensus on the environmental conditions that favor smut infection. Most reports indicate that rainy, humid weather is critical for infection, although some researchers have observed an association between dry weather and the prevalence of smut. Any environmental conditions (particularly drought) that result in poor pollination enhance the incidence of kernel galls on ears and may contribute to the conflicting obser-

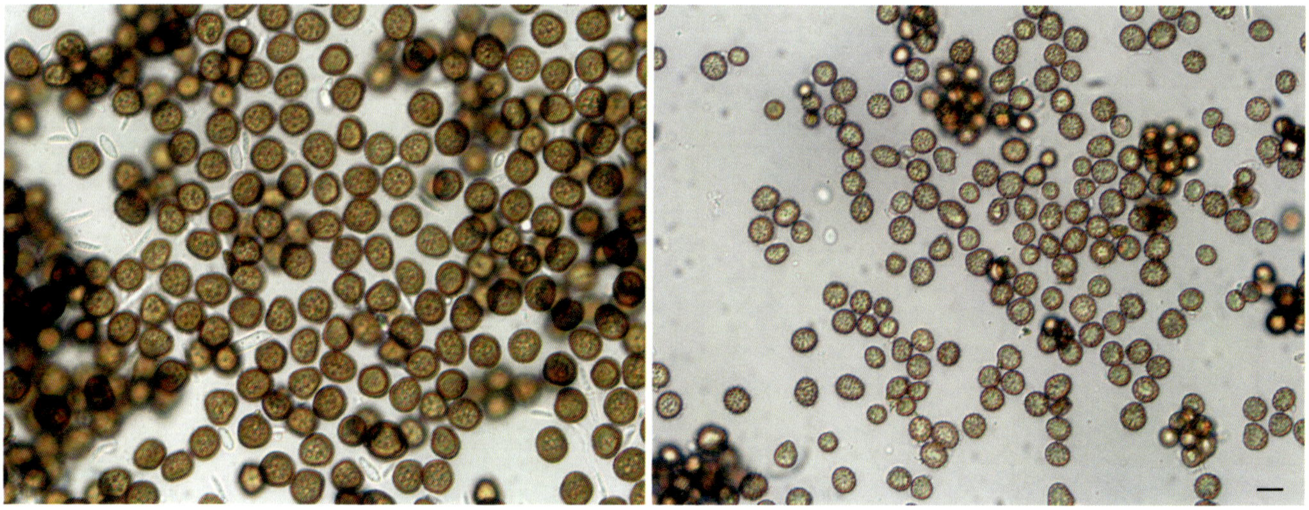

Fig. 125. Teliospores of *Sporisorium reilianum* (left) and *Ustilago maydis* (right). Bar (right) = 5 µm. (Courtesy D. Mayfield)

vations about conducive environmental factors. Several reports associate smut incidence with vigorously growing plants. Fertilizer with nitrogen has been reported to increase smut incidence, while phosphate fertilization decreased incidence.

During harvest, teliospores, naked or associated with infected plant tissue, become deposited into or onto the soil. They may also be aerially dispersed during harvest. Teliospores can contaminate the surface of healthy kernels that are harvested for grain or seed but they do not infect kernels. Teliospores do not pose a risk in grain storage, and the fungus is not seed transmitted.

Management

Many tactics for managing common smut have been recommended or evaluated, including crop rotation, sanitation, seed treatment, quarantines, application of foliar fungicides, modification of fertility, and biological control. In spite of these frequently mentioned management practices, the only practical and effective way to control common smut is to grow resistant hybrids.

Classification of corn as resistant or susceptible to common smut is based primarily on natural infection. Natural infection is also the primary selection pressure used to eliminate smut-susceptible lines in breeding programs. In spite of nearly 90 years of research, the genetics of resistance and the mechanisms of resistance to *U. maydis* are not fully understood. Early in the study of smut resistance, Griffiths demonstrated that no corn line is immune to infection when compatible sporidia are injected into susceptible host tissues. She concluded that resistance or susceptibility is largely a matter of the accessibility of susceptible plant parts to the invading fungus. Thus, resistance appears to be associated with morphological and phenological characteristics that allow corn to avoid infection. Some studies indicate that resistance is inherited polygenically and that resistant lines could be selected by evaluating the progeny of developmental materials test-crossed with susceptible lines. More recently, quantitative trail loci have been associated with reduced incidence of smut infection, but the function of those genes is unknown. It is likely that these quantitative trail loci are associated with avoidance mechanisms as suggested by Griffiths.

Variability in virulence, pathogenicity, and host specificity, has been reported from many collections of *U. maydis*. Nevertheless, factors determining resistance may be such that the genetic variability of *U. maydis* is relatively unimportant, and resistance may be effective against the broad array of genotypes that constitute a local population of *U. maydis*.

While quarantines and protectant seed treatments may be employed to prevent the introduction of *U. maydis* to the few areas

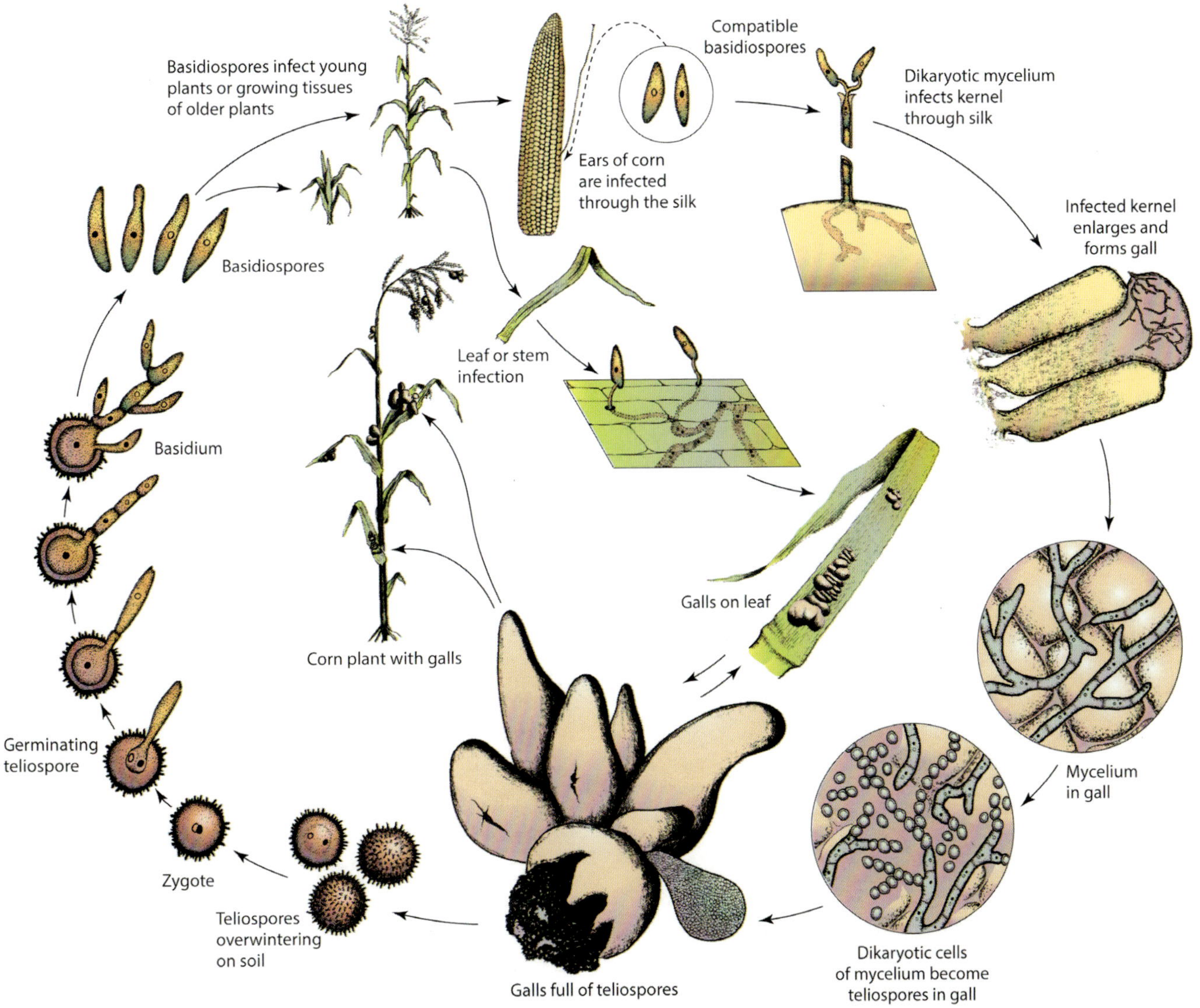

Fig. 126. Disease cycle of common smut, caused by *Ustilago maydis*. (Reproduced with permission of Elsevier, from Agrios, G. N. 2005. Plant Pathology, 5th ed. Elsevier Academic Press, New York; permission conveyed through Copyright Clearance Center)

where it does not already occur, it has long been recognized that contaminated seed is not an important source of inoculum.

Selected References

Christensen, J. J. 1963. Corn Smut Caused by *Ustilago maydis*. Monogr. 2. American Phytopathological Society, St. Paul, MN.

Kennedy, D. 1989. The Art of Mexican Cooking. Bantam, New York.

Pataky, J. K., and Chandler, M. A. 2003. Production of huitlacoche, *Ustilago maydis*: Timing inoculation and controlling pollination. Mycologia 95:1261-1270.

Snetselaar, K. M., and Mims, C. W. 1993. Infection of maize stigmas by *Ustilago maydis*: Light and electron microscopy. Phytopathology 83:843-850.

(Prepared by J. K. Pataky)

Head Smut

Head smut is found in North America in the deltas and intermountain valleys of the Pacific Northwest, California, Colorado, the high plains of Texas, North Dakota, South Dakota, Oklahoma, Kansas, Nebraska, and Minnesota (United States) and Ontario (Canada). It is also found in Argentina, Bolivia, Brazil, Chile, Colombia, Central America, Australia, China, Mexico, New Zealand, the former Soviet Union, South Africa, and southern Europe. Incidences of head smut as high as 80% have been observed.

Symptoms

Distinct symptoms of head smut first appear when ears and tassels form. Tassels and ears of infected plants often are replaced by smut sori. Each sorus is initially covered by a periderm, which eventually ruptures to reveal a dry mass of teliospores. Vascular elements remain relatively intact and appear as black, fibrous strands of tissue (Fig. 127). Rudimentary cob tissue may be evident internally, if galls are sliced or split open (Fig. 128). Tassel infection may be confined to individual spikelets, causing a shootlike growth, or the entire tassel may proliferate, forming bizarre structures (Fig. 129). If the tassel is infected, all ears on the plant will be smutted or aborted, with rudiments of leafy tissue replacing the normal ears (Fig. 130). Smutted ears are rounded and lack silks (Fig. 131). When the tassel is not infected, most of the ears

Fig. 127. Head smut symptoms on a corn ear. (Courtesy G. P. Munkvold)

Fig. 129. Head smut symptoms on a tassel. (Courtesy D. G. White)

Fig. 128. Head smut gall split open to reveal rudimentary cob tissue. (Courtesy G. P. Munkvold)

Fig. 130. Head smut symptoms on a corn ear, showing proliferation of vegetative tissue. (Courtesy G. P. Munkvold)

Fig. 131. Head smut symptoms on a corn ear. (Courtesy G. P. Munkvold)

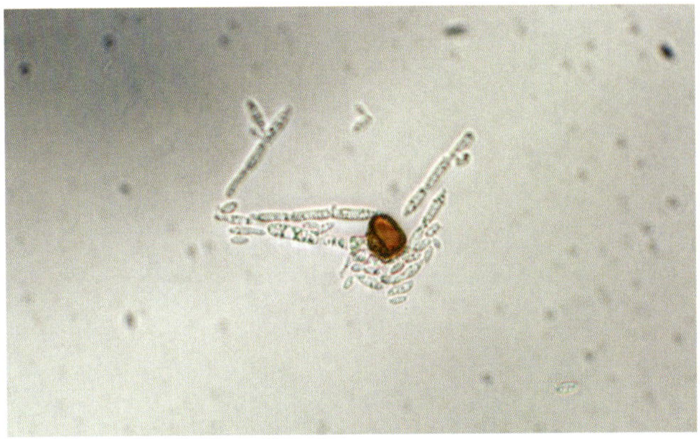

Fig. 132. Germinating teliospore of *Sporisorium reilianum* (syn. *Sphacelotheca reiliana*) with sporidia. (Courtesy D. Mayfield)

will be smutted but a few smut-free kernels may be produced. Infected plants with smutted tassels may be severely dwarfed. Excess tillering is common. Sori on the leaves develop as long, thin stripes.

Causal Organism

Head smut is caused by *Sporisorium reilianum* (syns. *Sorosporium reilianum*, *Sphacelotheca reiliana*, *Sporisorium holci-sorghi*, *Sporisorium reilianum* f. sp. *zeae*, and *Ustilago reiliana*). Teliospores are reddish brown to black, globose, abundantly and conspicuously spiny, and 9–14 µm in diameter. Although teliospores of the common smut and head smut pathogens have overlapping morphology, they can often be distinguished based on the larger size, thicker cell wall, finer echinulation, and darker pigmentation of *Sporisorium reilianum* teliospores compared with those of *U. maydis* (Fig. 125). Teliospores germinate and form basidia and lateral haploid sporidia that are small, hyaline, single celled, subglobose, and 7–15 µm in diameter (Fig. 132). Plus and minus mating types exist among the sporidia in approximately equal numbers. Infection occurs only from a sporidial mixture of mating types. Teliospores may germinate in soil, producing long, dikaryotic infection hyphae. Host-specific pathogenic specialization occurs in *Sporisorium reilianum*; one race group (or special form, if the *Sporisorium reilianum* f. sp. *zeae* nomenclature is used) is limited to corn and another to grain sorghums, sorgo, forage–sorghum hybrids, and some Sudan grass cultivars. Hybrids of *Sporisorium reilianum* that infect both corn and sorghum have been observed experimentally.

Disease Cycle and Epidemiology

The life cycle of *Sporisorium reilianum* differs from that of *U. maydis* in several important ways. As with *U. maydis*, teliospores can survive for several years in soil or crop residue and germinate to form sporidia. Rather than being dispersed to aboveground plant parts, *Sporisorium reilianum* attacks seedling roots, and unlike that of *U. maydis*, the mycelium develops systemically, invading undifferentiated floral tissues. Teliospores of *Sporisorium reilianum* may be externally seedborne and can be seed transmitted, but this is not an important source of inoculum.

Soil temperatures of 21–28°C and moderate to low soil moisture (15–25%, w/w) favor infection. The disease is exacerbated by nitrogen deficiency. Incidence decreases with applications of urea, ammonium sulfate, triple superphosphate, and calcium nitrate. Injuring seedlings has been an effective inoculation method in head smut nurseries and probably is associated with a higher incidence of natural infection.

Management

Resistance is available in corn hybrids. Inheritance of resistance appears to vary among sources of resistance. In some studies, resistance appeared to be partially dominant. The resistance of the dent corn line N6 was dominant in trials of uninjured plants but additive when seedlings were predisposed to infection by clipping. In studies done in China, resistance was quantitatively inherited, dominant, recessive, or controlled by the interaction of several nonallelic genes, depending on the parents. Hybrids with rapid seedling emergence may escape infection.

Seed treatment with systemic fungicides (e.g., tebuconazole) and in-furrow treatments with some fungicides have been effective management practices. Foliar applications have not been effective for head smut. Crop rotation is of questionable value since teliospores can survive in soil for prolonged periods. The disease can be avoided in areas where corn is planted and emerges at temperatures that are below those needed for teliospore germination.

Selected References

Ali, A., and Baggett, J. R. 1990. Inheritance of resistance to head smut disease in corn. J. Am. Soc. Hortic. Sci. 115:668-672.

Frederiksen, R. A. 1977. Head smut of corn and sorghum. Pages 89-105 in: Proc. Annu. Corn Sorghum Res. Conf., 32nd. American Seed Trade Association, Washington, DC.

Halisky, P. M. 1963. Head smut of sorghum, sudangrass, and corn, caused by *Sphacelotheca reiliana* (Kühn) Clint. Hilgardia 34:287-304.

(Prepared by J. K. Pataky)

False Smut

False smut has been reported worldwide. It is found on rice, corn, and other members of the family Gramineae but is of little economic importance.

Symptoms

False smut is characterized by galls (sclerotia) similar to those caused by ergot that replace the flowers on the tassels

Fig. 133. False smut. (Courtesy C. De Leon)

(Fig. 133). Usually only a few flowers are infected. The sclerotia superficially resemble smut sori.

Causal Organism

False smut is caused by *Ustilaginoidea virens* (syn. *U. oryzae*). Mature sclerotia are spherical, 4–15 mm in diameter, olive green to black, and velvety and have white interiors. The conidia (4–7 µm in diameter) are spherical to oval, echinulate, and olive green and are borne on short, sterigma-like projections from septate, yellowish green hyphae.

Disease Cycle and Epidemiology

The disease cycle is similar to that of ergot. False smut is favored during hot, wet weather.

Selected Reference

Haskell, R. J., and Diehl, W. W. 1929. False smut of maize, Ustilaginoidea. Phytopathology 19:589-592.

(Prepared by J. K. Pataky)

Rusts

Rust diseases are caused by fungi in the order Uredinales of the phylum Basidiomycota. Rusts have been recognized as serious threats to agriculture for millennia, causing devastating diseases on a wide range of crops, especially small grain cereals, legumes, and several trees, including pine and coffee. The economic importance of rusts is matched by few other groups of plant pathogens; outbreaks of rust in several crops have caused famines and resulted in massive shifts in agricultural production on numerous occasions. Rust diseases can produce a variety of symptoms and signs, but the most common and noticeable are pustules that break through the epidermis, releasing powdery spores that are usually rust colored, brown, orange, yellow, or white. Pustules usually appear on leaves or stems and the infections are local (with some exceptions). Rust fungi are obligate parasites, usually with very narrow host specialization. Within a morphologically uniform species of rust fungus, there can be numerous host-specific forms (*formae speciales*). Rusts have complicated life cycles, sometimes with as many as five different spore types, although some rusts produce only two spore types (teliospores and basidiospores) and are referred to as microcyclic. Rusts may be heteroecious (requiring two different hosts to complete their life cycle) or autoecious (able to complete their life cycle on a single host).

Within each *forma specialis* there can be numerous physiological races. Each race can attack only specific host genotypes, depending on the resistance genes (or lack thereof) in the plant. Resistance can sometimes be short lived since resistant crop cultivars are infected by previously unknown races. There are more than 5,000 known species of rust fungi (and even more *formae speciales*), but only three occur on corn. They differ in symptoms, temperature optima, and geographical distribution.

Selected References

Agrios, G. N. 2005. Plant Pathology, 5th ed. Elsevier Academic Press, Burlington, MA.
Cummins, G. B. 1941. Identity and distribution of three rusts of corn. Phytopathology 31:856-857.
Cummins, G. B. 1971. Rusts of Cereals, Grasses and Bamboos. Springer-Verlag, New York.
Hooker, A. L. 1985. Corn and sorghum rusts. Pages 207-236 in: The Cereal Rusts, Vol. 2, Diseases, Distribution, Epidemiology, and Control. A. P. Roelfs and W. R. Bushnell, eds. Academic Press, Orlando, FL.

(Prepared by G. P. Munkvold)

Common Rust

The common rust fungus was first described from plant tissue that was believed to be from sorghum. While the binomial *Puccinia sorghi* suggests sorghum as a host, in fact, the fungus does not infect *Sorghum* spp. As corn cultivation spread throughout the world from Central America, *P. sorghi* spread with the crop, and common rust is now found wherever corn is grown. It is the most common rust disease in temperate climates, appearing nearly every year in many areas. Common rust reduces grain yield by absorbing plant photosynthate and reducing the photosynthetic capacity of infected leaf tissue. Estimates of losses vary but are associated with the percentage of the leaf area infected at different host growth stages. Estimates of reductions in grain weight range from about 3–8% for each 10% of the total leaf area infected.

Symptoms

Uredinia are abundant on leaves, rupturing the leaf epidermis and creating blisterlike pustules that are easily recognized (Figs. 134 and 135). Urediniospores form in uredinia 7 days after infection on both leaf surfaces, distinguishing common rust from southern rust, which has little or no pustule development on the lower leaf surface. Uredinia frequently appear in bands as a result of infection that took place when the leaf tissue formed a whorl. The circular to elongate, golden brown to cinnamon brown pustules are sparsely scattered over both leaf surfaces and become brownish black as the infected leaf tissues mature and teliospores develop in place of urediniospores (Fig. 136A). When infection is severe, necrotic areas may develop surrounding pustules, scattered across the leaves, or in bands (Fig. 137); these can result in chlorosis and death of entire leaves and leaf sheaths, sometimes resembling the symptoms of other diseases.

Causal Organism

Puccinia sorghi causes common rust. Urediniospores (21–30 × 24–33 µm) are cinnamon brown and mostly spherical to broadly ellipsoidal. Spore walls are golden or cinnamon brown, 1.5–2 µm thick, and moderately echinulate with three to four equatorial germ pores (Fig. 138A). Each spore is binucleate, as is the mycelium that develops upon germination. Teliospores (14–25 × 28–46 µm) replace urediniospores in the pustules and are chestnut brown or golden brown to black, smooth, oblong to ellipsoid or obovoid, two celled with slight constrictions at the septa, and attached to pale yellowish to brownish pedicels up to 80 µm long (once or twice the length of the spores) (Figs. 136B and 138A). Aeciospores (9–13 × 18–6 µm) are pale yellow, verrucose, and mostly spherical or ellipsoid and occur in "cluster cups" on *Oxalis* spp.

Disease Cycle and Epidemiology

In certain areas of the world, teliospores germinate (usually in the spring) to form basidia on which small, thin-walled, hyaline, haploid basidiospores are produced. These spores cannot infect corn but germinate and penetrate the leaves of *Oxalis* spp., forming spermagonia with minute spermatia on the upper leaf surface. The spermatia fuse with flexuous (receptive) hyphae of the opposite mating type, initiating the aecial stage on the lower leaf surface. The binucleate aeciospores in the cluster

Fig. 134. Uredinia of the common rust pathogen, *Puccinia sorghi.* (Courtesy G. P. Munkvold)

Fig. 135. Uredinia of the common rust pathogen, *Puccinia sorghi.* (Courtesy G. P. Munkvold)

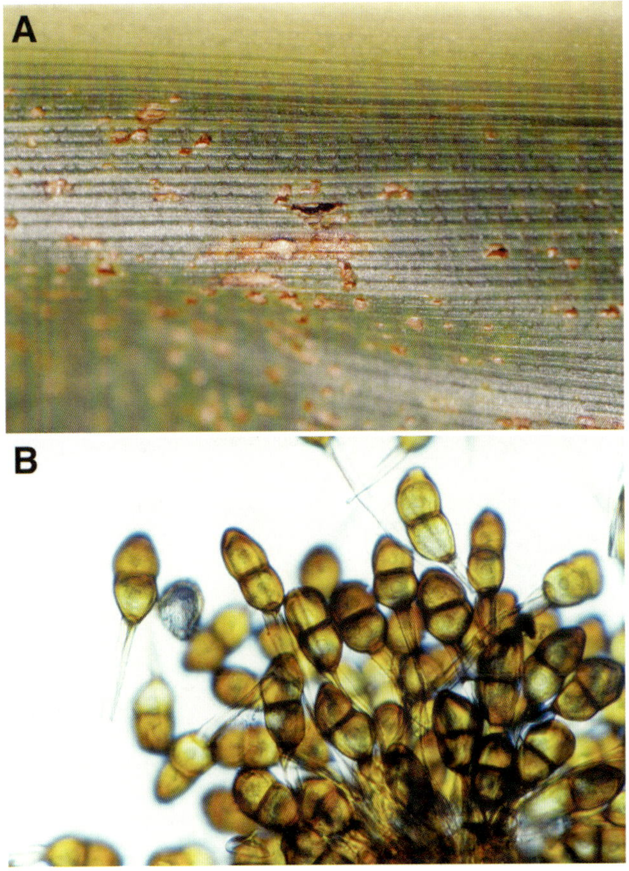

Fig. 136. Common rust pathogen, *Puccinia sorghi.* **A,** Uredinia and telia; and **B,** teliospores. (Courtesy G. P. Munkvold)

Fig. 137. Necrotic leaf lesions resulting from common rust infection. (Courtesy G. P. Munkvold)

cups are disseminated by the wind and infect corn leaves. These infections give rise to uredinia and urediniospores, the repeating stage of the fungus. Basidiospore infection and formation of aecia are found infrequently in temperate regions of Europe, the former Soviet Union, the United States, Mexico, South Africa, India, and Nepal on *Oxalis* spp. Uredial and telial stages appear worldwide where corn or teosinte is grown.

In most temperate areas of the world, including the North American Corn Belt, *P. sorghi* does not survive the winter on *Oxalis* spp.; hence, aeciospores do not serve as primary inocula for corn. Initial infection of corn coincides with the arrival of wind-blown urediniospores from nearby tropical and subtropical regions, where the aecial stage appear on *Oxalis* spp. Urediniospores form as secondary inocula on corn. In the midwestern United States, arrival of urediniospores and initial infection usually are in late May to early June. Moderate temperatures (16–25°C), high relative humidity (greater than 95%), and rainy weather favor rust development and spread. About 6 h of moisture is required for urediniospores to germinate and infect. Hence, leaves can often be infected even under relatively dry conditions because tissue in the leaf whorl is moist for the minimal period required for infection. Severity of infection of individual leaves is affected by the vegetative phase change in corn. Younger leaf tissue is more susceptible to infection than are mature, adult leaves, possibly because of differences in juvenile and adult leaf morphology, e.g., the presence of waxes and trichomes on mature leaves. Numerous physiological races of *P. sorghi* can be identified from reactions on corn lines with specific genes for resistance.

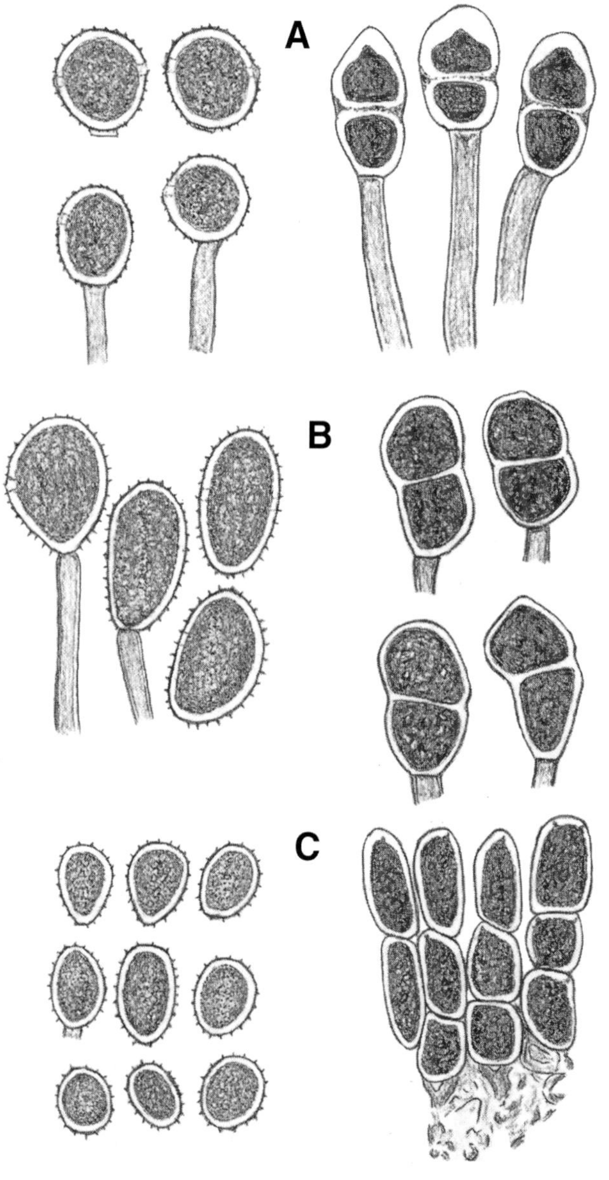

Fig. 138. Urediniospores and teliospores of three species of rust fungi. **A,** *Puccinia sorghi*; **B,** *Puccinia polysora*; and **C,** *Phakopsora zeae* (syn. *Physopella zeae*).

In most cases, dent corn hybrids grown in the North American Corn Belt avoid severe levels of rust inoculum when plants are at more susceptible juvenile stages with leaf whorls that serve as conducive sites for infection. By the time inoculum is more abundant, plants have passed through the vegetative stages, and whorls are no longer present once tassels have emerged. However, common rust can be severe on dent corn hybrids when planting is delayed and crops are at younger, more vulnerable growth stages later in the season, when inoculum is abundant.

P. sorghi does not survive winters on *Oxalis* spp. in the U.S. Corn Belt; therefore, viable inocula of *P. sorghi* typically do not overwinter. However, under unique circumstances, urediniospores may be present at the beginning of the planting season. In late-planted sweet corn crops harvested in the fall (e.g., late September to early October), teliospores have not yet replaced urediniospores in pustules. If this infected leaf tissue survives the winter, particularly under snow cover, where it is protected from several cycles of wetting and drying, viable urediniospores may be present to infect corn crops planted the following spring.

Management

Common rust is controlled effectively by growing resistant hybrids and cultivars or by timely applications of fungicides.

Resistance to common rust is clearly of two complementary types: specific and general. Specific resistance, sometimes called *Rp* resistance, is expressed as hypersensitive reactions with distinct infection types. Specific resistance is inherited simply and usually is dominant. Specific resistance has been identified at several gene regions on different chromosomes. Several individual *Rp* genes confer high levels of rust resistance to plants grown in the continental United States, but biotypes of *P. sorghi* with virulence against each of the *Rp* genes have been identified from various regions of the world. The distal end of the short arm of chromosome 10 is an interesting region, where resistance to the rust fungi *P. sorghi* and *P. polysora* may have evolved through repeated duplication with subsequent structural modifications of individual genes. Compound *Rp* genes, which carry one or more closely linked genes in the *Rp1* region of chromosome 10, confer high levels of resistance. Some compound genes may confer non-race-specific resistance. The *Rp3* region also consists of multiple, linked resistance genes similar to *Rp1*. Specific resistance to *P. sorghi* is also expressed by teosinte and *Tripsacum* spp.

General resistance ultimately results in reduced severity of infection, although distinct infection types are not recognized. One form of general resistance, which results from the increased resistance of mature leaves compared with that of juvenile leaves, is effective in all corn and may be associated with the rate at which a plant forms mature leaves. Other forms of general resistance reduce the number and size of uredinia or increase the length of the latent period. The degree of expression of various forms of general resistance appears to be dependent upon the environment in which the resistance is evaluated. Sources of resistance that effectively limit rust development in marginally favorable environments, such as the U.S. Corn Belt, may be relatively ineffective in environments that are extremely conducive to infection, such as Hawaii or other tropical areas.

Fungicidal control of common rust, mainly using fungicides represented in the strobilurin and triazole classes, is possible if initial applications are made while there is a limited amount of secondary inoculum. Usually, this requires an action threshold of only a few pustules per leaf, although the threshold can vary depending on host growth stage, level of general resistance, environmental conditions, and efficacy of the fungicide. Protectant and locally systemic fungicides are used routinely to control rust on seed corn and sweet corn crops, but they are rarely warranted for common rust control in hybrid grain production.

Selected References

Groth, J. V., Pataky, J. K., and Gingera, G. R. 1992. Virulence in eastern North American populations of *Puccinia sorghi* to *Rp* resistance genes in corn. Plant Dis. 76:1140-1144.

Hooker, A. L., and Yarwood, C. E. 1966. Culture of *Puccinia sorghi* on detached leaves of corn and *Oxalis corniculata*. Phytopathology 56:536-539.

Pataky, J. K., and Headrick, J. M. 1989. Management of common rust on sweet corn with resistance and fungicides. J. Prod. Agric. 2:362-369.

Schieber, E. 1971. Distribution of *Puccinia polysora* and *P. sorghi* in Africa and their pathogenicity on Latin American maize germ plasm. FAO Plant Prot. Bull. 19:25-31.

(Prepared by J. K. Pataky)

Southern Rust

Southern rust is primarily tropical or subtropical in distribution, mainly because of the temperature requirements for the disease; but it also is found in temperate regions. Southern rust is found in Africa, Southeast Asia, the Philippines, Australia, southern and central North America, Central America, South America, and the West Indies. In the United States, southern rust can be severe in the southeastern states and Texas; in the Corn Belt, it appears regularly but usually later in the season and its severity has been lower compared with that in the southeastern states. In recent years, southern rust has been found in the U.S. Corn Belt more frequently and earlier in the growing season, sometimes causing considerable economic damage. The disease has been reported as far north as southern Wisconsin and Minnesota (United States) and Ontario (Canada). Hosts include corn, plume grasses, teosinte, and gama grasses. As corn production has increased in Mexico and the southern United States, southern rust has become a more important disease in North America because of a larger contribution to overall corn production in these disease-conducive areas, as well as the capacity for increased production of inoculum that spreads northward.

Symptoms

Symptoms of southern rust resemble those of common rust but with very important, subtle differences. Uredinia of *Puccinia polysora* are light cinnamon brown to orange, circular to oval, 0.2–2.0 mm long, and densely scattered on the upper leaf surface. In nature, uredinia develop abundantly on upper leaf surfaces and rarely on lower leaf surfaces. Uredinia frequently occur on husk leaves of ears, ear shanks, stalks, and leaf sheaths (Figs. 139 and 140). In severe outbreaks, leaves are blighted by the development of numerous pustules (Fig. 141). Telia are circular to elongate, 0.2–0.5 mm in diameter, and chocolate brown to black and remain covered by the epidermis longer than those of *P. sorghi*. Telia of *P. polysora* often appear in circles around the uredinial pustules (Fig. 142).

Although the latent period of *P. polysora* is longer than that of *P. sorghi*, the temporal increase in southern rust severity often seems to be more rapid than that of common rust. This may be the result of the density of uredinia of *P. polysora* in infected tissues. Also uredinia of *P. polysora* form on leaf sheaths, husk

Fig. 140. Uredinia of the southern rust pathogen, *Puccinia polysora*, on a leaf sheath. (Courtesy G. P. Munkvold)

Fig. 139. Uredinia of the southern rust pathogen, *Puccinia polysora*. (Courtesy G. P. Munkvold)

Fig. 141. Leaf blighting resulting from severe southern rust infection. (Courtesy G. P. Munkvold)

leaves, and inner leaves of ear husks, where *P. sorghi* uredinia are found infrequently.

Causal Organism

Southern rust is caused by *Puccinia polysora*. Urediniospores (20–29 × 29–40 μm) are yellowish to golden and mostly ellipsoid or obovoid. The golden or yellowish walls are minutely and sparsely echinulate and 1–1.5 μm thick with four or five equatorial pores. Teliospores (18–27 × 29–41 μm) are chestnut brown, smooth, angularly ellipsoid or oblong, and rounded at both ends. They are very brittle, usually two celled, constricted at the septum, and borne on short (10–30 μm), persistent, yellow or brownish pedicels one-fourth or less the length of the spores (Fig. 138B).

Disease Cycle and Epidemiology

The role of *P. polysora* teliospores in the disease cycle of southern rust is not known. Although an alternate host for *P. polysora* has not been found, various combinations of virulence (i.e., races) occur in response to simply inherited host resistance, thus suggesting that sexual recombination in the pathogen is not infrequent. Urediniospores constitute both primary and secondary inocula and are carried to corn plants by wind or on infected plant tissue. Southern rust development is favored by warm, humid environments. In Africa, *P. polysora* is prevalent at low altitudes and *P. sorghi* at higher altitudes, probably because of temperature differences. In North America, *P. polysora* is endemic in southern states, while in the central United States, it generally is found later in the growing season and is associated with high humidity, warm night weather, and weather patterns that favor long-distance dispersal of urediniospores from southern states.

Management

Resistance is the primary management practice. Eleven genes controlling specific resistance to *P. polysora*, designated *Rpp1* to *Rpp11*, have been identified. The *Rpp9* gene on the short arm of chromosome 10 is closely linked to the *Rp1* region conveying *P. sorghi* resistance. Like the *Rp1* region, *Rpp9* may contain numerous alleles or may actually be very closely linked genes. Many of the other previously identified *Rpp* genes have not been maintained in breeding collections nor have the sources of these genes. Physiologic races are known. Races virulent against the *Rpp9* gene were found in Georgia (United States) in 2008.

Mass selection by native farmers for general resistance to southern rust is probably the major reason many local cultivars grown in East and West Africa developed resistance before hypersensitive resistance was widely distributed. In an interesting observation, Van der Plank noted that resistance to *P. polysora* must be deleterious to plants in areas where resistance is not needed and cited as evidence the lack of resistance in corn from areas of the world where *P. polysora* does not exist. Excellent sources of resistance to *P. polysora* are found among tropical and subtropical maize lines, such as those developed in Hawaii by Brewbaker.

Fungicides that control common rust are usually effective against southern rust. Timing of applications to achieve maximum benefit may differ between the two rust diseases. The timing of fungicide applications can be facilitated by information available through IPIPE (Integrated Pest Information Platform for Extension and Education), a public Internet resource in which U.S. scientists monitor the occurrence of southern corn rust. IPIPE also reports the distribution of certain races of *P. polysora*.

Selected References

Brewbaker, J. L., Kim, S. K., So, Y. S., Logrono, M., Moon, H. G., Ming, R., Lu, X. W., and Josue, A. D. 2011. General resistance in maize to southern rust (*Puccinia polysora*). Crop Sci. 51:1393-1409.

Cammack, R. H. 1959. Studies on *Puccinia polysora* Underw. III. Description and life cycle of *P. polysora* in West Africa. Trans. Br. Mycol. Soc. 42:55-58.

Hulbert, S. H. 1997. Structure and evolution of the *rp1 complex* conferring rust resistance in maize. Annu. Rev. Phytopathol. 35:293-310.

Schieber, E. 1971. Distribution of *Puccinia polysora* and *P. sorghi* in Africa and their pathogenicity on Latin American maize germ plasm. FAO Plant Prot. Bull. 19:25-31.

(Prepared by J. K. Pataky)

Tropical Rust

Tropical rust appears to be limited to warm, humid areas of Central and South America and the Caribbean. It has not been reported in the United States. Teosinte and *Tripsacum* spp. are also hosts.

Symptoms

Yellow pustules occur in small groups mostly 0.3–1.0 mm long on the upper leaf surfaces and are covered by the epidermis, except for a small pore or slit. The pustules later develop into dark purplish, circular or oblong blotches with creamy centers. The dark, chocolate brown to blackish telia are usually scattered in groups around the uredinia (Figs. 143 and 144).

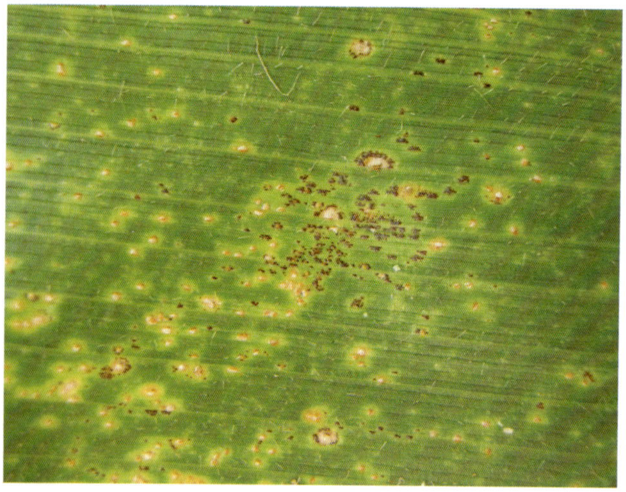

Fig. 142. Uredinia and telia of the southern rust pathogen, *Puccinia polysora*. (Courtesy G. P. Munkvold)

Fig. 143. Uredinia and telia of the tropical rust pathogen, *Phakopsora zeae* (syn. *Physopella zeae*). (Courtesy B. M. Anderson)

Fig. 144. Uredinia and telia of the tropical rust pathogen, *Phakopsora zeae* (syn. *Physopella zeae*). (Courtesy C. De Leon)

Causal Organism

Tropical rust is caused by *Phakopsora zeae* (syns. *Angiopsora zeae, Physopella zeae*). The urediniospores (15–22 × 22–33 μm) are sessile, ellipsoid or obovoid, and colorless to pale yellow with nearly hyaline, moderately echinulate walls 1.5–2 μm thick. The teliospores (12–18 × 16–36 μm) are cylindrical, golden brown to chestnut brown, one celled, and found in chains of two to three (Fig. 138C). The oblong to angularly ellipsoid teliospores are sessile with smooth, golden brown walls 1.5–2 μm thick at the sides and 2.5 μm at the apex of apical spores.

Disease Cycle and Epidemiology

The disease cycle of tropical rust is similar to that of southern rust. Aecia are unknown. No alternate host has been reported.

The disease is favored by warm to hot, humid weather and is found at low altitudes. At least two races of *Phakopsora zeae* are known.

Management

Sources of resistance have been identified, including resistance expressed as chlorotic flecks and small uredinia. Segregation for reaction type has been observed in some Guatemalan lines.

Selected Reference

Schieber, E., and Dickson, J. G. 1963. Comparative pathology of three tropical corn rusts. Phytopathology 53:517-521.

(Prepared by J. K. Pataky)

Identifying Rusts

Puccinia sorghi can be differentiated from *P. polysora* by its larger, less densely appearing, elongated, brick red-colored uredinia (Fig. 145); darker, rounder urediniospores (Fig. 146); erumpent telia; and teliospores with thicker walls, longer pedicels, and bullet-shaped terminal cells. Uredinia of *P. sorghi* are common on both leaf surfaces; those of *P. polysora* are much more numerous on the upper leaf surface. Uredinia of *P. polysora* are more common on husk leaves of ears, ear shanks, stalks, and leaf sheaths than are those of *P. sorghi*. The telia of *P. polysora* often remain covered and frequently are found in rings around uredinia, whereas telia of *P. sorghi* replace the uredinia. *Phakopsora zeae* is identified by hyaline urediniospores; covered, blackish telia; and sessile teliospores in short chains (Fig. 138).

Fig. 145. Common rust (left) and southern rust (right). (Courtesy D. G. White)

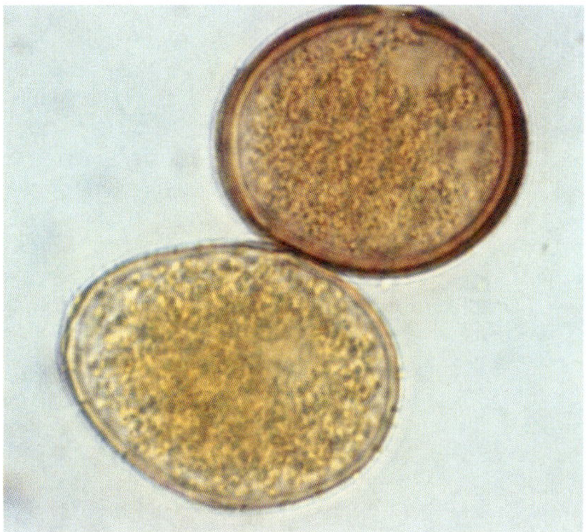

Fig. 146. Urediniospores of *Puccinia sorghi* (top) and *Puccinia polysora* (bottom). (Courtesy D. G. White)

Selected References

Cummins, G. B. 1941. Identity and distribution of three rusts of corn. Phytopathology 31:856-857.
Schieber, E., and Dickson, J. G. 1963. Comparative pathology of three tropical corn rusts. Phytopathology 53:517-521.

(Prepared by J. K. Pataky)

Fungal Stalk Rots

Stalk rots continue to be widespread and serious diseases of corn, although significant improvements in hybrid resistance continue to be made. The incidence and severity of stalk rot varies greatly from year to year and from field to field, although some stalk rot is found in every field every year. The term "stalk rot" is often used to include stalk breakage, stalk lodging, premature death of the plant, and occasionally even root lodging. For purposes of this section, however, stalk rot refers to the decay of the internal pith tissues of the stalk.

The earliest symptom of stalk rot may be plant wilting (Fig. 147), which eventually leads to premature death. Over the course of several days, leaves may become gray or streaked (resembling sunscald, Fig. 148), the ear droops, and the outer rind of the lower stalk turns brown (Fig. 149). When the outer stalk tissue is brown, the pith tissue in the lowest internode is rotted and separated from the rind (Fig 150A); the vascular bundles may be intact at first (Fig. 150B), and then the pith may become hollowed out (Fig. 150C). As the rotting pith tissue pulls away from the rind, the structural integrity of the plant changes from that of a solid rod to a hollow tube. With the structural integrity destroyed, the tube more readily lodges. Plants with rotted stalks almost always have rotted roots. In fact, high levels of root rot before flowering predispose the plant to stalk rotting later in the season. The stalk rot complex is difficult to characterize because a number of fungi and sometimes bacteria are involved in the decay of the pith. This is further complicated by the fact that stalk rot is generally regarded as a disease of senescence and factors involved in the acceleration of senescence may alter the susceptibility of the plant.

Stalk rot causes losses in several different ways. Yield losses result from premature plant death (Fig. 151), which stops normal grain fill. Yield losses also occur when plants with rotted stalks lodge (Fig. 152) and grain cannot be harvested with mechanical equipment. If stalk rot is severe, harvest is slowed and valuable time is lost (Fig. 153). Another loss occurs when ears on lodged plants touch the soil, enhancing the development of ear rots that reduce grain quality and possibly cause dockage when the grain is marketed. Because losses caused by stalk rot

Fig. 147. Healthy plant (left) compared with a wilted plant as a result of stalk rot (right). (Courtesy G. P. Munkvold)

Fig. 148. Symptoms of scalding on corn leaves caused by stalk rot. (Courtesy G. P. Munkvold)

Fig. 149. Healthy plant (left) compared with a prematurely dead plant as a result of stalk rot (right). (Courtesy D. G. White)

Fig. 150. Stalk rot symptoms in lower stalk. **A,** Pith deterioration; **B,** pith disintegration with vascular bundles intact; and **C,** lower stalk hollowed out as a result of stalk rot. (Courtesy G. P. Munkvold)

occur in several different ways, precise yield loss estimates are difficult to obtain. No method is available to establish plots that are completely free of stalk rot.

Several techniques have been used in attempts to estimate losses caused by stalk rot. Yields of hybrids during years with little natural stalk rot have been compared with yields during years when stalk rot is severe. However, these comparisons are difficult to make because of year-to-year variation in average yield. Another technique is to compare yields from inoculated plots with those of uninoculated plots. The problems with this method are that natural infection can occur in uninoculated plots and stalk rot resulting from inoculation does not exactly duplicate natural stalk rot. A third technique involves paired plant comparisons, in which grain yields of adjacent diseased and healthy plants are compared. However, the number of kernels is often higher on diseased plants than on healthy plants. The final problem is that in a number of studies, losses have been estimated on the basis of hand-harvested yield, and losses resulting from lodging, slowed harvest, or ear rots are not taken into account.

Considerable research has been done on the factors that influence stalk rot. Obviously, the most important factor is the inherent susceptibility of the hybrid. However, a hybrid that is considered susceptible may have stalk rot in some locations and/or in some years but not in others. Several factors, including weather conditions, moisture availability, grain fill and kernel number, cultural practices, plant densities, severity of foliar disease, physical leaf damage, cloud cover, and insect damage, may alter the susceptibility of the hybrid. Factors that affect stalk rot development have been reviewed by J. L. Dodd, who proposed a photosynthetic stress-translocation balance concept of predisposition of corn to stalk rot. In general, when growing conditions are favorable early in the season, corn plants develop a large number of kernels. After flowering, the kernels are a sink for carbohydrates within the plant. When the plant does not produce enough carbohydrates to fill grain and maintain all tissues, the grain sink (ear) has priority over other tissues. Carbohydrate levels are not maintained in root and lower stem tissues, which have low priority in the intraplant allocation of carbohydrates. Cells in these tissues senesce and become more susceptible to colonization by various stalk-rotting organisms. Many of the stalk rot pathogens are opportunistic invaders of senescing tissue. Unlike most other corn diseases, in which a specific organism causes a specific disease, stalk rot can be caused by whichever stalk-rotting organism happens to be present in or near the plant as root and stalk tissues begin to senesce. More often than not, several different fungi can be isolated from rotted stalks.

A review of all the factors involved in plant stress and their relationship to stalk rot would be too lengthy to cover here. Therefore, three factors that can be manipulated to manage stalk rots are discussed: soil fertility, tillage, and plant density.

In general, stalk rot is more severe and the incidence is greater with increased fertility. A number of studies have evaluated the effects of various nutrients on stalk rot. The most widely studied nutrient is nitrogen (N), which is reported to both increase and decrease stalk rot severity. The variable effect of nitrogen on stalk rot appears to be influenced by factors such as the stalk rot pathogen, susceptibility of the plant, source of nitrogen, and relative availability of other nutrients. Most research indicates that, when there is an excess of nitrogen in relation to potassium, stalk rot severity increases. Other research indicates that, if nitrogen levels are high early in the growing season and nitrogen is then lost through denitrification and leaching, stalk rot severity also increases. This finding implies a need for a balanced and continuous nitrogen supply throughout the growing season to maintain living cells in the pith for a longer period of time. A balanced and continuous nitrogen supply helps to explain the reduction in stalk rot that occurs when nitrification inhibitors, such as 2-chloro-6(trichloromethyl)pyridine(nitrapyrin), are mixed with anhydrous ammonia. Therefore, the effect of ni-

Fig. 151. Premature death of a corn plant caused by stalk rot. (Courtesy G. P. Munkvold)

Fig. 152. Lodging of corn plants caused by stalk rot. (Courtesy G. P. Munkvold)

Fig. 153. Reel attachment on corn combine to gather lodged stalks for harvest. (Courtesy G. P. Munkvold)

trogen may not be dependent upon how much is applied at the beginning of the season but on whether it is available throughout the season.

Potassium (K), which is involved in stomatal functions as well as metabolic pathways, also has been related to stalk rot. When plants are deficient in potassium, the rate of photosynthesis is lower, perhaps resulting in more rapid pith senescence. Also, pith cells in plants grown with adequate to high levels of potassium have thicker walls and may not rot as quickly as cells with thinner walls. When adequate potassium is present, stalk rot severity is reduced.

The response of corn to phosphorus (P) varies with the season, the hybrid, and the pathogen. While phosphorus does not decrease stalk rot severity when applied at high levels, adequate phosphorus fertilization seems to afford some protection against stalk rot.

A number of studies have related tillage practices to the occurrence of plant diseases. In general, debris that is left on the surface of the soil favors the development of many foliar diseases because many pathogenic fungi overwinter in debris on the soil surface, where they survive well and are easily disseminated to the foliage of the next crop. Foliar diseases increase the predisposition of plants to stalk rots by reducing the total photosynthetic area of the leaf, thereby decreasing carbohydrate production and accelerating the senescence of pith cells. However, in the absence of leaf blights, conservation tillage has been shown to actually reduce the levels of several stalk rot diseases, perhaps because surface residues reduce evaporation of soil moisture, resulting in less stress to the plant. With anthracnose stalk rot, conservation tillage may result in more symptoms on the surface of the stem; however, the level of internal rot may not be increased because stress may be reduced.

As plant populations increase, the incidence and severity of stalk rot also increase, especially in susceptible hybrids. If cloudy weather occurs, foliar diseases are present, or fertility is unbalanced in a field, plant populations become an even more critical contributing factor. Stalk rot levels also increase when irrigation ceases and plants are stressed before grain fill is complete.

Another factor that may influence the development of corn stalk rots is the use of insect-resistance genes introduced through genetic engineering. European corn borers, *Ostrinia nubilalis*, and other lepidopteran insects bore into the corn stem and make wounds that allow for stalk-rotting fungi to enter (Fig. 154). Transgenes derived from the bacterium *Bacillus thuringiensis* (*Bt* genes) code for the expression of insecticidal proteins that control European corn borers and several other insects. Several studies have found a lower incidence of stalk rot in hybrids with *Bt* genes compared with their respective non-*Bt* near-isogenic lines. However, this is not consistent across all studies and hybrids, which is possibly because stalk-rotting fungi penetrate in many different ways.

Fig. 154. Stalk rot initiated by European corn borer injury. (Courtesy G. P. Munkvold)

Selected References

Dodd, J. L. 1977. A photosynthetic stress-translocation balance concept of corn stalk rot. Pages 122-130 in: Proc. Annu. Corn Sorghum Res. Conf., 32nd. H. D. Loden and D. Wilkinson, eds. American Seed Trade Association, Washington, DC.

Dodd, J. L. 1980. The role of plant stresses in development of corn stalk rots. Plant Dis. 64:533-537.

Dodd, J. L. 1980. Grain sink size and predisposition of *Zea mays* to stalk rot. Phytopathology 70:534-535.

Christensen, J. J., and Wilcoxson, R. D. 1966. Stalk Rot of Corn. Monogr. 3. American Phytopathological Society, St. Paul, MN.

Doupnik, B., Jr., and Boosalis, M. G. 1980. Ecofallow—A reduced tillage system—And plant diseases. Plant Dis. 64:31-35.

Gatch, E. W., Hellmich, R. L., and Munkvold, G. P. 2002. A comparison of maize stalk rot occurrence in Bt and non-Bt hybrids. Plant Dis. 86:1149-1155.

Hooker, A. L., and Britton, M. P. 1962. The effects of stalk rot on corn yields in Illinois. Plant Dis. Rep. 46:9-13.

Hooker, A. L., and White, D. G. 1976. Prevalence of corn stalk rot fungi in Illinois. Plant Dis. Rep. 60:1032-1034.

Keller, N. P., Bergstrom, G. C., and Carruthers, R. I. 1986. Potential yield reductions in maize associated with an anthracnose/European corn borer pest complex in New York. Phytopathology 76:586-589.

Parker, D. T., and Burrows, W. C. 1959. Root and stalk rot in corn as affected by fertilizer and tillage treatment. Agron. J. 51:414-417.

Warren, H. L., Huber, D. M., Nelson, D. W., and Mann, O. W. 1975. Stalk rot incidence and yield of corn as affected by inhibiting nitrification of fall-applied ammonium. Agron. J. 67:655-660.

White, D. G., Hoeft, R. G., and Touchton, J. T. 1978. Effect of nitrogen and nitrapyrin on stalk rot, stalk diameter, and yield of corn. Phytopathology 68:811-814.

Wysong, D. S., and Kerr, E. 1969. Hybrid corn yield reduction by stalk rot and root rot in Nebraska. Plant Dis. Rep. 53:326-329.

(Prepared by D. G. White)

Gibberella Stalk Rot

Symptoms

Symptoms of Gibberella stalk rot are similar to those of other stalk rots. Affected plants wilt, the leaves change from light to dull green, and the lower stalks become straw colored. The internal pith tissue disintegrates, leaving only the vascular bundles. Plants often lodge at the nodes. A diagnostic sign of Gibberella stalk rot is the reddish discoloration that occurs inside the stalk (Fig. 155), although it is not always present (Fig. 156). Another characteristic is the presence of superficial, small, round, black perithecia, often at the internode, that can be easily scraped from the stalk surface (Fig. 157).

Causal Organism

Fusarium graminearum (syn. *Gibberella zeae*) causes Gibberella stalk rot. Perithecia are bluish black and superficial (Fig. 158) and contain eight ascospores arranged obliquely in one row. The ascospores (3–5 × 10–30 μm) are hyaline, have three septa, and are slightly curved and taper uniformly at the ends. Perithecia are often seen in the fall, but usually spores contained in the perithecia are not mature until the following spring or summer. *F. graminearum* produces macroconidia (4–6 × 30–60 μm) that are hyaline, are slightly curved with pointed ends, and have three to five septa (Figs. 159 and 160). Some isolates of the fungus produce chlamydospores. *F. graminearum* has been divided into at least nine cryptic species, based on DNA sequence data and other characteristics. It is not yet clear how many of the described species are associated with corn diseases, but *F. graminearum sensu stricto* is the species found on corn in North America.

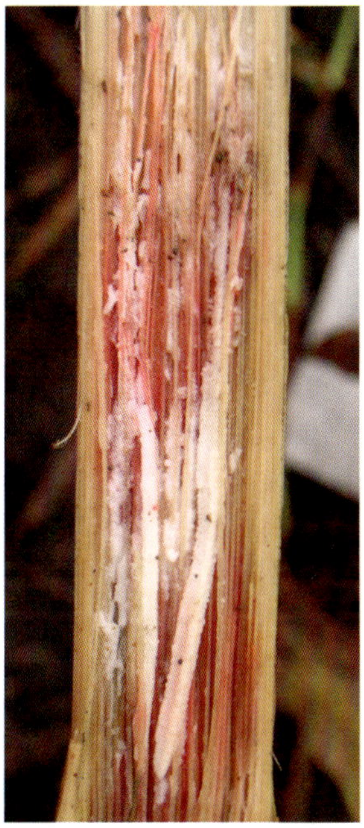

Fig. 155. Pink discoloration in stalk pith caused by Gibberella stalk rot. (Courtesy G. P. Munkvold)

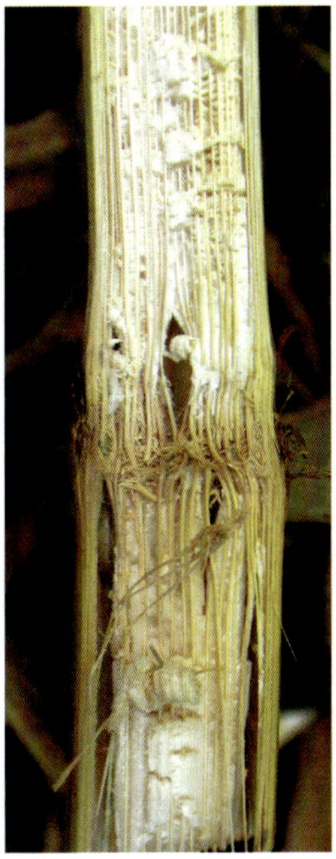

Fig. 156. Gibberella stalk rot without pink discoloration. (Courtesy L. Ruttman, DuPont Pioneer)

Fig. 157. Gibberella stalk rot showing perithecia that are easily scraped off by a thumbnail. (Courtesy D. G. White)

Fig. 158. Perithecia of *Gibberella zeae* on a leaf sheath. (Courtesy G. P. Munkvold)

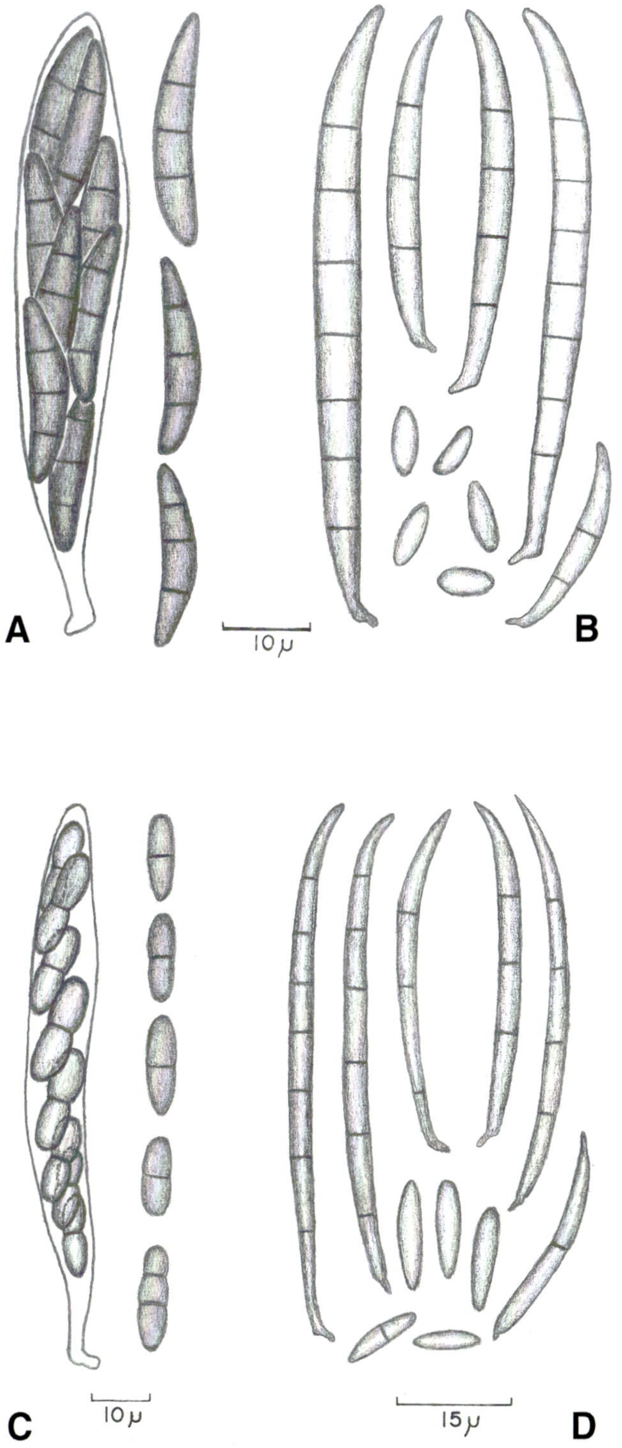

Fig. 159. A, Ascus and ascospores of *Fusarium graminearum* (syn. *Gibberella zeae*). **B,** Conidia of *Fusarium graminearum*. **C,** Ascus and ascospores of *Fusarium verticillioides* (syn. *Gibberella moniliformis*). **D,** Macroconidia and microconidia of *Fusarium verticillioides*.

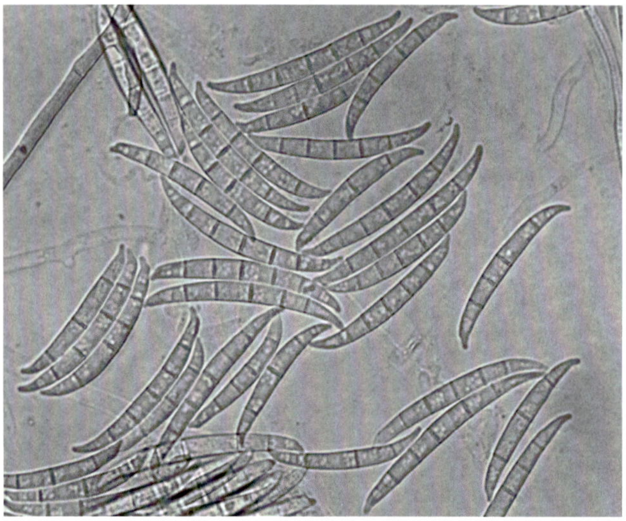

Fig. 160. Macroconidia of *Fusarium graminearum*. (Courtesy K. Seifert of the Department of Agriculture and Agri-Food, Government of Canada; © Her Majesty the Queen in Right of Canada, as represented by the Minister of Agriculture and Agri-Food Canada)

Disease Cycle and Epidemiology

Under warm, wet conditions, perithecia produce ascospores that are disseminated by wind and may infect corn plants. Inoculum may also be produced as conidia during the summer. Stalk infections usually occur shortly after pollination, developing at the origin of the leaf sheaths, through aborted ear shoots, or around the brace roots. The fungus also may enter through the roots and grow up into the lower stem or through stalk wounds made by insects or other means. *F. graminearum* also infects wheat, barley, oat, and rye, causing scab (also called Fusarium head blight) and seedling blights. The fungus is common, and the incidence and severity of Gibberella stalk rot are not limited by inoculum in most corn-growing areas.

Management

Even though stalk rot and lodging continue to be serious problems, considerable gains have been made in the development of resistant hybrids. Resistance genes against *F. graminearum* have been reported, and inbred lines have been identified as effective sources of resistance. Resistance to stalk rot and lodging is a major consideration in the marketing of a corn hybrid, and seed companies devote considerable resources to selecting hybrids with this resistance. However, one of the most important characteristics of a hybrid is yield. Genotypes with higher yields tend to have bigger ears, which act as large sinks for carbohydrates in the plant. The larger ears result in reduced carbohydrate levels in the lower stem, predisposing the plant to stalk rot. Therefore, the balance between breeding for resistance to stalk rot and breeding for high yield is delicate. Breeding for resistance to stalk rot also is greatly complicated by the relatively large number of fungi that can cause stalk rots and by the effects of various environmental factors that predispose plants to stalk rot. Also, a hybrid can be resistant to pith decay but have unacceptable lodging tendencies caused by poor rind strength, poor root strength, or both. Cultural practices, such as balanced fertility, proper plant population, and any other practice that lessens plant stress, can help reduce the amount of stalk rot. Control of stalk-boring insects is also helpful for stalk rot management.

Selected References

Michaelson, M. E. 1957. Factors affecting development of stalk rot of corn caused by *Diplodia zeae* and *Gibberella zeae*. Phytopathology 47:499-503.

O'Donnell, K., Ward, T. J., Geiser, D. M., Kistler, H. C., and Aoki, T. 2004. Genealogical concordance between the mating type locus and seven other nuclear genes supports formal recognition of nine phylogenetically distinct species within the *Fusarium graminearum* clade. Fungal Genet. Biol. 41:600-623.

Santiago, R., Reid, L. M., Zhu, X., Butron, A., and Malvar, R. A. 2010. Gibberella stalk rot (*Fusarium graminearum*) resistance of maize inbreds and their F1 hybrids and their potential for use in resistance breeding programs. Plant Breed. 129:454-456.

Thayer, P., and Williams, L. E. 1960. Effect of nitrogen, phosphorus, and potassium concentrations on the development of Gibberella stalk- and root-rot of corn. Phytopathology 50:212-214.

Yang, D. E., Zhang, C. L., Zhang, D. S., Jin, D. M., Weng, M. L., Chen, S. J., Nguyen, H., and. Wang, B. 2004. Genetic analysis and molecular mapping of maize (*Zea mays* L.) stalk rot resistant gene *Rfg1*. Theor. Appl. Genet. 108:706-711.

(Prepared by D. G. White)

Diplodia (Stenocarpella) Stalk Rot

Diplodia stalk rot was considered to be the most important stalk rot of corn in the midwestern United States until the late 1960s and early 1970s. At that time, the incidence and severity of Diplodia stalk rot declined and the incidence and severity of Gibberella stalk rot and anthracnose stalk rot increased. The occurrence of Diplodia stalk rot began to increase again in the 1990s with the increased use of conservation tillage, and *Stenocarpella maydis* is now found relatively commonly in rotted stalk tissues.

Symptoms

Symptoms of Diplodia stalk rot are similar to those of Gibberella stalk rot, but Diplodia stalk rot does not include pink or red discoloration of the tissue. A characteristic sign of Diplodia stalk rot is the presence of subepidermal, minute, dark brown to black pycnidia embedded in the rind tissue of the lower stem (Fig. 161).

Causal Organism

Diplodia stalk rot is caused by *Stenocarpella maydis* (syns. *Diplodia maydis* and *D. zeae*). The fungus produces globose or flask-shaped pycnidia containing olive to brown, elliptical, two-celled, straight to slightly curved conidia (5–6 × 25–30 µm) (Fig. 162). Conidia are produced in a sticky matrix, and under wet conditions, pycnidia may produce dark spore tendrils of numerous spores sticking together. Less commonly, pycnidia contain colorless, long, narrow, threadlike spores that are 1–2 × 25–35 µm. No sexual stage of the fungus is known.

Fig. 161. Diplodia stalk rot with pycnidia embedded in stalk tissue. (Courtesy G. P. Munkvold)

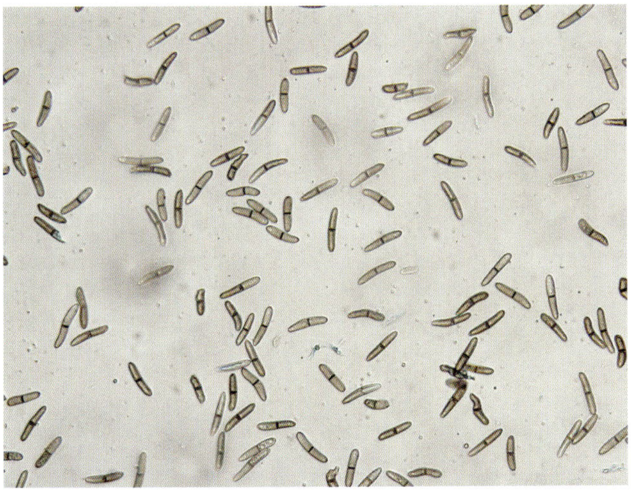

Fig. 162. Conidia of *Stenocarpella maydis*. (Courtesy M. P. Romero Luna)

Disease Cycle and Epidemiology

S. maydis overwinters in stalk debris that is either buried or on the soil surface. Under warm, moist conditions, the spores are extruded from the pycnidia on stalks on the soil surface and disseminated by rain and wind. Infection of plants takes place mainly through the crown, mesocotyl, or roots. Occasional infections take place at the nodes between the crown and the ear. The fungus may also enter through wounds caused by insects or other means.

Corn is the only host for *S. maydis*, and the disease is most common where debris from a previous crop is on the soil surface. This may help to explain the reduced incidence of Diplodia stalk rot in the midwestern United States during the 1960s and 1970s, when fall tillage was widely practiced. As conservation tillage has become more widely practiced, the incidence of Diplodia stalk rot has increased, and *Diplodia*-infected stalk tissue that remains on the soil surface is an important source of inoculum for Diplodia ear rot the next season.

Management

Genetic resistance to Diplodia stalk rot is highly correlated with resistance to Gibberella stalk rot, and management practices are similar.

Selected Reference

Michaelson, M. E. 1957. Factors affecting development of stalk rot of corn caused by *Diplodia zeae* and *Gibberella zeae*. Phytopathology 47:499-503.

(Prepared by D. G. White)

Anthracnose Stalk Rot

Since 1855, anthracnose stalk rot has been known to be found in the United States. However, the disease was not considered to be important until sometime after 1970. The prevalence and severity of the disease has increased, and it is now considered to be one of the most important stalk rot diseases.

Symptoms

Stalk infections become evident at various stages of growth, depending on the susceptibility of the plant. Early symptoms may appear as dark elliptical lesions near the stalk base (Fig. 163). The lower stalk tissues of acutely susceptible inbred lines (e.g., C123 and Mo940) and hybrids with very susceptible inbred parents may be so severely rotted during early growth stages that plants are killed before pollen production. On most inbred lines and hybrids, however, symptoms are not visible until just prior to normal senescence. Unlike other common stalk rot diseases, the pathogen may rot several internodes of the plant. In some cases, portions of the plant above the ear die prematurely while the lower portions of the plant remain green (top dieback) (Fig. 164). Later in the season, the prematurely killed top portions of the plant lodge and fall off. Anthracnose stalk rot is easily recognized late in the season by the shiny black color on the outer stalk (Figs. 165 and 166). This discoloration may be uniform or blotchy and may appear anywhere on the stalk. Occasionally, discoloration may be seen only near

Fig. 163. Anthracnose stalk rot lesion at the base of a stalk. (Courtesy G. P. Munkvold)

Fig. 164. Anthracnose top dieback. (Courtesy G. P. Munkvold)

the nodal plate. The black color results from what appear to be masses of darkly pigmented fungal mycelium just under the rind surface. The internal stalk tissues are often blackened or discolored, and the stalk can be easily crushed between the thumb and forefinger. With many plants, the outer stalk surface is discolored but internal tissues are not discolored or rotted. Conversely, stalks that are systemically colonized may have pith discoloration while the rinds remain green and unblemished.

Causal Organism

Colletotrichum graminicola (syn. *Glomerella graminicola*), the cause of foliar anthracnose, also causes anthracnose stalk rot (see the Anthracnose Leaf Blight section under Foliar Diseases above).

Disease Cycle and Epidemiology

The mechanisms of penetration and infection of stalk tissues are not completely understood. *C. graminicola* is the only common stalk rot pathogen that also causes leaf blight, and some researchers postulate that spores produced in lesions on the leaves are washed behind the leaf sheaths and penetrate the stalks. However, the appearance of the leaf blight and stalk rot in the same field during the same season is not well correlated. Unlike most other stalk rot pathogens, *C. graminicola* can infect directly through the stalk epidermis and can cause vascular wilt symptoms. The fungus also enters the stalks through insect feeding sites and other wounds. Anthracnose leaf blight and anthracnose stalk rot appear to be more severe where corn residues are on the soil surface, and anthracnose stalk rot incidence decreases with increased distance from crop residue. In fields without apparent anthracnose leaf blight, infection likely takes place from penetration of the roots by buried inoculum from previously diseased plants.

Fig. 165. Anthracnose stalk rot. (Courtesy G. P. Munkvold)

Fig. 166. Anthracnose stalk rot. (Courtesy G. P. Munkvold)

Management

The use of resistant hybrids is the most suitable method for managing anthracnose stalk rot. When the disease became a problem during the early 1970s, certain hybrids, particularly those with inbred line C123, were identified as extremely susceptible. These hybrids and susceptible related lines are no longer utilized. Currently used commercial hybrids are usually not susceptible until after flowering, when senescence of pith tissue takes place. Numerous sources of resistance have been identified, and resistance has been incorporated into many commercial hybrids. In many sources, the resistance is complexly inherited. Two sources, LB31 and MP305, are known to have resistance genes that have been incorporated by backcrossing into a few commercial hybrids. A gene from MP305, identified as *Rcg1*, has been incorporated into commercial hybrids and provides a high level of resistance to anthracnose stalk rot. It should be noted that resistance to the stalk rot phase of the disease is not highly correlated with resistance to the leaf blight phase. Cultural practices such as balanced fertility and reduction of stress are also useful for reducing the incidence of anthracnose stalk rot. Control of the corn borer can be important because the fungus is known to enter through its feeding sites.

Selected References

Badu-Apraku, B., Gracen, V. E., and Bergstrom, G. C. 1987. A major gene for resistance to anthracnose stalk rot in maize. Phytopathology 77:957-959.

Frey, T. J., Weldekidan, T., Colbert, T., Wolters, P. J. C. C., and Hawk, J. A. 2011. Fitness evaluation of *Rcg1*, a locus that confers resistance to *Colletotrichum graminicola* (Ces.) G.W. Wils. using near-isogenic maize hybrids. Crop Sci. 51:1551-1563.

Jung, M., Weldekidan, T., Schaff, D., Paterson, A., Tingey, S., and Hawk, J. 1994. Generation-means analysis and quantitative trait locus mapping of anthracnose stalk rot genes in maize. Theor. Appl. Genet. 89:413-418.

Keller, N. P., Bergstrom, G. C., and Carruthers, R. I. 1986. Potential yield reductions in maize associated with an anthracnose/European corn borer pest complex in New York. Phytopathology 76:586-589.

Lipps, P. E. 1985. Influence of inoculum from buried and surface corn residues on the incidence of corn anthracnose. Phytopathology 75:1212-1216.

Lipps, P. E. 1988. Spread of corn anthracnose from surface residues in continuous corn and corn-soybean rotation plots. Phytopathology 78:756-761.

Perkins, J. M., and Hooker, A. L. 1979. The effects of anthracnose stalk rot on corn yields in Illinois. Plant Dis. Rep. 63:26-30.

Toman, J., Jr., and White, D. G. 1993. Inheritance of resistance to anthracnose stalk rot of corn. Phytopathology 83:981-986.

Warren, H. L., and Nicholson, R. L. 1975. Kernel infection, seedling blight, and wilt of maize caused by *Colletotrichum graminicola*. Phytopathology 65:620-623.

Williams, L. E., and Willis, G. M. 1963. Disease of corn caused by *Colletotrichum graminicolum*. Phytopathology 53:364-365.

(Prepared by D. G. White)

Charcoal Rot

Charcoal rot is a common stalk rot disease in hot, dry areas or years. It is found in most other areas during years with drought conditions. Initial symptoms are similar to those of Diplodia stalk rot, Gibberella stalk rot, and Fusarium stalk rot. A characteristic sign of charcoal rot is the presence of numerous, minute, black sclerotia, particularly on the vascular bundles and inside the rind of the stalk, causing the stalk to appear gray-black (Figs. 167 and 168). The name of the disease is derived from this symptom.

Causal Organism

Charcoal rot is caused by *Macrophomina phaseolina* (syns. *Macrophomina phaseoli*, *Botryodiplodia phaseoli*, and *Sclerotium bataticola*). Some strains of the fungus produce conidia; however, isolates from corn are sterile and do not form spores. Sclerotia (0.05–0.22 mm) are generally smooth and black.

Disease Cycle and Epidemiology

Charcoal rot development is favored by soil temperatures of 30–42°C and low soil moisture. *M. phaseolina* overwinters as sclerotia and may penetrate roots and lower stems during the growing season. The fungus has a broad host range, including sorghum and soybean, and the disease is more common on soybean than on corn in much of the midwestern United States. The survival of the fungus on crops, such as soybeans, and weed host plants provides a source of inoculum that is capable of infecting senescing corn stalks during conditions of drought. This may explain why the disease is widespread on corn during years that are hot and dry.

Management

Limited information is available on the inheritance of resistance to charcoal rot. The fungus is believed to infect mostly senescing tissue; hybrids that are resistant to Gibberella stalk rot and Diplodia stalk rot also have resistance to charcoal rot. Selection for the "stay-green" trait (ability to avert premature senescence) has been used as an approach to improve charcoal rot resistance. As with most other stalk rot diseases, any cultural practice that helps avoid the early senescence of stalk tissue helps reduce the severity of this disease.

Selected Reference

Christensen, J. J., and Wilcoxson, R. D. 1966. Stalk Rot of Corn. Monogr. 3. American Phytopathological Society, St. Paul, MN.

(Prepared by D. G. White)

Fig. 167. Charcoal rot and microsclerotia of *Macrophomina phaseolina*. (Courtesy G. P. Munkvold)

Fig. 168. Charcoal rot. (Courtesy G. P. Munkvold)

Fusarium Stalk Rot

Fusarium stalk rot is more important in warm, dry areas, even though the causal organisms may be found wherever corn is grown. The pathogens can be isolated from both rotted and healthy stalks near the end of the growing season. Although *Fusarium* spp. can be isolated from symptomless stalks and do not always cause symptoms, these species have been shown to cause severe stalk rot in plants inoculated by the toothpick method, and under conducive conditions in warm, dry climates, Fusarium stalk rot can cause significant losses.

Symptoms

The symptoms of Fusarium stalk rot are not easily distinguished from those of Diplodia stalk rot or Gibberella stalk rot, and there rarely are signs of the fungus that make it easily recognizable. Stalks rotted by *Fusarium* spp. may display white to pink mycelium in the internally rotted tissues (Fig. 169), but this sign is not reliable for definitive identification. In fact, Fusarium stalk rot may be diagnosed, in part, by the absence of signs of Gibberella stalk rot, Diplodia stalk rot, anthracnose stalk rot, or charcoal rot.

Causal Organisms

Fusarium stalk rot is caused by *Fusarium verticillioides* (syns. *F. moniliforme*, *Gibberella moniliformis*, and *G. fujikuroi* mating population A); *F. proliferatum* (syns. *G. intermedia* and *G. fujikuroi* mating population D); and *F. subglutinans* (syns. *G. subglutinans* and *G. fujikuroi* mating population E). The perfect stage is rarely seen. Macroconidia (2.5–5 × 15–60 μm) are hyaline, are curved near the tips, and have three to five septa (Fig. 159). Microconidia (2–3 × 5–12 μm) are abundant, single celled, and borne in chains (*F. verticillioides* and *F. proliferatum*) (Fig. 170) or false heads (*F. verticillioides*, *F. proliferatum*, and *F. subglutinans*). Conidiophores of *F. proliferatum* and *F. subglutinans* are unbranched and branched polyphialides and monophialides (Fig. 171), whereas *F. verticillioides* has only unbranched and branched monophialides. A few other *Fusarium* spp. also may be associated with corn stalk rot.

Taxonomic note: Prior to 1976, the causal organism of Fusarium stalk rot and Fusarium ear rot was usually reported as

Fig. 169. Fusarium stalk rot. (Courtesy G. P. Munkvold)

Fig. 170. Microconidia of *Fusarium verticillioides*. (Courtesy K. Seifert of the Department of Agriculture and Agri-Food, Government of Canada; © Her Majesty the Queen in Right of Canada, as represented by the Minister of Agriculture and Agri-Food Canada)

F. moniliforme. *F. moniliforme* var. *subglutinans* had been recognized as a variety of *F. moniliforme* since 1925. However, many authors did not make this distinction because of the similar appearance of cultures. *F. subglutinans* was described as a separate species in 1983; *F. proliferatum* was separated from *F. moniliforme* in 1976, but many researchers did not begin using the name until after 1983. Because its morphology in culture is similar in appearance to that of *F. moniliforme* (now *F. verticillioides*), some authors still identify *F. proliferatum* as *F. moniliforme* or *F. verticillioides*. The current nomenclature indicates that the causal organisms of Fusarium stalk rot and Fusarium ear rot are *F. verticillioides*, *F. proliferatum*, *F. subglutinans*, and *F. temperatum*. Cultures of these four fungi have very similar appearances, although they can be distinguished by microscopic characters. Routine identification of large numbers of isolates can be difficult because careful microscopic observation is necessary. The most reliable species identification is through gene sequence analysis of several well-studied fungal genes.

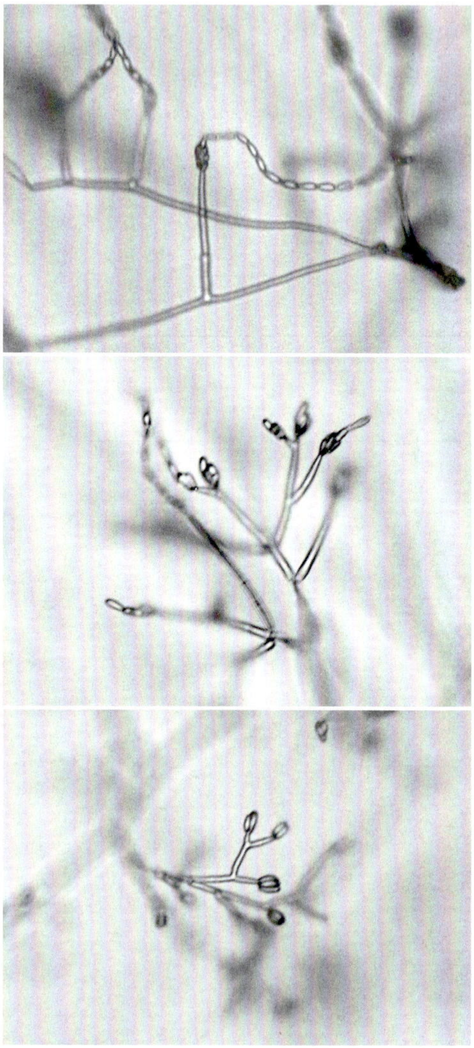

Fig. 171. Conidiophores of *Fusarium* spp. *Fusarium verticillioides* (top) has only monophialides. *Fusarium proliferatum* (middle) and *Fusarium subglutinans* (bottom) have both monophialides and polyphialides. (Courtesy B. Summerell)

Disease Cycle and Epidemiology

The fungi survive on crop residue in the soil or on the soil surface. Under favorable conditions, they may infect roots or stalks. All four species (*F. verticillioides*, *F. proliferatum*, *F. subglutinans*, and *F. temperatum*) may be present throughout the life cycle of the plant, some possibly originating from infected seed.

Management

In general, hybrids resistant to other stalk rot diseases are also resistant to Fusarium stalk rot, and cultural practices that reduce the incidence of other stalk rot diseases should be effective. Researchers in one study reported that resistance to Fusarium stalk rot in inbred line 61C is attributable to two genes. However, this resistance has not been utilized in commercial corn hybrids in the United States.

Selected References

Christensen, J. J., and Wilcoxson, R. D. 1966. Stalk Rot of Corn. Monogr. 3. American Phytopathological Society, St. Paul, MN.

Hooker, A. L., and White, D. G. 1976. Prevalence of corn stalk rot fungi in Illinois. Plant Dis. Rep. 60:1032-1034.

(Prepared by D. G. White)

Pythium Stalk Rot

Pythium stalk rot develops during extended hot and wet conditions. It is not considered to be of economic importance in most of the United States because it is found on individual plants in localized areas and does not become severe over extensive areas.

Symptoms

Unlike many other stalk rot diseases, Pythium stalk rot may be found prior to flowering. The disease usually is confined to the first internode above the soil line. Stalks are not completely broken off, and the diseased area is brown, water-soaked, soft, and collapsed (Fig. 172). The stalks may become twisted and distorted as the pith tissue completely rots. Tops of infected plants may remain green for several weeks because the vascular bundles remain intact.

Fig. 172. Pythium stalk rot. (Courtesy D. G. White)

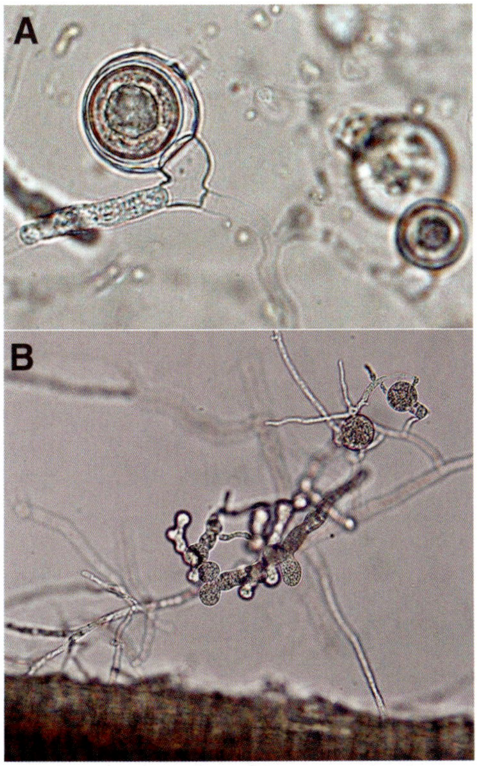

Fig. 173. *Pythium aphanidermatum.* **A,** Oospore inside an oogonium with an antheridium; and **B,** inflated filamentous sporangium. (Courtesy Z. G. Abad)

Causal Organism

Pythium stalk rot is caused by *Pythium aphanidermatum* (syn. *P. butleri*). The hyphae are hyaline and nonseptate. Sporangia (4–20 × 50–1,000 μm) are inflated filamentous and branched or unbranched and usually form complexes (Fig. 173). Zoospores (7.5 × 12 μm) are reniform and laterally biciliate. Oogonia (22–27 μm in diameter) are spherical, terminal, and rarely intercalary. Antheridia are usually monoclinous and typically intercalary, although often terminal. Oospores (17–19 μm in diameter) are aplerotic and single (Fig. 173).

Disease Cycle and Epidemiology

Very little research has been conducted on Pythium stalk rot. It is suspected that the organism survives as oospores in soil for several years and maintains its population on other hosts. The disease develops primarily during extended periods of hot (32°C), wet or very humid weather. It is seen only where soil drainage is poor and humidity is high.

Management

Very little information on management is available. Any practice that improves soil drainage could be somewhat effective.

(Prepared by D. G. White)

Other Fungal Stalk Rot Diseases

Black bundle disease, caused by *Sarocladium strictum* (syn. *Cephalosporium acremonium*), was reported in several studies conducted from 1920 to 1930 but is not currently considered to be a problem in the U.S. Corn Belt. It is important to note that, when it was reported, hybrid corn was not yet used and it appears that it was found on open-pollinated varieties. It is entirely possible that the susceptible open-pollinated varieties were eliminated from breeding efforts as inbred lines were being developed for hybrid corn. Symptoms include reddening of the leaves and stalk at the dough stage and a blackening of the vascular bundles that extends through several internodes. Infected plants may be barren, have excessive tillers, or have multiple ears at one node.

Nigrospora stalk rot, caused by *Nigrospora oryzae*, is also found occasionally on corn. The symptoms include black streaking on the exterior of the stalk with black discoloration of the pith (Fig. 174). The fungus is more often found in association with cobs and grain and is not considered to be an aggressive stalk invader.

Red root rot, caused by *Setophoma terrestris* (syn. *Phoma terrestris*), may be associated with shallow, dark brown, blotchy

Fig. 174. Nigrospora stalk rot. (Courtesy G. P. Munkvold)

lesions at or just below the soil surface. Lesions are reddish on dead stalks and may be confused with those caused by Gibberella stalk rot. The fungus is also commonly associated with roots and causes reddening of the brace roots (see the Red Root Rot section under Root Rots above).

Phaeocytostroma ambiguum (syn. *Phaeocytosporella zeae*) is associated with lesions on maturing stalks at or below ground level. The fungus produces pycnidia resembling those of *Stenocarpella maydis*, except that they are elongated rather than round. The organism is not known to cause economic damage as a stalk-rotting pathogen.

Other fungi associated with stalks are usually regarded as secondary stalk invaders. They may cause decay of pith tissue when stalks are inoculated. Some of the more common genera include *Alternaria*, *Aspergillus*, *Bipolaris*, *Cladosporium*, *Curvularia*, *Mucor*, *Penicillium*, *Rhizoctonia*, *Rhizopus*, and *Trichoderma*.

Selected Reference

Christensen, J. J., and Wilcoxson, R. D. 1966. Stalk Rot of Corn. Monogr. 3. American Phytopathological Society, St. Paul, MN.

(Prepared by D. G. White)

Late Wilt

Late wilt was first reported in Egypt as a vascular wilt disease of corn and is now found in all corn-growing areas of the country. Ninety percent of the plants of susceptible cultivars may be infected, and yield losses may approach 40%. The disease also is well established in India, although differences are apparent. Late wilt has also been reported in Hungary, Portugal, and Spain. Unconfirmed reports include other countries in Africa, Europe, and the Middle East. In Egypt, the fungus infects plants only through the roots; in India, late wilt symptoms can be induced by stalk inoculation. Isolates of the causal fungus from Egypt and India also differ slightly in morphology, pathogenicity, and sensitivity to fungicides.

Symptoms

Initial infection takes place through the roots, and root tip discoloration can be observed. Aboveground symptoms first become apparent during the tasseling stage. The initial symptom is a rapid wilting of lower leaves that progresses upward. Leaves assume a bleached or scorched appearance, and the entire plant may die prematurely (Fig. 175). Yellowish to reddish brown streaks appear on the basal internodes (Fig. 176). Stalks begin to dry up and have a shrunken and hollow appearance. Pith and vascular bundles become dark yellow to brownish (Figs. 177 and 178). Stalk symptoms may be modified by the invasion of secondary organisms around the nodes, which usually results in stalk rot. If kernels form, they are poorly developed. Seedborne infection or high inoculum levels in soil can result in stunted roots, poor emergence and vigor, or damping-off.

Causal Organism

Late wilt is caused by *Magnaporthiopsis maydis* (syns. *Cephalosporium maydis, Harpophora maydis*). Four clonal lineages, differing in virulence on corn, have been described from Egypt. Lineages of isolates from other areas are not clearly defined. Originally described as *C. maydis*, this species has over the years been transferred to *Acremonium*, *Harpophora*, and, most recently, *Magnaporthiopsis*. It is allied with other grass pathogens, such as *M. poae*, which causes summer patch of turfgrass.

The fungus grows best on potato dextrose agar plus 0.2% yeast extract at 29–30°C, with little or no growth below 10°C or above 36°C. Colonies have a felty appearance and are white but become gray or black with age. The margin of the colony has a characteristic rhizoidal appearance. Branched or unbranched conidiophores (60–250 µm or longer) are terminal or lateral, bearing conidia on phialides in heads. Conidia are hyaline, single celled, and oblong to oval, varying in size according to geographical origin: 1.2–3.4 × 2.3–13.8 µm (Egypt); 3–4 × 7–11 µm (India); and 3–3.6 × 6–19 µm (Hungary). They germinate by one, two, or (rarely) three germ tubes. Anastomosis of

Fig. 175. Premature plant death and bleaching caused by late wilt. (Courtesy G. P. Munkvold)

Fig. 176. Late wilt external stalk symptoms. (Courtesy G. P. Munkvold)

young hyphae and germ tubes is common. Small, sclerotium-like structures may appear in culture.

Disease Cycle and Epidemiology

M. maydis is soilborne or seedborne and infects corn seedlings through the roots or mesocotyl. Initially, the fungus grows superficially on roots, producing hyphae with short, thick-walled, and swollen, hyphopodia-like cells. After penetration, *M. maydis* grows intra- and intercellularly to colonize xylem tissue and then grows or is translocated to upper parts of the plant. By flowering, when symptoms usually first appear, the fungus has colonized the entire stalk and vascular tissue is blocked with hyphae and gumlike deposits, resulting in wilt symptoms. When plants are infected early and infections are severe, ears can be colonized by 12–13 weeks after planting and the fungus then can move through pedicels to colonize pericarp, endosperm, and embryo tissues. Kernel infection results in seedborne dissemination and also causes seed rot and damping-off. Seedlings are most susceptible to initial infection; as plants mature, susceptibility declines and there is no further infection from about 50 days after planting.

Optimal temperatures for late wilt development are 21–27°C, but the disease progresses rapidly at temperatures up to 32°C. Temperatures above 35°C limit the growth of *M. maydis*. Optimal soil pH for the fungus is 6.5, but it grows under a wide range of pH levels, from 4.5 to 10.0. In Egypt, late wilt is associated with low soil moisture levels, but in India, disease incidence is increased when rainfall is above average.

Fig. 177. Stalk shriveling caused by late wilt. (Courtesy G. P. Munkvold)

Fig. 178. Vascular discoloration caused by late wilt. (Courtesy G. P. Munkvold)

M. maydis can remain viable in the soil as sclerotia for several years in the absence of a host, usually in the top 20 cm of the soil profile. Although it is a weakly competitive saprophyte, the production of sclerotia in infested host debris ensures its long-term survival. The pathogen is spread through the movement of soil, crop residue, or infected seed. Tillage operations can spread the pathogen within fields and inoculum can be spread among fields by tillage equipment. Lupine, the only alternate host, can play a role in the parasitic survival of the pathogen. Saturated soils lessen sclerotial survival and the incidence of late wilt.

Seedborne infection appears to be common but has a minor role in the epidemiology of the disease. Infection incidence in seed from naturally infected plants has reportedly ranged from 1 to 11%. Planting infected seed can lead to disease transmission and soil infestation; therefore, infected seed is a potential means for disease spread. However, field outbreaks in areas where the disease is established are believed to be caused by soilborne inoculum.

Management

Resistant hybrids have been identified and they are recommended where the disease is a problem. Many sources of resistance have been identified in Egypt, and these have significantly reduced the impact of late wilt in that country. Other sources of resistance have been identified in proprietary germplasm that could be deployed elsewhere in the world. Resistance is usually under polygenic control, but major resistance genes have been reported. Use of disease-free seed could limit the risk of disease spread. *M. maydis* can be detected in seed using culture-plate methods, and species-specific polymerase chain reaction primers have been developed. Hot-water seed treatment can reduce seed transmission. In India, systemic fungicide seed treatments reduced the incidence of late wilt and increased yields, but seed treatments have not been effective in Egypt. Tillage to bury infested crop residue and *M. maydis* sclerotia can reduce pathogen survival and subsequent infection. Crop rotation involving paddy-cultivated rice and balanced soil fertility provide some control. Irrigation should be adequate, but not excessive, particularly during tasseling. Soil solarization can effectively reduce pathogen survival in areas where conditions of sunlight duration and intensity are sufficient.

Selected References

El-Shafey, H. A., El-Shorbagy, F. A., Khalil, M. A., Ikbal, I., and El-Assiuty, E. M. 1988. Additional sources of resistance to late wilt disease of maize caused by *Cephalosporium maydis*. Agric. Res. Rev. (Egypt) 66:221-230.

Gams, W. 2000. *Phialophora* and some similar morphologically little-differentiated anamorphs of divergent ascomycetes. Stud. Mycol. 45:187-200.

Molinero-Ruiz, M. L., Melero-Vara, J. M., and Mateos, A. 2010. *Cephalosporium maydis*, the cause of late wilt in maize, a pathogen new to Portugal and Spain. Plant Dis. 94:379.

Pecsi, S., and Nemeth, L. 1998. Appearance of *Cephalosporium maydis* Samra, Sabet and Hingorani in Hungary. Meded. - Fac. Landbouwkd. Toegepaste Biol. Wet. (Univ. Gent) 63(3a):873-877.

Sabet, K. A., Zaher, A. M., Samra, A. S., and Mansour, I. M. 1970. Pathogenic behavior of *Cephalosporium maydis* and *C. acremonium*. Ann. Appl. Biol. 66:257-263.

Saleh, A. A., and Leslie, J. F. 2004. *Cephalosporium maydis* is a distinct species in the *Gaeumannomyces-Harpophora* species complex. Mycologia 96:1294-1305.

Zeller, K. A., Ismael, A.-S. M., El-Assiuty, E. M., Fahmy, Z. M., Bekheet, F. M., and Leslie, J. F. 2002. Relative competitiveness and virulence of four clonal lineages of *Cephalosporium maydis* from Egypt toward greenhouse-grown maize. Plant Dis. 86:373-378.

(Prepared by H. A. El-Shafey and L. E. Claflin;
Revised by G. P. Munkvold)

Ear and Kernel Rots

Fungi cause a number of ear and kernel rots of corn. Some may be severe and result in losses in yield and grain quality. Perhaps more importantly, many of the ear rot fungi produce mycotoxins that can affect the safety, feed value, and marketability of grain (see the Mycotoxins and Mycotoxicoses section below). Mycotoxin contamination of grain used for ethanol production also is very important because of the use of coproducts of ethanol production (distillers' grains) as livestock feed supplements. Mycotoxins generally are enriched in the distillers' grains, which are an important revenue source for ethanol production facilities. Colonization of the grain by ear rot fungi is enhanced by insect, bird, or hail damage and by stalk lodging, which often results in the ears touching the ground. These conditions often lead to mixed infections of ear rot pathogens. Several of the ear rot pathogens also cause premature sprouting of kernels in the ear, which reduces grain quality and storability. Infection of kernels in the field can also lead to rot and the accumulation of mycotoxins during storage. The major ear rot diseases are Aspergillus ear rot, Diplodia (Stenocarpella) ear rot, Fusarium ear rot or pink ear rot, Gibberella ear rot or red ear rot, Penicillium ear rot and blue eye, and Nigrospora ear and cob rot. Some of the less common ear and kernel rots are listed in Table 3 (Figs. 179–182).

Fig. 179. Trichoderma ear rot. (Courtesy G. P. Munkvold)

TABLE 3. Other Corn Ear Rots

Disease Causal Organism(s)	Fruiting Structures	Symptoms and Other Relevant Information
Trichoderma ear rot *Trichoderma viride*	Mature fungal growth is dark green to dark bluish green in culture. Tufts of conidiophores are spread almost evenly on the surface of the colony. Conidia are rough walled, obovoid to broad ellipsoidal, and 2.8–4.5 µm in diameter.	Dark green to dark bluish green fungal growth is found on and between kernels, often covering the entire ear (Fig. 179). The disease is associated with injury to the developing ear and is usually found on broadly distributed plants. It is much more common in tropical areas and is reported to be a serious problem in southern China.
Gray ear rot *Botryosphaeria zeae* (syns. *Physalospora zeae* and *Macrophoma zeae*	Black perithecia contain clavate to cylindrical asci. Ascospores (8–11 × 21–33 µm) are hyaline to greenish, oval to ellipsoidal, and one celled. Pycnidia are submerged, black, globose, and protruding and have short necks. Conidia (6.5–8.5 × 17–31 µm) are unicellular and ellipsoidal.	Grayish white mold is found on and between kernels. Ears infected early have tightly adherent and bleached husks. Symptoms resemble those of Diplodia ear rot, except the ears at harvest are slate gray instead of grayish brown. Small, black sclerotia are often scattered throughout the cob. Perithecia and pycnidia may appear in large lesions on corn leaves. The disease was very common in seed production fields when double-cross hybrids were being produced from susceptible inbred lines but has virtually disappeared since then.
Black kernel rot *Lasiodiplodia theobromae* (syn. *Botryodiplodia theobromae*)	Conidia (10.5–17.25 × 21.5–31.5 µm) are oval, brown, and have one septum and are borne on white, aerial mycelium. Black, subglobose to globose pycnidia may appear singly or in clumps.	During early stages of infection, kernels show discoloration. As disease progresses, kernels become black with small eruptions that may contain pycnidia. A dry rot develops later. The disease is very rare.
Cladosporium rot *Cladosporium herbarum* and *Cladosporium cladosporioides* (syn. *Hormodendrum cladosporioides*)	Conidiophores are erect, up to 250 µm long, smooth, sparsely branched, and pale olive brown to brown. Conidia often appear in chains of two or three, have one to three septa, and are pale olive brown, subcylindrical, and mostly 4–6 × 8–15 µm.	Dark, greenish black, blotched or streaked kernels are scattered over the ear (Figs. 180 and 181). Discoloration appears first where the kernels are attached to the cob. Later, blotches progress upward but seldom reach the crown. When completely colonized, ears are dark and lightweight. The disease is often associated with insect injury on kernels and may also appear after frost damage. It is commonly associated with other ear rot diseases.
Rhizoctonia ear rot *Waitea circinata* (syn. *Rhizoctonia zeae*)	No spores are produced. Sclerotia are mostly superficial and are 0.1–0.5 mm in diameter. Hyphae are typically constricted at right angles.	A salmon pink mold first appears on the ear. Later, infected ears become dull gray, and numerous white to salmon-colored to dark brown or black sclerotia develop on the outer husks. The disease usually follows extended periods of warm weather. This disease is very rare.
Rhizopus ear rot *Rhizopus stolonifer*	Sporangiophores are 1–2 mm tall, globose sporangia are 100–150 µm in diameter, and sporangiospores are pale brown, ellipsoidal or subglobose, and 5–7 × 7–12 µm.	Coarse, pale gray mycelium densely covers part or all of the ear (Fig. 182). Development is favored by warm, wet conditions and physical injury to the ear. Dark sporangia are scattered throughout the mycelium.

Fig. 180. Cladosporium ear rot. (Courtesy D. G. White)

Fig. 181. Cladosporium ear rot. (Courtesy G. P. Munkvold)

Fig. 182. Rhizopus ear rot. (Courtesy G. P. Munkvold)

Selected References

Bezuidenhout, H., and Marasas, F. W. O. 1978. *Botryosphaeria zeae*: The cause of grey ear rot of maize (*Zea mays*) in South Africa. Phytophylactia 10:21-24.

Del Rosario, M. S. E. 1954. The fungus causing black kernel rot of corn. Philipp. J. Agric. 37:623-631.

McGee, D. C. 1988. Maize Diseases: A Reference Source for Seed Technologists. American Phytopathological Society, St. Paul, MN.

Munkvold, G. P. 2003. Cultural and genetic approaches to managing mycotoxins in maize. Annu. Rev. Phytopathol. 41:99-116.

Munkvold, G. P. 2003. Mycotoxins in corn—Occurrence, impacts, and management. Pages 811-881 in: Corn: Chemistry and Technology, 2nd ed. P. J. White and L. A. Johnson, eds. AACC International, St. Paul, MN.

Punithalingam, E. 1976. *Botryodiplodia theobromae*. Descriptions of Pathogenic Fungi and Bacteria, no. 519. Commonwealth Mycological Institute, Kew, England.

(Prepared by D. G. White)

Aspergillus Ear Rot

Aspergillus ear rot causes little damage to the ears during most years because colonization is often restricted to a few kernels. However, under drought-stress conditions, or when there is extensive physical damage to kernels from insects or other causes, colonization of the ears can be severe. It is most serious in geographical locations with high temperatures and drought environments. The disease is of concern because *Aspergillus flavus* and a related species, *A. parasiticus*, produce aflatoxins, which are potent carcinogens. *Aspergillus niger* and other species in *Aspergillus* section *Nigri* can cause ear rot symptoms in the field; species in this group do not produce aflatoxins, but some strains can produce another mycotoxin, fumonisin B_2. These fungi, as well as other *Aspergillus* spp., also cause storage rots.

Symptoms

Usually, only a few kernels on an ear display symptoms (Fig. 183), but severe colonization can occur (Fig. 184). Masses of

Fig. 183. *Aspergillus flavus* sporulating on a corn ear. (Courtesy G. P. Munkvold)

Fig. 184. Severe Aspergillus ear rot. (Courtesy G. P. Munkvold)

yellow-green or olive green spores often are found on and between diseased kernels. Symptoms of *A. niger* infection are similar to those of *A. flavus* or *A. parasiticus* but are distinct because the spore masses are dark brown to black (Fig. 185). Older colonies of *A. flavus* and *A. parasiticus* may turn dark green to brown but retain more yellow color than the colonies of other ear rot fungi, such as *Penicillium* or *Trichoderma* spp. Although any part of the ear may be infected with *Aspergillus* spp., the tip is the most common site. Sporulation is evident on kernels that are injured, especially around sites damaged by insects (Fig. 186). However, the fungus may be present in kernels with no visible sporulation (Fig. 187), although these infected kernels often appear dull and discolored. Kernels infected by *Aspergillus* spp. may fluoresce under ultraviolet light (Fig. 188) because of the production of kojic acid. Because so-called "bright green-yellow fluorescence" (BGYF) sometimes is associated with the presence of aflatoxins, ultraviolet light has been used as a quick screen for potential aflatoxin contamination. However, this approach is not a direct method for detecting aflatoxins and is not recommended for this purpose.

Fig. 185. *Aspergillus niger* sporulating on a corn ear. (Courtesy G. P. Munkvold)

Fig. 186. Infection of insect-damaged kernels by both a *Penicillium* sp. and a *Fusarium* sp. (left) and by *Aspergillus flavus* (right). (Courtesy G. P. Munkvold)

Causal Organisms

Aspergillus flavus and *A. parasiticus* are morphologically similar and are the only species that produce aflatoxins. *A. flavus* is the predominant species on corn. Conidial heads are globose to radiate to columnar and light yellow-green to olive brown (Fig. 189). Vesicles are globose or subglobose, and the larger vesicles have both metulae and phialides. Conidia are globose to subglobose, smooth to slightly rough, and 3–7 µm (mostly 3.5–4.5 µm) in diameter. Sclerotia, when present, are dark red to black; globose, subglobose, or vertically elongate; and 200–700 µm in diameter. There are two morphological variants within *A. flavus*, known as S strains and L strains, according the typical size of their sclerotia. S strains produce

Fig. 187. *Aspergillus flavus*-infected kernels under white light. (Courtesy D. G. White)

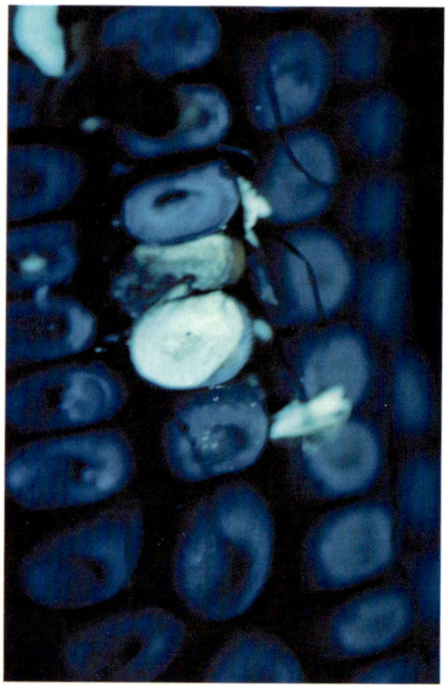

Fig. 188. *Aspergillus flavus*-infected kernels under ultraviolet light. (Courtesy D. G. White)

many small sclerotia (less than 400 µm), fewer conidia, and high levels of aflatoxins. L strains produce fewer, larger sclerotia, more numerous conidia, and variable amounts of aflatoxins. *A. flavus* can also be differentiated into two phylogenetic groups (Groups I and II); Group I includes L and S strains that produce only B-type aflatoxins; whereas Group II consists of S strains, some of which produce both B- and G-type aflatoxins, while others produce only B-type aflatoxins. Other *Aspergillus* spp. that may be found include *A. niger* and *A. oryzae*, but they can be distinguished from *A. flavus* and *A. parasiticus*. In the case of *A. niger*, conidia are dark brown to black (Fig. 190); *A. oryzae* strains have variable morphology, but a key difference is that neither of these species produces aflatoxins.

Disease Cycle and Epidemiology

A. flavus is a thermotolerant fungus; thus, it is more likely than many other fungi and bacteria to survive at high temperatures and under dry conditions. The fungus overwinters in soil and debris, and its populations greatly increase during hot, dry weather. Conidia of the fungus are spread to silks by wind or insects. Hyphae may then grow on the silk tissue and down the silks into the kernels. Insect damage predisposes the kernels to fungal penetration, and insects may carry the spores directly to the damaged kernels. Larvae of lepidopteran insects, such as corn earworm (*Helicoverpa zea*), fall armyworm (*Spodoptera frugiperda*), and European corn borer (*Ostrinia nubilalis*), as well as some adult beetles, such as *Carpophilus* spp., have been associated with enhanced *A. flavus* infection and elevated aflatoxin levels in corn. Plants that are stressed, particularly by drought, are more susceptible to infection by *A. flavus* and to the production of aflatoxin.

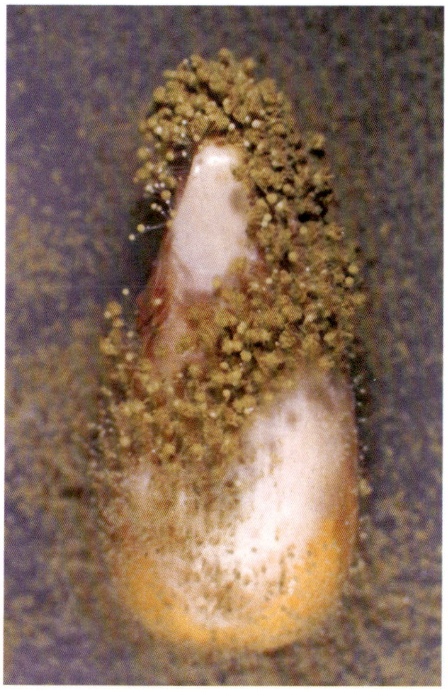

Fig. 189. Conidial heads of *Aspergillus flavus* on a corn kernel. (Courtesy M. da Silva)

Fig. 190. Conidial heads of *Aspergillus niger* on a corn kernel. (Courtesy M. da Silva)

Management

Use of resistant hybrids is the best and most economical management practice. Several commercial hybrids with some resistance have been identified, and some are used in the southern United States where drought conditions frequently favor infection by *A flavus*. Most hybrids used in the midwestern United States are considered to be susceptible and are at risk during years with drought conditions. Even resistant hybrids may not have a level of resistance high enough to escape damage during severe droughts. A number of inbred lines with resistance have been identified, and molecular markers associated with genes for resistance are known. Unfortunately, most inbred lines with resistance are very late maturing, have low yield in hybrid combination, and have serious agronomic deficiencies that make them unusable in modern hybrids. The most likely solution involves the incorporation of genes associated with resistance into commercially usable inbred lines. Advances in corn genomics and proteomics have facilitated the identification of key genes and proteins involved in resistance. Resistant genotypes under drought stress often have a higher expression of stress-related genes compared with that from more susceptible genotypes. Genomics studies of *A. flavus* may result in eventual opportunities for genetically engineered resistance. Strains of *A. flavus* that do not produce aflatoxins have shown to reduce aflatoxin accumulation in grain by competitively excluding aflatoxin producers. Use of atoxigenic strains has a possibility of becoming a component of aflatoxin management programs, and commercial products are now available and have shown some promise. Some control may be achieved by irrigation to avoid stress. Insecticides or transgenic resistance to control insects that feed on ears can be partially effective. Reducing moisture stress during and after pollination can have significant value for decreasing the risk of infection and aflatoxin accumulation. This can be accomplished by adjusting planting date, proper plant density, and adequate, timely irrigation. Timely harvest can help to avoid additional fungal growth and aflatoxin production during drydown. Rapid drying of harvested grain can significantly aid in limiting further development of *A. flavus* in grain bins. Crop rotation and tillage have little or no effect, except when better plant growth with less stress results. Proper storage of kernels after harvest prevents the additional accumulation of aflatoxins.

Selected References

Campbell, K. W., and White, D. G. 1995. Inheritance of resistance to Aspergillus ear rot and aflatoxin in corn genotypes. Phytopathology 85:886-896.

Campbell, K. W., and White, D. G. 1995. Evaluation of corn genotypes for resistance to Aspergillus ear rot, kernel infection, and aflatoxin production. Plant Dis. 79:1039-1045.

Campbell, K. W., Hamblin, A. M., and White, D. G. 1997. Inheritance of resistance to aflatoxin production in the cross between corn inbreds B73 and LB31. Phytopathology 87:1144-1147.

Dorner, J. W. 2010. Efficacy of a biopesticide for control of aflatoxins in corn. J. Food Prot. 73:495-499.

Flett, B. C., McLaren, N. W., and Wehner, F. C. 1998. Incidence of ear rot pathogens under alternating corn tillage practices. Plant Dis. 82:781-784.

Raper, K. B., and Fennell, D. I. 1965. The Genus *Aspergillus*. Williams & Wilkins, Baltimore.

(Prepared by D. G. White)

Diplodia (Stenocarpella) Ear Rot

Diplodia ear rot was the most prevalent ear rot of midwestern-grown corn in the first half of the twentieth century, but this disease decreased in importance during the 1960s and 1970s. Subsequently, Diplodia ear rot has re-emerged with increased incidence and severity in the midwestern United States along with the use of conservation tillage. The disease is most severe when corn follows corn and there is wet weather at and shortly after silking. In North America, this disease has not been associated with mycotoxin production or mycotoxicoses; however, in the Southern Hemisphere, diplodiosis is a well-documented toxicosis associated with *Stenocarpella maydis*.

Symptoms

Ears are most susceptible early in their formation. Symptoms usually first appear at the base of the ear (Fig. 191). Early infection of ears usually results in mummification, in which the entire ear may be shrunken and turn grayish brown (Fig. 192). Infected kernels appear to be "glued" to the husk by the white mycelium of the fungus. Often, the leaf sheath is rotted and ear leaves are killed (Fig. 193). On ears, the fungus is visible as a conspicuous, coarse, white to grayish brown mycelial growth over the husk and kernels. Infected kernels are dull gray to brown. Infected ears are very light and may be totally rotted. Black pycnidia are produced late in the season on the husks, kernels, cobs, and rotted stalks (Fig. 194). If infection takes place several weeks after flowering, the ears may show no external symptoms. Kernels from these ears may be asymptomatic, but the ears often have white fungal mycelium between the kernels. Infected kernels with no symptoms of rot can present a serious problem with hybrid seed production. When asymptomatic infected seed is planted, the fungus may rot the mesocotyl of the seedling and plants will be stunted and low in yield. Some isolates of the fungus may cause premature germination of the kernels.

Causal Organisms

Diplodia ear rot is caused by *Stenocarpella maydis* (syns. *Diplodia maydis* and *D. zeae*) and *Stenocarpella macrospora* (syn. *D. macrospora*). Other diseases caused by these fungi are Diplodia leaf streak, Diplodia (Stenocarpella) stalk rot, and Diplodia leaf spot.

Fig. 191. Diplodia (Stenocarpella) ear rot initial symptoms at the base of an ear. (Courtesy G. P. Munkvold)

Fig. 192. Diplodia (Stenocarpella) ear rot. (Courtesy G. P. Munkvold)

Fig. 193. Dead husk leaves resulting from Diplodia (Stenocarpella) ear rot. (Courtesy G. P. Munkvold)

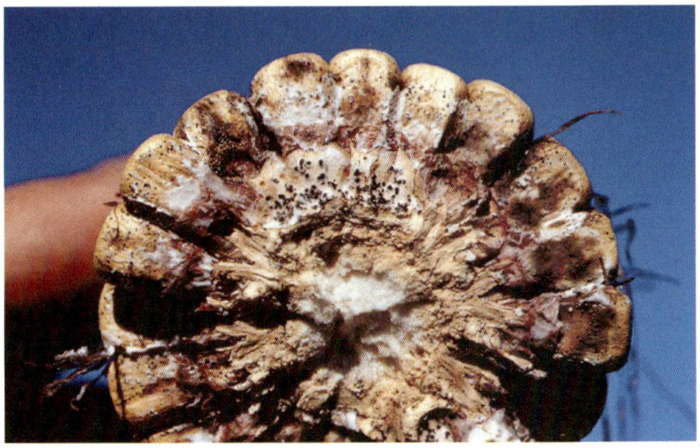

Fig. 194. Diplodia (Stenocarpella) ear rot and pycnidia of *Stenocarpella maydis*. (Courtesy D. G. White)

Disease Cycle and Epidemiology

Corn is the only host of *S. maydis*. Consequently, the most important source of Diplodia ear rot inoculum is infected corn stalk residue from a previous crop. Pycnidia embedded in stalks on the soil surface produce conidia during wet weather. The conidia are disseminated in splashing water and usually penetrate the husk at the base of the ear, and hyphae grow into the ear. With husk infection, the fungus appears to grow between the ear shoot and the stalk or between the ear shoot and the sheath of the ear leaf. Conidia may also be deposited on silks, and the fungus then grows down the silks into the ear. Ears are most susceptible at flowering until 3 weeks later.

Management

Hybrids differ in their susceptibility to Diplodia ear rot, and commercial hybrids are screened for resistance. A few studies have been done to identify the most susceptible commercial hybrids. Little has been on the genetics of resistance. Crop rotation and fall tillage of corn residue reduce the disease levels. Certain fungicides sprayed at flowering have been shown to be effective in the control of Diplodia ear rot in seed production fields but are not likely to provide an economical return for the control of this disease in commercial grain production.

Selected References

Dorrance, A. E., Hinkelmann, K. H., and Warren, H. L. 1998. Diallel analysis of Diplodia ear rot resistance in maize. Plant Dis. 82:699-703.

Flett, B. C., and McLaren, N. W. 1994. Optimum disease potential for evaluating resistance to *Stenocarpella maydis* ear rot in corn hybrids. Plant Dis. 78:587-589.

Flett, B. C., McLaren, N. W., and Wehner, F. C. 1998. Incidence of ear rot pathogens under alternating corn tillage practices. Plant Dis. 82:781-784.

Kleinschmidt, C. E., and White, D. G. 2003. Evaluation of Quadris and Tilt for control of Diplodia ear rot in corn, 2002. Fungicide and Nematicide Tests 58:FC046. American Phytopathological Society, St. Paul, MN.

(Prepared by D. G. White)

Fusarium Ear Rot or Pink Ear Rot

Fusarium ear rot is found wherever corn is grown. It is most severe when drought stress occurs, followed by high humidity at and immediately after flowering. These conditions are mostly common along coastal areas, where the disease is more severe. Recognition of the importance of this disease greatly increased when it was determined that the fumonisin mycotoxins are produced by the causal organisms.

Symptoms

Symptoms vary greatly depending upon genotype, environment, and disease severity. Usually, individual or groups of infected kernels are scattered randomly on the entire ear. White, pale pink, salmon-colored, or lavender fungal growth on kernels, silks, or both is typical (Fig. 195). Infected kernels may also exhibit a starburst symptom, i.e., white streaks radiating from the point of silk attachment at the cap of the kernel or from the base (Figs. 196 and 197). These streaks are corroded channels within the pericarp caused by fungal growth. Air in the channels breaks the transparency of the pericarp so that the yellow aleurone beneath cannot be seen. Fungal growth is frequently found at the tip of the ear, where infection is associated with insect and other types of damage (Fig. 198), or symptoms may appear anywhere on the ear where insects have fed (Fig. 199). It has also been seen where seed coats are broken as a result of the noninfectious conditions known as silk cut and

Fig. 195. Fusarium ear rot. (Courtesy G. P. Munkvold)

Fig. 196. Starburst symptom of Fusarium ear rot. (Courtesy G. P. Munkvold)

Fig. 197. Fusarium ear rot. (Courtesy G. P. Munkvold)

popped kernel (Fig. 200). When disease is very severe, the entire ear may be pale and appear weathered with fungal growth on and between kernels. Rarely, the entire ear can be mummified (Fig. 201). *Fusarium verticillioides*, *F. proliferatum*, and *F. subglutinans* are commonly isolated from asymptomatic, apparently sound kernels, and it is difficult to find grain or seed lots in which one or more of these fungi are not associated with at least a small percentage of kernels. Even when symptoms are not evident, fumonisins may be found in grain.

Causal Organisms

Fusarium kernel rot is caused by several *Fusarium* spp.: primarily *Fusarium verticillioides* (syns. *F. moniliforme*, *Gibberella moniliformis*, and *G. fujikuroi* mating population A); *F. proliferatum* (syns. *G. intermedia*, *G. fujikuroi* var. *intermedia*, and *G. fujikuroi* mating population D); *F. subglutinans* (syns. *G. subglutinans* and *G. fujikuroi* mating population E); and *F. temperatum*. Less common are *F. oxysporum*, *F. poae*, *F. sporotrichioides*, and *F. tricinctum*. Details about these causal organisms are in the Fusarium Stalk Rot disease section.

Disease Cycle and Epidemiology

The *Fusarium* spp. overwinter in debris, where they produce macroconidia and microconidia under warm, humid conditions. Microconidia are the most common propagules. Airborne conidia can colonize silks, and the fungus grows down the senescing silks and colonizes kernels, resulting in starburst symptoms, visible mold, or symptomless infection. Some species also systemically invade the stalk and reach the ears through the ear shank. This results either from seed transmission or from root or stalk infection. The fungus also commonly infects the kernels at the embryo end, rarely causing visible symptoms. Symptomless infection is common, especially following systemic colonization. Infection may also follow pericarp injury, e.g., that caused by silk cut. Insect damage is an important factor leading to infection and ear rot; and under some conditions, there is a close correlation between insect damage and ear rot severity. Several types of insects are associated with increased Fusarium ear rot severity: primarily larvae of the European corn borer, corn earworm, and other lepidopteran insects; beetles in the families Nitidulidae and Chrysomelidae (corn rootworms); and thrips (*Frankliniella* spp.). These insects can act as vectors for the conidia of the fungi and create pericarp wounds that are very susceptible to infection.

Management

Corn hybrids differ in their susceptibility to Fusarium ear rot and to accumulation of fumonisins in grain. Molecular markers associated with resistance have been identified and are being used to cross resistance into commercially used inbred lines. The use of insect resistance genes introduced through genetic engineering has been demonstrated to reduce Fusarium ear rot and result in lower fumonisin levels in corn grain. European corn borers (*Ostrinia nubilalis*) and other lepidopteran insects

Fig. 198. Fusarium ear rot at the ear tip resulting from insect damage. (Courtesy G. P. Munkvold)

Fig. 199. Fusarium ear rot resulting from insect damage. (Courtesy G. P. Munkvold)

Fig. 200. Fusarium ear rot symptoms on split kernels. (Courtesy G. P. Munkvold)

Fig. 201. Severe Fusarium ear rot. (Courtesy M. Mancl, DuPont Pioneer)

feed on kernels and cobs, making wounds that allow for increased colonization by *Fusarium* spp. Transgenes derived from the bacterium *Bacillus thuringiensis* (*Bt* genes) code for the expression of insecticidal proteins that control European corn borers and several other insects. Timing of planting to avoid drought stress and peak insect populations during early grain fill reduces the risk of severe disease and mycotoxin accumulation. Adequate irrigation and timely harvest can also reduce the risks. Proper storage of kernels after harvest prevents the additional accumulation of fumonisins.

Selected References

Clements, M. J., Campbell, K. W., Maragos, C. M., Pilcher, C., Headrick, J. M., Pataky, J. K., and White, D. G. 2003. Influence of Cry1Ab protein and hybrid genetics on fumonisin contamination and Fusarium ear rot of corn. Crop Sci. 43:1283-1293.

Clements, M. J., Maragos, C. M., Pataky, J. K., and White, D. G. 2004. Sources of resistance to fumonisin accumulation in grain and Fusarium ear and kernel rot of corn. Phytopathology 94:251-260.

Headrick, J. M., Pataky, J. K., and Juvik, J. A. 1990. Relationships among carbohydrate content of kernels, condition of silks after pollination, and the response of sweet corn inbred lines to infection of kernels by *Fusarium moniliforme*. Phytopathology 80:487-494.

Munkvold, G. P., McGee, D. C., and Carlton, W. M. 1997. Importance of different pathways for maize kernel infection by *Fusarium moniliforme*. Phytopathology 87:209-217.

Nelson, P. E., Toussoun, T. A., and Marasas, W. F. O. 1983. *Fusarium* Species: An Illustrated Manual for Identification. Pennsylvania State University Press, University Park.

Robertson, L. A., Kleinschmidt, C. E., White, D. G., Payne, G. A., Maragos, C. M., and Holland, J. B. 2006. Heritabilies and correlation of Fusarium ear rot resistance and fumonisin contamination resistance in two maize populations. Crop Sci. 46:353-361.

Schaafsma, A. W., Miller, J. D., Savard, M. E., and Ewing, R. J. 1993. Ear rot development and mycotoxin production in corn in relation to inoculation method, corn hybrid, and species of *Fusarium*. Can. J. Plant Pathol. 15:185-192.

Shelby, R. A., White, D. G., and Bauske, E. M. 1994. Differential fumonisin production in maize hybrids. Plant Dis. 78:582-584.

(Prepared by D. G. White)

Gibberella Ear Rot or Red Ear Rot

Gibberella ear rot is distributed worldwide but is more severe in cool, humid areas. The disease can cause serious problems because of the production of various mycotoxins, notably the trichothecenes and zearalenone, which are toxic to livestock and humans.

Symptoms

Gibberella ear rot, sometimes called red ear rot, typically is characterized by a pink to reddish mold that appears at the tip and grows down the ear (Fig. 202). Symptoms also may be associated with insect or hail damage to ears. In some cases, symptoms develop from the middle or base of the ear (Fig. 203). If infected early, the entire ear may rot and be covered with a pinkish mycelium that causes the husk to tightly adhere to the ear (Fig. 204). The husk leaves may die prematurely and superficial, blue-black perithecia may appear on the husks and ear shanks (Figs. 205 and 206).

Causal Organisms

Gibberella ear rot is caused by *Fusarium graminearum* (syn. *Gibberella zeae*) or occasionally *F. culmorum*, *F. boothii*, or *F. meridionale*. Details about *F. graminearum* are in the Gibberella Stalk Rot disease section.

Disease Cycle and Epidemiology

G. zeae overwinters in corn debris and can also be found associated with debris of numerous other hosts. Inoculum produced on wheat during epidemics of wheat scab (also called Fusarium head blight) may be important. Macroconidia are produced under moist conditions and moderate temperatures and are splash dispersed. Ascospores also are important propagules; they are forcibly discharged from perithecia on crop debris and can be dispersed long distances. The fungus infects silks, grows into the ear, and progresses down the ear during grain fill. Silks are most susceptible within the first week after silk emergence. Cool, wet weather during this period leads to high levels of infection. Infections may also take place at the base of the ear when there is heavy rainfall late in the season, especially on hybrids where the ears do not turn down and water collects between husks and kernels at the base. The disease may be more severe in fields previously planted to corn or wheat, especially if debris remains on the soil surface. Physical

Fig. 202. Gibberella ear rot initiated from the ear tip. (Courtesy G. P. Munkvold)

Fig. 203. Gibberella ear rot. (Courtesy G. P. Munkvold)

Fig. 204. Severe Gibberella ear rot. (Courtesy G. P. Munkvold)

Fig. 205. Dead husk leaves at the distal end of an ear resulting from Gibberella ear rot infection. (Courtesy G. P. Munkvold)

damage to the ears promotes more severe disease development, but Gibberella ear rot is not closely associated with insect damage.

Management

Corn lines differ in resistance to Gibberella ear rot; for example, the fungus does not grow as well on silk tissue of inbred lines CO272 and CO387 as on silks of susceptible inbred lines. Commercial hybrids are screened for resistance and there is a wide range of susceptibility levels among commercial hybrids.. Hybrids with extremely tight husks appear to be more severely affected than hybrids with loose husks. Also, hybrids with ears that remain upright have a tendency to be more susceptible. Major differences in resistance to *G. zeae* and to mycotoxin formation by the causal organisms have been reported among inbred lines and hybrids. Inbred lines derived from late-maturing southern white corn populations have a tendency to be very susceptible. Molecular markers associated with resistance have been reported for inbred line CO387 and eventually could be used to backcross resistance into commercially used inbred lines. Crop rotation (with noncereal crops) and tillage can reduce local inoculum levels and the risk of disease, but aerially dispersed ascospores limit the effectiveness of these tactics. Controlling insects has a limited impact on Gibberella ear rot and associated mycotoxins but can contribute to management. Proper storage of kernels after harvest prevents the additional accumulation of mycotoxins.

Fig. 206. Perithecia of *Gibberella zeae* on husk leaves. (Courtesy G. P. Munkvold)

Selected References

Ali, M. L., Taylor, J. H., Jie, L., Sun, G., William, M., Kasha, K. J., Reid, L. M., and Pauls, K. P. 2005. Molecular mapping of QTLs for resistance to *Gibberella* ear rot, in corn, caused by *Fusarium graminearum*. Genome 48:521-533.

Assabgui, R. A., Reid, L. M., Hamilton, R. I., and Arnason, J. T. 1993. Correlation of kernel (E)-ferulic acid content of maize with resistance to *Fusarium graminearum*. Phytopathology 83:949-953.

Flett, B. C., McLaren, N. W., and Wehner, F. C. 1998. Incidence of ear rot pathogens under alternating corn tillage practices. Plant Dis. 82:781-784.

Nelson, P. E., Toussoun, T. A., and Marasas, W. F. O. 1983. *Fusarium* Species: An Illustrated Manual for Identification. Pennsylvania State University Press, University Park.

Reid, L. M., Bolton, A. T., Hamilton, R. I., Woldemariam, T., and Mather, D. E. 1992. Effect of silk age on resistance of maize to *Fusarium graminearum*. Can. J. Plant Pathol. 14:293-298.

Reid, L. M., Mather, D. E., Hamilton, R. I., and Bolton, A. T. 1992. Diallel analysis of resistance in maize to *Fusarium graminearum* infection via the silk. Can. J. Plant Sci. 72:915-923.

Reid, L. M., Spaner, D., Mather, D. E., Bolton, A. T., and Hamilton, R. I. 1993. Resistance of maize hybrids and inbreds following silk inoculation with three isolates of *Fusarium graminearum*. Plant Dis. 77:1248-1251.

Schaafsma, A. W., Miller, J. D., Savard, M. E., and Ewing, R. J. 1993. Ear rot development and mycotoxin production in corn in relation to inoculation method, corn hybrid, and species of *Fusarium*. Can. J. Plant Pathol. 15:185-192.

Snijders, C. H. A. 1994. Breeding for resistance to *Fusarium* in wheat and maize. Pages 37-58 in: Mycotoxins in Grain: Compounds Other Than Aflatoxin. J. D. Miller and H. L. Trenholm, eds. AACC International, St. Paul, MN.

(Prepared by D. G. White)

Fig. 207. Nigrospora ear rot. (Courtesy D. G. White)

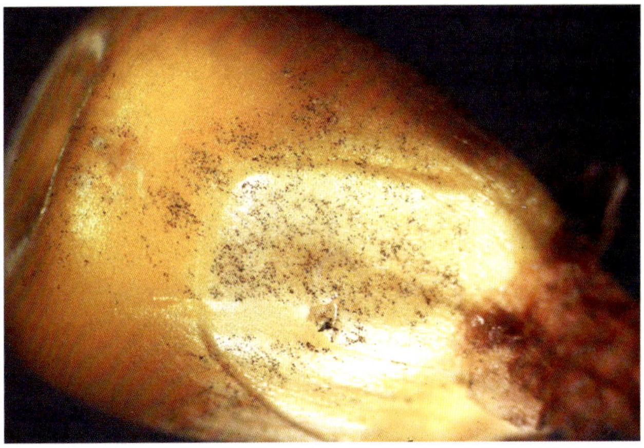

Fig. 209. *Nigrospora oryzae* spores on a corn kernel. (Courtesy G. P. Munkvold)

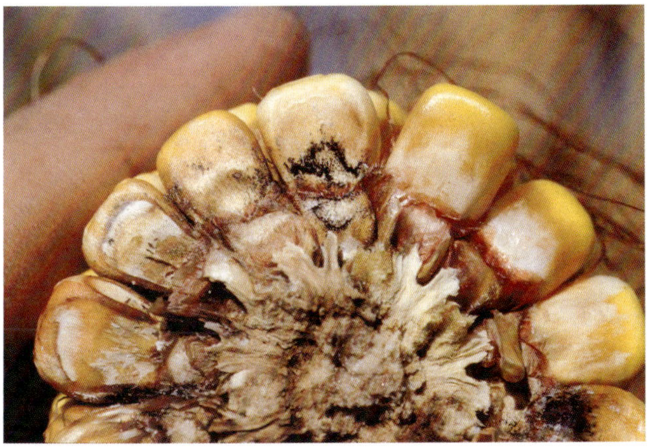

Fig. 208. Nigrospora ear rot. (Courtesy D. G. White)

Fig. 210. Severe Nigrospora ear rot. (Courtesy G. P. Munkvold)

Nigrospora Ear and Cob Rot

Nigrospora ear rot is uncommon, except on corn killed by early frost, stalk rot, or other factors. Sometimes it is found on ears rotted by *Stenocarpella* or *Fusarium* spp.

Symptoms

Symptoms are not conspicuous until the ears are harvested. The distinguishing feature is a shredding of the cob, usually at the butt end, when the ears are picked mechanically (Fig. 207). Cobb tissue may then be included in harvested grain, making drying and storage more difficult. There may be gray mycelium on the kernels, which are also "peppered" with masses of small, round, black spores (Figs. 208 and 209). In severe cases, dark gray mycelium may partially cover the base of the ear (Fig. 210). Affected ears are lightweight and chaffy, and kernels are slightly bleached or streaked, poorly finished, and easily pressed into the cob. In yellow and white hybrids, the chaff is brown and pale yellow or gray, respectively.

Causal Organism

Nigrospora ear rot is caused by *Nigrospora oryzae* (syns. *Khuskia oryzae* and *Basisporium gallarum*). Conidia are black, spherical to ovoid, 10–16 μm in diameter (mostly 12–14 μm), and borne singly on short, branched conidiophores 3–7 μm thick. Perithecia are globose and 200 μm in diameter. Ascospores (5–7 × 16–21 μm) are hyaline and two celled.

Disease Cycle and Epidemiology

N. oryzae survives on plant debris. It is a weak parasite and attacks ears only after plants are weakened or killed prematurely by frost, drought, root injury, leaf blights, stalk rot, or other ear rot pathogens. The fungus also can colonize stalk tissue but is not considered an aggressive stalk rot pathogen.

Corn grown in infertile soil appears to be more susceptible than that grown in fertile soil, possibly because poor nutrition results in premature death.

Management

The disease is rarely economically important, and very little information is available on the susceptibility of corn hybrids. Any cultural practice that promotes healthy plants or prevents damage to the ear reduces the incidence of the disease.

Selected Reference

Standen, J. H. 1945. *Nigrospora oryzae* (B. and Br.) petch on maize. Phytopathology 35:552-564.

(Prepared by D. G. White)

Penicillium Ear Rot and Blue Eye

Several *Penicillium* spp. can be found on kernels in the field or in storage and in silage. Damage in the field is usually minor, but in some cases, it can involve most of the ear. *Penicillium* spp. can be important because of their effects on grain quality, including mold damage and the presence of associated mycotoxins. The most important mycotoxin associated with *Penicillium* spp. is ochratoxin A, produced by *P. verrucosum*, but several others occasionally are found.

Fig. 211. Penicillium ear rot. (Courtesy D. G. White)

Fig. 212. Penicillium ear rot. (Courtesy D. G. White)

Symptoms

Penicillium ear rot is found primarily on ears with mechanical or insect injuries. The characteristic sign is the powdery green or blue-green mold on and between the kernels, usually at the ear tip (Figs. 211 and 212). Kernels with fungal growth normally become bleached and streaked. If the grain is stored at high moisture levels (higher than 18%), the fungus may invade the grain and discolor the embryo. "Blue eye" is a blue discoloration of the embryo resulting from the presence of the blue-green spores of the fungus (Fig. 213). Blue eye may also be caused during storage by *Aspergillus glaucus*, which can grow at grain moisture contents as low as 14.5%. Penicillium ear rot may be confused with Trichoderma ear rot because of similarity of the color of the conidia of the causal organisms.

Causal Organisms

Several *Penicillium* spp. may be isolated from kernels at harvest, but *P. oxalicum* is the only one that causes a definite ear rot after inoculation. Numerous other species, including *P. chrysogenum*, *P. cyclopium*, *P. funiculosum*, *P. expansum* (syn. *P. glaucum*), and *P. viridicatum*, also commonly are found either in the field or in storage. The conidia of *Penicillium* spp. are borne on long conidiophores that branch in a broomlike fashion (Fig. 214). They are one celled and globose to ovoid, and under the microscope, they resemble glass beads.

Fig. 213. Blue eye symptom and *Penicillium* sp. sporulating on pedicels. (Courtesy D. G. White)

Fig. 214. Conidiophores and conidial heads of a *Penicillium* sp. (Courtesy M. da Silva)

Disease Cycle and Epidemiology

P. oxalicum and other *Penicillium* spp. are commonly found on plant debris. Conidia are produced abundantly and are common in the air. These fungi usually enter the ear through wounds caused by insects or birds.

Management

No information is available on the susceptibility of corn hybrids. Any practice that prevents damage to the ear or grain following harvest provides some control. Proper storage of kernels after harvest prevents the additional growth of *Penicillium* spp. in grain.

(Prepared by D. G. White)

Storage Molds

Storage molds can be found anywhere corn grain is stored, but they are particularly important in developing countries where there are often high humidities and temperatures and sophisticated storage facilities may not be available. Fungi that cause decay of corn grain have traditionally been classified as field fungi or storage molds, although these groups may overlap. Field fungi are usually characterized as those that infect kernels in the field and require grain moistures of approximately 20% or more for growth in kernel tissues. They are not active at lower moisture content in storage and eventually die out. These include most of the fungi discussed in the previous section. Storage molds are primarily saprophytic fungi that may externally contaminate but typically do not invade kernels in the field. They may become associated with kernels during harvest, transport, or storage. They can grow at lower grain moisture content, between 13 and 20%, and under a wide range of temperature conditions, although higher temperatures typically result in greater damage by storage molds. They can survive for long periods in stored grain at low moisture content. Storage molds reduce the dry matter and nutritional quality of grain and can produce mycotoxins that can lead to economic impacts in livestock production. Other losses occur because of price discounts associated with mold damage or there is a complete loss of marketability. Grain drying and aeration costs are primarily incurred to prevent storage mold development. Grain molds also pose a safety hazard to workers because of inhalation exposure and the risk of engulfment and asphyxiation when entering grain bins to break up cakes or crusts formed by molded grain.

Storage molds also can be found in silage and high-moisture corn storage if anaerobic conditions are not maintained.

Symptoms

Grain affected by storage mold becomes discolored and a musty odor usually develops. Eventually, the grain shows signs of mycelial growth or sporulation. This often appears blue-green or olive green because of the presence of *Penicillium* spp. or *Aspergillus* spp.; however, white, pink, black, or other colors are possible. The grain may cake or form a surface crust, preventing the grain from flowing during transfer. Grain heating and moisture accumulation may be noticed.

Causal Organisms

The most common storage molds are fungi in the genera *Aspergillus* and *Penicillium* (Table 4); however, a wide range of microorganisms also may be involved, including other filamentous fungi, yeasts, and bacteria. If high levels of moisture exist in storage, ear rot pathogens and other field fungi, such as *Alternaria* spp., *Cladosporium* spp., *Fusarium* spp., *Nigrospora oryzae*, *Rhizopus stolonifer*, and *Mucor* spp., can colonize stored grain.

Disease Cycle and Epidemiology

The most important factors affecting the severity of mold damage in storage are grain moisture, temperature, amount of broken grain and foreign material, amount of initial inoculum present in the grain or bin, activity of arthropods, and length of storage time. Storage molds are ubiquitous in the agricultural environment, surviving in soil, crop and weed residue, organic material that collects in harvesting and transport equipment, and storage facilities. Most of them produce abundant airborne spores that can nearly always be detected in the air around storage facilities. Propagules are often carried into storage along with the grain, either as conidia on the grain surface or as mycelium or conidia associated with broken kernels, weed seeds, or other foreign material. If storage bins are not cleaned, propagules may be present before the grain is deposited in storage. Fungal growth and colonization depends on grain moisture and temperature. The average grain moisture may not be predictive of the risk, because grain moisture is not uniform and the wettest part of the bin determines the risk of mold development. The various organisms differ in their temperature and moisture optima (Table 4), and a succession of species can be found as grain becomes warmer and moister because of fungal respiration. As molds grow, both temperature and moisture content rise. Water vapor follows temperature gradients and increases the moisture content of the cooler parts of the grain mass. Moisture may also accumulate because of leaks and moisture migration resulting from convective air currents. This can lead to wet grain at the top surface, followed by mold growth and crusting. If insects or mites are active in the grain, they increase the risk of storage molds by physically damaging the grain and by releasing heat and moisture through respiration. Broken kernels are much more susceptible to invasion by storage fungi.

Grain inexorably deteriorates over time, and models have been developed to predict the maximum safe storage time for grain, depending on grain moisture content and temperature. These relationships have been developed empirically for different cereal grains, based on the length of time to reach 0.5% dry matter loss (Table 5).

Management

Storage mold prevention depends primarily on limiting the moisture content of grain entering storage and managing the

TABLE 4. Temperature and Relative Humidity Conditions for Growth of Selected Storage and Field Fungi in Corn

	Temperature, °C			Relative Humidity	Grain Moisture
Fungal Species	Lower Limit	Optimum	Upper Limit	Lower Limit	Lower Limit
Aspergillus candidus				72	14.5–15.0
Aspergillus flavus	10–15	37–39	45–48	82	16.0–16.5
Aspergillus fumigatus			82	16.0–16.5	
Aspergillus glaucus	0–5	30–35	40–45	73	14.0–14.5
Aspergillus ochraceus			80	15.0–15.5	
Aspergillus restrictus	5–10	30–35	40–45	70	13.5–14.0
Aspergillus versicolor			75	14.5–15.0	
Fusarium graminearum (syn. *Gibberella zeae*)	4	28	32	94	20.0–21.0
Fusarium verticillioides	4	28–30	36	91	19.0–20.0
Penicillium brevi-compactum	–2	20–24	30–32	81	16.0–16.5
Penicillium chrysogenum				78	14.5–15.5
Penicillium citrinum				80	15.0–15.5
Penicillium cyclopium	–2	20–24	30–32	81	16.0–16.5
Penicillium funiculosum	8	31–33	35–37	91	19.0–20.0
Penicillium oxalicum	8	31–33	35–37	86	17.0
Penicillium viridicatum	–2	20–24	34–36	81	16.0–16.5
Rhizopus and *Mucor* spp.	...	...	...	92–94	20.0–21.0
Trichothecium roseum	...	...	...	90	19.0–19.5

TABLE 5. Safe Storage Time (Months) for Corn Grain According to Temperature and Grain Moisture Conditions[a]

Temperature (°C)	Corn Moisture Content (%)						
	13	14	15	16	17	20	24
5	150	58	28	14	9.0	3.0	1.3
10	84	34	16	9	5.3	1.6	0.5
15	47	21	10	5	3.2	1.0	0.3
20	30	13	6	3	2.0	0.6	0.2
25	18	8	4	2	1.2	0.3	0.1

[a] Values are based on predicted time to reach 0.5% dry matter loss.

environmental conditions in storage. Extensive literature exists describing the proper management of stored corn. Most grain in developed countries is stored in closed bins with aeration systems, which provide the best management potential. Other open or closed structures are often used, especially in developing countries, and their tendency to promote mold development depends on the extent to which grain moisture content and temperature can be managed. Inadequate storage facilities are a major cause of mold and mycotoxin problems in developing countries. Even in developed countries, corn grain is often stored in the open for a limited duration during the cold months. Grain bins and handling systems should be kept clean and free from debris and accumulation of fines. Storage insects should be controlled at all times. Grain entering storage should be free from broken kernels and foreign matter to the best extent possible. Grain that has low test weight, is noticeably damaged, or is molded from the field should not be stored for long periods. Corn grain dried to ≤15% moisture content is safe for storage over the winter months, and corn grain dried to ≤13% is safe for long storage periods. Grain that goes into storage at greater than 15% is at a high risk for spoilage by storage molds (Tables 4 and 5). If fines accumulate during bin loading, they should be removed immediately. Grain should be cooled to ≤4.5°C as soon as possible; temperature should be monitored closely and maintained through aeration. Grain in storage must be inspected frequently for signs of mold. If pockets of mold or crusts form, they should be removed as soon as possible. If a problem is suspected, grain samples can be collected for mycotoxin testing. Grain monitoring systems are available that can track and record moisture, temperature, and carbon dioxide conditions in bins, providing an early warning for the development of storage fungi.

Antifungal agents can be applied to grain to suppress mold growth in storage. These products, such as propionic acid, do not kill the mold already present or reduce toxins already formed. Treated grain may not be suitable for all end uses, so antifungal agents should not be used on stored grain unless it is certain that the grain can be marketed after treatment.

Selected References

Reed, C. R. 2006. Managing Stored Grain to Preserve Quality and Value. AACC International, St. Paul, MN.

Sauer, D. B., Meronuck, R. A., and Christensen, C. M. 1992. Microflora. Pages 313-340 in: Storage of Cereal Grains and Their Products, 4th ed. D. B. Sauer, ed. AACC International, St. Paul, MN.

Watson, S. A., 1987. Measurement and maintenance of quality. Pages 125-184 in: Corn: Chemistry and Technology. S. A. Watson and P. E. Ramstad, eds. AACC International, St. Paul, MN.

(Prepared by G. P. Munkvold)

Mycotoxins and Mycotoxicoses

Mycotoxins are toxic compounds produced by fungi. When consumed by animals and humans, they cause diseases called mycotoxicoses. Symptoms of mycotoxicoses differ greatly depending on the mycotoxin consumed and the animal species affected and include feed refusal, reduced weight gain, liver tumors, kidney necrosis, hyperestrogenism, and increased susceptibility to other diseases. Diagnosis of mycotoxicosis may be difficult in many cases because clinical symptoms of the disease may not be diagnostic. This is particularly true for cases of subacute or chronic mycotoxicoses and in situations in which more than one mycotoxin is present. If mycotoxicosis is suspected, the feed source should be analyzed for mycotoxins. However, tracing mycotoxicosis back to a contaminated feed source may be difficult because the tainted feed may have been entirely consumed before the symptoms appeared.

Mycotoxins in corn are produced by fungi growing in the grain. Although many fungi attack corn kernels, *Aspergillus*, *Fusarium*, and *Penicillium* spp. are the most important mycotoxin producers. Species of these genera are capable of producing a number of mycotoxins and are well adapted for growing on corn kernels with relatively low water contents.

Mycotoxins can be produced in the field, during storage, or both, although most mycotoxin contamination occurs in the field. It is highly dependent on environmental conditions and is often related to other factors, such as damage caused by insects and birds. Mycotoxin contamination in storage is almost always associated with improper drying or storage conditions. For proper grain storage, the corn must be dry (15% moisture content or below for less than 1 year, and 13% or below for longer-term storage) and free of insects, and the air movement must be properly regulated to avoid the accumulation of moisture.

Sampling is the most critical step in determining mycotoxin contamination. In many cases, there may be relatively few highly contaminated kernels in a lot, but these kernels can have mycotoxin concentrations sufficient to cause mycotoxicosis. In corn, for example, one kernel may contain an aflatoxin level of 400,000 parts per billion (ppb). One such kernel would be sufficient to contaminate 5 lb of grain at a level unacceptable for commercial grain trade. Sampling of grain for accurate mycotoxin determination is always problematic because mycotoxin-contaminated grain is not distributed uniformly throughout the load of harvested grain or in a storage bin. Storage bins present a particular sampling problem. "Hot spots" for mycotoxins often are found in a bin as a result of localized areas of high moisture content or insect activity. Care should be taken to follow established sampling procedures. For adequate sampling, it is recommended that a 5-kg sample be removed from the grain lot and coarsely ground. A smaller subsample should be finely ground (to approximately the size of ground coffee) for analysis. In general, the analysis can be done on 25- to 50-g samples. Several types of analyses are available for quantification of mycotoxins. The continuum of available methods ranges from those that are quick and convenient (and less precise) to those that are extremely precise (and require more time and instrumentation). Kits are available for on-site analysis and these are usually based on enzyme-linked immunosorbent assays.

More sophisticated chromatography and spectroscopy methods are used for precise quantification of diverse mycotoxins.

Although many mycotoxins have been reported on corn grain, some have been reported only infrequently. The more commonly found mycotoxins are discussed here.

Selected References

Cannavan, A., Whitaker, T., Maestroni, B., Doko, B., and Slate, A. 2010. Sampling Procedures to Detect Mycotoxins in Agricultural Commodities. Springer, New York.

Council for Agricultural Science and Technology (CAST). 2003. Mycotoxins: Risks in Plant, Animal and Human Systems. Task Force Report 139. CAST, Ames, IA.

Diaz, D., ed. 2005. The Mycotoxin Blue Book. Nottingham University Press.

Leslie, J. F., Bandyopadhyay, R., and Visconti, A. 2008. Mycotoxins: Detection Methods, Management, Public Health and Agricultural Trade. CABI, Cromwell Press, Trowbridge, United Kingdom.

Magan, N., Medina, A., and Aldred, D. 2011. Possible climate-change effects on mycotoxin contamination of food crops pre- and postharvest. Plant Pathol. 60:150-163.

Munkvold, G. P. 2003. Mycotoxins in corn—Occurrence, impacts, and management. Pages 811-881 in: Corn: Chemistry and Technology, 2nd ed. P. J. White and L. A. Johnson, eds. AACC International, St. Paul, MN.

Sinha, K. K., and Bhatnagar, D. 1998. Mycotoxins in Agriculture and Food Safety. Marcel Dekker, New York.

(Prepared by G. A. Payne)

Aflatoxins and Aflatoxicosis

The term "aflatoxins" refers to a family of toxic and carcinogenic compounds produced by two *Aspergillus* spp. (*A. flavus* and *A. parasiticus*). The four major forms of aflatoxins found in grain are B_1, B_2, G_1, and G_2. Aflatoxins are the most highly toxic of the known mycotoxins. The most biologically active of this group, aflatoxin B_1, is a potent, naturally occurring carcinogen that primarily affects the liver, although other organs can also be affected. Because of their carcinogenicity, aflatoxins in products intended for human consumption are a major concern. Two other forms of aflatoxin, M_1 and M_2, can be found in foods, particularly in milk after animal consumption of contaminated grain. The U.S. Food and Drug Administration (FDA) has established action levels of 20 parts per billion (ppb) for aflatoxin B_1 in grain and feed products and 0.5 ppb for aflatoxin M_1 in milk (Table 6). Grain or milk containing higher amounts cannot be moved via interstate commerce. Allowable limits for aflatoxin contamination in the European Union and some other importing countries are even lower (Table 6), and intentionally mixing aflatoxin-contaminated grain with sound grain is illegal.

In livestock, the acute toxicity of aflatoxins is of more concern than their carcinogenicity. While the mechanism of carcinogenicity is becoming clear, less is understood about the toxic effects of aflatoxin. The clinical symptoms of aflatoxicosis may not be distinctive, but they can be grouped into three classes: (i) primary acute, (ii) chronic, and (iii) secondary. Primary acute aflatoxicosis results in death. The liver appears to be most affected by acute aflatoxicosis, and the 50% lethal doses (LD_{50}) range from 0.34 mg/kg for ducks to 15–18 mg/kg for chickens. Rarely do animals die of acute aflatoxicosis in the United States. More often, animals show chronic effects of aflatoxicosis, including reduced weight gain, hemorrhages, liver damage, and capillary fragility. The symptoms of secondary aflatoxicosis are decreased phagocytic efficiency and cell-mediated immunity in animals. In some cases, this results in death or poor health because the animals are more susceptible to other diseases.

Aflatoxins are the mycotoxins with the biggest impact on human health because of their occurrence in tropical and subtropical areas and the high proportion of corn in the diets of people living in some of these high-risk areas. Acute effects of aflatoxins on humans include death, reproductive problems, anemia, immune system suppression, and jaundice. Chronic effects include growth stunting and liver cancer. Aflatoxins have a major impact on the incidence of liver cancer in some parts of the world, and there is an interaction between aflatoxin exposure and hepatitis C infection that greatly increases the probability of liver cancer development.

Aflatoxins are produced by two morphologically similar fungi, *A. flavus* and *A. parasiticus*. These two fungi can grow on a number of food and feed sources, but they are of most concern on corn, cottonseed, peanuts, and tree nuts. *A. flavus* is much more common on corn than is *A. parasiticus*. Growth of the fungus and aflatoxin contamination can occur in the field or after harvest, but the major concern is infection in the field. Aflatoxin contamination can develop under a range of conditions, but it is more prevalent during years when the corn plants suffer from environmental stress. Contamination is often associated with drought and temperature stress, nitrogen deficiency, and significant insect damage.

The frequency of aflatoxin contamination is greater in the southern regions of the United States than in other parts of the country. In the Midwest, contamination is most commonly associated with improper storage, except during years with above-average temperature and below-average rainfall, when preharvest aflatoxin contamination can be a serious problem.

Several methods for detecting aflatoxin contamination are available. In the past, grain was screened using a black light, which makes *Aspergillus*-infected kernels glow with a bright green-yellow fluorescence. However, this test does not directly measure aflatoxin contamination and is not reliable. Some states have laboratories that analyze samples for aflatoxin contamination. The most commonly used assays in the grain trade are based on antibodies to aflatoxin. Several kits are available.

TABLE 6. Recommendations and Regulations for Safe Limits on Mycotoxin Concentrations in Grain in the United States and European Union

	Grain for Human Food[a]		Grain for Animal Feed[a]	
Mycotoxin	USA[b]	EU[c]	USA[b]	EU[c]
Aflatoxins	20 ppb	2–4 ppb[d]	20–300 ppb[e]	10–50 ppb[e]
Deoxynivalenol	1,000 ppb	750 ppb	5,000–10,000 ppb[e]	1,750 ppb
Fumonisins	2,000–4,000 ppb[d]	1,000 ppb	5,000–100,000 ppb[e]	4,000 ppb
Zearalenone	No guidance levels, case-by-case basis	75–100 ppb[d]	No guidance levels, case-by-case basis	100–350 ppb[e]

[a] Measured in parts per billion (ppb).
[b] Munkvold, 2003.
[c] Commission Regulation (EC) No. 1126/2007.
[d] Varies among specific food items.
[e] Varies among livestock species.

Aflatoxin contamination can be reduced by treating corn with ammonia. Ammoniation has been approved in some states, but it has not been approved by the FDA for use on corn destined for interstate trade. The following guidelines have been established for feeding contaminated corn to animals: dairy, 20 ppb; mature breeding beef cattle, swine, and poultry, 100 ppb; finishing swine, 200 ppb; and finishing beef, 300 ppb. The effects of aflatoxin depend greatly on the general health of the animals. Aflatoxins are metabolized and excreted from animals, and thus, the meat is not contaminated. The only animal product of concern is milk, in which another form of aflatoxin, M_1, can accumulate when lactating cows are given contaminated feed.

Selected References

Council for Agricultural Science and Technology (CAST). 2003. Mycotoxins: Risks in Plant, Animal and Human Systems. Task Force Report 139. CAST, Ames, IA.

Eaton, D. L., and Groopman, J. D., eds. 1994. The Toxicology of Aflatoxins: Human Health, Veterinary, and Agricultural Significance. Academic Press, New York.

Georgianna, D. R., and Payne, G. A. 2009. Genetic regulation of aflatoxin biosynthesis: From gene to genome. Fungal Genet. and Biol. 46:113-125.

Park, D. L. 1993. Perspectives on mycotoxin decontamination procedures. Food Addit. Contam. 10:49-60.

Payne, G. A. 1998. Process of contamination by aflatoxin-producing fungi and their impact on crops. Pages 279-306 in: Mycotoxins in Agriculture and Food Safety. D. B. K. K. Sinha, ed. Marcel Dekker, New York.

Reddy, K. R. N., Salleh, B., Saad, B., Abbas, H. K., Abel, C. A., and Shier, W. T. 2010. An overview of mycotoxin contamination in foods and its implications for human health. Toxin Rev. 29(1):3-26.

Rodricks, J. V., Hesseltine, C. W., and Mehlman, M. A. 1977. Mycotoxins in Human and Animal Health. Pathotox Publishers, Park Forest South, IL.

Whitiker, T. B. 1990. Problems associated with testing agricultural commodities for aflatoxin: Errors in sampling, sample preparation, and analysis. Pages 1-11 in: A Perspective on Aflatoxin in Field Crops and Animal Food Products in the United States. U.S. Department of Agriculture, Washington, DC.

(Prepared by G. A. Payne)

Fumonisins

Fumonisins are a group of mycotoxins described in the late 1980s. There are at least 28 fumonisin analogs, but fumonisins B_1, B_2, and B_3 are the most common. Their toxicity is a result of the disruption of sphingolipid metabolism in cells, which can lead to a variety of diseases affecting different organs. Fumonisins are toxic and carcinogenic to the liver and kidneys in laboratory studies. Among livestock species, horses and swine appear to be the most sensitive to fumonisins. In horses, donkeys, and mules, fumonisins cause equine leukoencephalomalacia (blind staggers). In swine, they cause pulmonary edema. Both of these diseases are usually fatal. Chronic fumonisin exposure leads to delayed growth and development in swine, which can be a significant economic problem. In humans, fumonisins are associated with increased incidence of esophageal cancer in South Africa and China. Fumonisin exposure during pregnancy has been linked to neural tube birth defects in specific human populations with high proportions of corn in the diet. This connection has been confirmed in laboratory studies showing that fumonisin consumption interferes with folate uptake and this, in turn, affects neural tube development in the fetus.

Fumonisins are produced by several *Fusarium* spp. The most commonly found species on corn are *F. verticillioides* (syns. *F. moniliforme* and *Gibberella moniliformis*) and *F. proliferatum* (syn. *Gibberella intermedia*). Both of these fungi are pathogens on corn worldwide, but *F. verticillioides* is the one most commonly associated with mycotoxicoses in horses and swine. *F. proliferatum*, however, includes the highest fumonisin-producing isolates and can produce several other mycotoxins, including beauvericins, fusaproliferin, and moniliformin. It is also becoming more frequently isolated from corn in Europe. *F. proliferatum* and *F. verticillioides* often are found together on corn, where they can infect roots, stems, ears, and kernels.

These fungi can grow over a wide range of conditions, but the contamination of grain with fumonisins is associated with environmental conditions unfavorable for the plant, such as heat and drought stress. Drought stress early in the season increases the incidence of infection and systemic colonization by *F. verticillioides*. Fumonisins may continue to be produced in infected grain that is poorly stored.

Little is known about the resistance of corn lines to these two fungi, but breeding lines do differ in susceptibility to fumonisin accumulation. U.S. Food and Drug Administration (FDA) guidelines for the levels of fumonisin in corn and corn byproducts are no higher than 5 parts per billion (ppb) for horses, 20 ppm for swine, and 60 ppm for cattle (Table 6). Commercial test kits are available for fumonisin detection.

Selected References

Castelo, M. M., Sumner, S. S., and Bullerman, L. B. 1998. Occurrence of fumonisins in corn-based food products. J. Food Prot. 61:704-707.

Council for Agricultural Science and Technology (CAST). 2003. Mycotoxins: Risks in Plant, Animal and Human Systems. Task Force Report 139. CAST, Ames, IA.

Desjardins, A. E. 2006. *Fusarium* Mycotoxins: Chemistry Genetics, and Biology. American Phytopathological Society, St. Paul, MN.

Miller, J. D., and Trenholm, T. L. 1994. Mycotoxins in Grain: Compounds Other Than Aflatoxin. AACC International, St. Paul, MN.

Munkvold, G. P., and Desjardins, A. E. 1997. Fumonisins in maize: Can we reduce their occurrence? Plant Dis. 81:556-565.

Parsons, M. W., and Munkvold, G. P. 2010. Relationships of immature and adult thrips with silk-cut, Fusarium ear rot and fumonisin B1 contamination of maize in California and Hawaii. Plant Pathol. 59:1099-1106.

(Prepared by G. A. Payne)

Deoxynivalenol and Other Trichothecenes

Trichothecenes comprise a large family of mycotoxins (at least 148 have been identified) that are protein synthesis inhibitors and potent immunosuppressants. Syndromes caused by trichothecenes include alimentary toxic aleukia, fusariotoxicoses, and moldy corn toxicoses. Swine and other monogastric animals (including humans) show the greatest sensitivity to trichothecenes. Because many molecular species of trichothecenes can be produced by the same fungi and many of these toxic species may be found in feed at the same time, it is not always possible to identify the causative toxin. Deoxynivalenol (DON) is the most frequently found trichothecene in corn, but there are several other biologically active trichothecenes produced by *Fusarium* spp., including nivalenol, T-2 toxin, HT-2 toxin, and diacetoxyscirpenol (DAS). In general, poultry are more resistant to trichothecenes than are swine, but T-2 and DAS appear to be more toxic to chickens than is DON. Both of these toxins can cause oral lesions in broiler chickens at levels as low as 1 part per million (ppm), and reduced food intake and retarded growth can occur at 4 ppm. Ruminants appear to be relatively resistant to trichothecenes.

Acute trichothecene toxicosis is possible. However, the more common symptoms, such as poor growth of the affected animal, are subtle and nonspecific. In swine, the first symptom of DON toxicosis is often decreased consumption of the contaminated feed and reduced weight gain. If the animal is maintained on contaminated feed or if the levels of the mycotoxin increase, clinical symptoms may develop, including diarrhea, lethargy, blanched skin color, dermal irritation, hypothermia and shivering, intestinal hemorrhage, and eventual complete feed refusal. Because trichothecenes are immunosuppressants, animals may also be more susceptible to other diseases.

Trichothecenes are produced by several *Fusarium* spp., including *F. graminearum* (syn. *Gibberella zeae*), *F. culmorum*, *F. cerealis* (syn. *F. crookwellense*), *F. poae*, *F. sporotrichioides*, and *F. equiseti*. *F. graminearum*, the major DON producer in corn, causes Gibberella ear rot of corn and head scab of wheat. *F. graminearum* has been subdivided into at least nine cryptic species (see the Gibberella Stalk Rot section) and all are capable of trichothecene mycotoxin production. *F. graminearum* strains differ in their capacity to produce mycotoxins; some strains produce DON and its acetylated derivative 15-ADON, while others produce DON and 3-ADON, and others primarily produce nivalenol. Metabolic profiles (chemotypes) of *F. graminearum* strains may be related to their pathogenic aggressiveness.

DON can be produced in the field and in storage. In the field, infection by *F. graminearum* and the production of DON are favored by cool, wet summers. DON accumulation in stored corn is favored when grain moisture content is 20.5% or more and temperatures are 21–29°C. Grain containing DON may also contain smaller amounts of other trichothecenes and the mycotoxin zearalenone. Production of DON is halted if grain is dried to moisture contents of 15% or lower.

DON causes reduced weight gain and reduced animal feeding, especially in swine, at levels as low as 1–3 ppm. Vomiting and total feed refusal does not occur until DON concentrations are much higher (greater than 10 ppm). The U.S. Food and Drug Administration (FDA) has recommended that DON levels not exceed 1 ppm for human food. The following limits have been recommended: 10 ppm for cattle and chickens, not to exceed 50% of the diet; 5 ppm for swine, not to exceed 20% of the diet; and 5 ppm for all other animals, not to exceed 40% of the diet. The prognosis for recovery from DON mycotoxicosis is excellent once the animals are removed from the contaminated food source. DON is rapidly metabolized by the animals, and residues in meat do not accumulate to levels that pose a hazard to human health.

Selected References

Council for Agricultural Science and Technology (CAST). 2003. Mycotoxins: Risks in Plant, Animal and Human Systems. Task Force Report 139. CAST, Ames, IA.

Desjardins, A. E. 2006. *Fusarium* Mycotoxins: Chemistry Genetics, and Biology. American Phytopathological Society, St. Paul, MN.

Miller, J. D., and Trenholm, T. L. 1994. Mycotoxins in Grain: Compounds Other Than Aflatoxin. AACC International, St. Paul, MN.

Munkvold, G. P. 2003. Mycotoxins in corn—Occurrence, impacts, and management. Pages 811-881 in: Corn: Chemistry and Technology, 2nd ed. P. J. White and L. A. Johnson, eds. AACC International, St. Paul, MN.

Park, D. L. 1993. Perspectives on mycotoxin decontamination procedures. Food Addit. Contam. 10:49-60.

Robertson, A. E., Munkvold, G. P., Hurburgh, C. R., and Ensley, S. 2011. Effects of natural hail damage on ear rots, mycotoxins, and grain quality characteristics of corn. Agron. J. 103:193-199.

(Prepared by G. A. Payne)

Zearalenone

Zearalenone, once referred to as F-2 toxin, is a nonsteroidal fungal toxin with estrogenic properties. Swine appear to be the most sensitive farm animals. Symptoms of the mycotoxicosis vary, but in severe cases, they include enlargement or swelling and reddening of the vulva in gilts and sows and, in some cases, vaginal and rectal prolapses. Other signs may include swelling of the mammary glands and atrophy of the ovaries. In young males, the toxin can cause feminization, which includes swelling of the prepuce, testicular atrophy, and enlargement of the mammary glands. The main concern in swine, however, is delayed estrus in breeder sows. Although swine are the most sensitive, field cases of zearalenone toxicity in cattle and sheep have been well documented. Except at high toxin concentrations, poultry do not seem to be significantly affected.

Zearalenone has been consistently reported from *Fusarium graminearum* (syn. *Gibberella zeae*) and the related species *F. cerealis* (syn. *F. crookwellense*) and *F. culmorum*. Production of the toxin in the field is related to excessive rainfall during the autumn. High levels of zearalenone may accumulate in corn that has a high moisture content at harvest and is then placed in storage. Maximum levels of the mycotoxins are produced when temperatures are low or when there is fluctuation between low and moderate temperatures. Symptoms of mycotoxicosis often appear in the spring after swine are fed stored corn.

The major clinical effects of zearalenone are estrogenic, and concentrations of 1–5 parts per million (ppm) produce clinical signs of vulva edema in young gilts. Zearalenone is rapidly metabolized in animals and humans and eliminated mainly as water-soluble glucuronides. Thus, zearalenone does not accumulate in animal meat.

Selected References

Desjardins, A. E. 2006. *Fusarium* Mycotoxins: Chemistry Genetics, and Biology. American Phytopathological Society, St. Paul, MN.

Munkvold, G. P. 2003. Mycotoxins in corn—Occurrence, impacts, and management. Pages 811-881 in: Corn: Chemistry and Technology, 2nd ed. P. J. White and L. A. Johnson, eds. AACC International, St. Paul, MN.

(Prepared by G. A. Payne)

Ochratoxins

Ochratoxins are produced by several *Aspergillus* and *Penicillium* spp. The most commonly found ochratoxin-producing species on corn in temperate regions is *Penicillium verrucosum*, while *Aspergillus alutaceus* var. *alutaceus* (syn. *A. ochraceus*) is the species most frequently producing ochratoxins in the tropics and subtropics. Ochratoxin A is the most studied ochratoxin. Swine, ducklings, and chickens can be affected, but the toxin is of major concern in swine and is the principal cause of porcine nephropathy. The kidneys are most affected, and the effects of ochratoxin are more visible at the organ level than as overt symptoms. There is evidence for both acute and chronic forms of ochratoxin toxicosis in livestock and poultry. In poultry, the symptoms include weakness, possible anemia, decreased feed consumption, decreased egg production, and reduced growth. The effects on cattle and other ruminants are not well characterized.

Ochratoxins have been found in grain in North America and most European countries. They have been implicated as a factor in human nephropathy and urinary tract tumors in the Balkan countries. The mycotoxin citrinin also causes kidney damage in laboratory animals and may interact synergistically with ochratoxin A. In the United States, there are no U.S. Food and Drug Administration (FDA) guidelines for ochratoxin A levels, although they are regulated in the European Union and in other countries.

Ochratoxin-producing fungi colonize grain mainly in storage. The fungi may grow superficially on the kernel surface or within the germ area beneath the pericarp. The greenish blue mold growth in this area has led to the name "blue eye" to describe this corn disease.

In temperate regions, ochratoxin is not produced in crops before harvest, and improper storage in these areas is the principal cause of ochratoxin A contamination. Storage of corn below 16% moisture content prevents the accumulation of ochratoxin.

Selected Reference

Bayman, P., and Baker, J. L. 2006. Ochratoxins: A global perspective. Mycopathologia 162:215-223.

(Prepared by G. A. Payne)

Diplodiosis and Diplodiatoxin

Consumption of maize infected with *Stenocarpella maydis* (syns. *Diplodia maydis* and *D. zeae*) has been associated with a nervous condition characterized by ataxia and paralysis in sheep and cattle. This syndrome has been described in southern Africa, Argentina, Australia, Brazil, and New Zealand in sheep and cattle grazing in cornfields after harvest. The disease, referred to as diplodiosis, has been attributed to several potentially toxic compounds, including diplodiatoxin and diplonine. Feeding studies with these compounds in cattle have not been reported. Rats fed purified diplodiatoxin exhibit tremors, convulsions, reduced feed intake, and a loss in body weight. Diplonine induced neurotoxic signs in guinea pigs that were fed the purified compound. Diplodiatoxin has been isolated from naturally infected maize in Illinois (United States), along with two other bioreactive compounds.

Selected Reference

Wicklow, D. T., Rogers, K. D., Dowd, P. F., and Gloer, J. B. 2011. Bioactive metabolites from *Stenocarpella maydis*, a stalk and ear rot pathogen of maize. Fungal Biol. 115:113-142.

(Prepared by G. A. Payne)

Other Mycotoxins

The mycotoxins citrinin, patulin, mycophenolic acid, cyclopiazonic acid, and roquefortine C can be produced by *Penicillium* spp. found in corn. They can be found in kernels and in freshly harvested or ensiled silage. Cyclopiazonic acid and sterigmatocystin also can be produced by *Aspergillus flavus* and *A. parasiticus*, but they are found infrequently in corn. Mycotoxins under watch for increased incidence in corn are the beauvericins, fusaproliferin, and moniliformin. As mentioned in the Fumonisins section, *Fusarium proliferatum* can produce all three of these toxins. Also, *F. subglutinans* (syn. *Gibberella subglutinans*) can produce beauvericins and fusaproliferin.

Selected References

Desjardins, A. E. 2006. *Fusarium* Mycotoxins: Chemistry Genetics, and Biology. American Phytopathological Society, St. Paul, MN.

Munkvold, G. P. 2003. Mycotoxins in corn—Occurrence, impacts, and management. Pages 811-881 in: Corn: Chemistry and Technology, 2nd ed. P. J. White and L. A. Johnson, eds. AACC International, St. Paul, MN.

Munkvold, G. P., Logrieco, A., Moretti, A., Ferrancane, R., and Ritieni, A. 2009. Dominance of Group 2 and fusaproliferin production by *Fusarium subglutinans* from Iowa maize. Food Addit. Contam. 26:388-394.

(Prepared by G. A. Payne)

Diseases Caused by Viruses

Viruses cause important diseases of corn worldwide. Some viruses are widespread (e.g., *Maize dwarf mosaic virus* and *Maize mosaic virus*); many others have a limited incidence and are localized (e.g., *Maize fine streak virus*, *Maize leaf fleck virus*, *Maize necrotic streak virus*, and *Maize white line mosaic virus*).

Yield losses from some corn virus diseases are sporadic and devastating (e.g., maize streak). However, losses from others (e.g., maize dwarf mosaic) may occur each year, but the losses are relatively minor because of highly resistant or tolerant germplasm, good control of weeds that harbor viruses, or infection that takes place late in the growing season.

Properties of corn viruses vary. They may have either an RNA or DNA genome. The RNA is either single or double stranded. The particles (virions) are rod shaped, sphere shaped, or bacilliform. Many corn viruses are transmitted by arthropod vectors. A few are soilborne (e.g., *Maize white line mosaic virus* and *Maize necrotic streak virus*). Some corn viruses with aphid, mite, thrips, or beetle vectors can be experimentally transmitted by rub inoculation with infected sap. In addition, many arthropod-transmitted viruses (including the leafhopper- and planthopper-transmitted viruses *Maize chlorotic dwarf virus*, *Maize rayado fino virus*, *Maize streak virus*, *Maize mosaic virus*, *Maize rough dwarf virus*, and *Maize stripe virus*) are also mechanically transmissible to seeds by vascular puncture inoculation.

Virus diseases are often first detected by the symptoms they cause. Foliar symptoms commonly consist of changes in coloration. The resulting patterns may be mosaics of yellow and green, or there may be a general chlorosis or limited chlorotic spots, streaks, or rings. Sometimes the leaves are greener than normal. Occasionally, the leaves are reddish or purplish. Symptoms on leaves may also involve morphological aberrations. Leaves may be puckered or their margins torn because of unequal growth, or they may feel rough to the touch because of enations or swollen veins. Some leaves are flaccid or brittle when infected; they may fail to unroll and resemble a rat's tail or appear twisted when unfurled. Stunting of infected plants is also common, but the degree of stunting often depends upon the growth stage at which the plant is infected and the host genotype. Root symptoms are limited. With some viral infections (e.g., *Maize rough dwarf virus*), roots are stubby and have fissures in the cortical tissues. Generally, the types of symptoms observed on infected plants depend upon the interaction of the virus and the host at a particular time, with the most severe symptoms often observed on plants infected early in development. Leaves of corn infected with *Maize dwarf mosaic virus* usually exhibit mosaic patterns. However, in some corn inbred lines or hybrids (e.g., Ky61:2335), the symptoms consist of limited chlorotic streaks. Others (e.g., Wf9 × Oh51A) may express a recovery response in which emerging leaves of a plant have very little or no mosaic symptoms. Grain yield losses may result from sterility, kernel abortion, reduction in ear size, or lack of ear formation. Occasionally, multiple ears also occur. In addition to losses in grain yield, there is a loss in total biomass. In some cases (e.g., *Maize dwarf mosaic virus* and *Sugarcane mosaic virus*), virus-infected plants may be more severely infected by fungal pathogens (e.g., southern corn leaf blight, northern corn leaf spot, gray leaf spot, Diploida leaf streak) than plants that are not virus infected.

Accuracy in disease identification is increased by a long working experience with the disease, by knowledge of symptom variations associated with various corn genotypes or various virus isolates, and by awareness of other pathogens in a particular area that might coinfect corn. The geographic range of a particular virus disease is another helpful tool in disease identification. For example, *Maize streak virus* is currently not found in the United States, and *Maize chlorotic dwarf virus* is not found in Africa. This approach, however, is not suitable for viruses that are widespread (e.g., *Maize mosaic virus* or *Maize stripe virus*). Infections by other organisms, genetic and nutritional disorders, and noxious substances such as herbicides and insect toxins can also cause virus-like symptoms. Since symptoms can vary with host, environment, and virus combination, biological, serological, and molecular assays along with symptom observation lead to the most accurate disease identification. Useful assays for identifying viral diseases include serological assays, such as enzyme-linked immunosorbent assays (ELISA), where the etiological agent is known and sensitive and discriminating antibodies have been developed, and polymerase chain reaction (PCR)-based assays, which are sensitive but rely on some knowledge of the virus' nucleic acid sequence. Virus chips are also being developed for some crops for rapid screening and identification of disease-causing agents. As next-generation sequencing becomes more affordable, it may become a useful tool for virus discovery and diagnosis since no previous sequence information is required. Several viruses are seedborne and seed transmitted in corn, and seed health tests for these viruses are important management tools. Seed health testing methods for viruses are based on ELISA or PCR, and they must be developed and validated specifically for use on seeds.

The discovery or emergence of new corn virus diseases has not abated. In 1996, a new virus disease of corn was reported in the High Plains of the western United States. A new agent causing maize necrotic streak was found in Arizona in 1998, and another new virus causing maize fine streak was described from Georgia in 2002. Serious outbreaks of maize lethal necrosis were recently reported in China (2010) and Kenya (2011). Maize lethal necrosis has subsequently been reported throughout eastern Africa. Although there are many viruses that infect corn, only a selected few are described here. Some less common diseases are listed in Table 7 (Figs. 215–218).

Many factors influence virus disease emergence and the impact of virus diseases. For a virus to cause disease requires a combination of susceptible host genotypes, a mechanism of virus dissemination, and reservoirs (via alternate hosts or continuous cropping) to preserve populations of the virus, the vector, or both. Global movement of plants, insects, and viruses over time impacts the distribution of diseases. Changes in cultural practices, the genetic composition of resistance in currently available genotypes, the prevalence of vector species or vector adaptation, and the habitats of virus reservoirs all play roles in disease development. Adaptation of viruses or vectors from native monocot species to corn may also play a role in the emergence of new diseases. To date, however, the origin of many virus diseases is not clear. But as cultural practices and the distribution of plants, insects, and viruses adapt and change over time, the continued emergence of corn viruses is to be expected.

Strong genetic resistance is the most effective means of virus disease management. Highly resistant maize germplasm

Fig. 216. Maize fine streak. (Courtesy R. Louie)

Fig. 217. Maize necrotic streak. (Courtesy R. Louie)

Fig. 215. *Brome mosaic virus.* (Courtesy R. Louie)

Fig. 218. Maize white line mosaic. (Courtesy R. Louie)

TABLE 7. Other Selected Corn Virus Diseases

Disease	Distribution	Hosts	Primary Symptoms	Causal Agent[a]
Barley yellow dwarf	Worldwide distribution	Corn, barley, oats, wheat, other grasses	Chlorosis of leaf tips and margins; dark red, purple, or orange leaf color, on older leaves first; slight stunting	*Barley yellow dwarf virus*-PAV (BYDV-PAV) and *Barley yellow dwarf virus*-MAV (BYDV-MAV), *Luteovirus*; *Cereal yellow dwarf virus*-GPV (CYDV-GPV) and *Cereal yellow dwarf virus*-RPV (CYDV-RPV), *Polerovirus*; possibly other *Luteorviridae*. Virions are icosahedral particles, approximately ~25 nm in diameter, containing a monopartite ssRNA genome.
Brome mosaic	Worldwide distribution	Wide host range, including corn, wheat, oats, barley, rye, other grasses	Symptom severity varies with host genotype and virus isolate, from chlorotic streaking to local necrotic lesions to systemic necrosis (Fig. 215)	*Brome mosaic virus* (BMV), *Bromovirus*, *Bromoviridae*. Virions are icosahedral particles, 28 nm in diameter, containing the viral tripartite ssRNA genome.
Cereal chlorotic mottle	Australia, Morocco	Corn, barley, wheat, oats, other grasses	Initial chlorotic stripes followed by mild chlorotic mottle	*Cereal chlorotic mottle virus* (CCMV), *Nucleorhabdovirus*, *Rhabdoviridae*. Virions are bacilliform, 65 × 240 nm, containing a monopartite ssRNA genome.
Cucumber mosaic	Worldwide distribution	Wide host range, including monocots and dicots	Chlorotic elliptical spots that form leaf stripes; severely infected plants are stunted with necrotic leaf spots or death at seedling stage	*Cucumber mosaic virus*, (CMV), *Cucumovirus*, *Bromoviridae*. Virions are icosahedral, 29 nm in diameter, containing tripartite ssRNA genome.
Maize fine streak	United States (Georgia)	Corn, wheat, oats, rye, barley, foxtail, annual ryegrass, quack grass	Short, fine chlorotic streaks along veins (Fig. 216)	*Maize fine streak virus* (MFSV), *Nucleorhabdovirus*, *Rhabdoviridae*. Virions are bacilliform, 231 × 71 nm, membrane-bound particles containing ssRNA monopartite genome.
Maize necrotic streak	United States (Arizona)	Corn	Yellow-white spots and streaks, 1–2 mm, develop on emerging leaves, becoming spindle shaped as lamina expands; chlorotic bands merge over time with necrotic edges (Fig. 217)	*Maize necrotic streak virus* (MNeSV), *Zeavirus*, *Tombusviridae*. Virions are isometric particles, approximately 32 nm in diameter, containing a ssRNA monopartite genome.
Maize white line mosaic	United States, France, Italy	Corn	High-contrast mosaics of white, yellow, and yellow-green in rectangular blocks develop on infected leaves (Fig. 218)	*Maize white line mosaic virus* (MWLMV), *Aureusvirus*, *Tombusviridae*. Isometric particles are ssRNA and 35 nm in diameter; a satellitelike virus, 17-nm isometric particle, is associated with a larger virus particle.
Rice stripe	Japan	Corn, rice, barley, wheat, millets	Chlorotic leaf streaks; stunting of plants	*Rice stripe virus* (RSV), *Tenuivirus*, unassigned family. Virions are flexuous filaments of different lengths, 8 nm in diameter, containing four negative or ambisense ssRNA genome segments.

[a] ssRNA = single-stranded RNA.

[b] + = transmissible by this method, and – = not transmissible by this method. VPI = vascular puncture inoculation of seed.

[c] 1. Benavente, L. M., and Scofield, S. R. 2011. A new tool for functional genomics in maize. New Phytol. 189:363-365.
2. Beuve, M., Naibo, B., Foulgocq, L., and Lapierre, H. 1999. Irrigated hybrid maize crop yield losses due to barley yellow dwarf virus-PAV luteovirus. Crop Sci. 39:1830-1834.
3. Bocktahler, L. E., and Kaesberg, P. 1962. The molecular weight and other biophysical properties of Bromegrass mosaic virus. Biophys. J. 2:1-9.
4. Boothroyd, C. W. 1981. Virus diseases of sweet corn. Pages 103-109 in: Viruses and viruslike disease of maize in the United States. Southern Coop. Ser. Bull. 247. D. T. Gordon, J. K. Knoke, and G. E. Scott, eds. Ohio Agricultural Research and Development Center, Wooster, OH.
5. Boothroyd, C. W., and Israel, H. W. 1980. A new mosaic disease of corn. Plant Dis. 64:218-219.
6. Conti, M., Luisoni, E., Caciagli, P., and Milne, R. G. 1981. A new isometric maize virus in northeastern Italy. Pages 9-13 in: Proc. Conf. Virus Dis. Gramineae Eur., 3rd. Rothamsted Experimental Station, Herpenden, United Kingdom.
7. Ding, X. S., Schneider, W. L., Chaluvadi, S. R., Mian, M. A. R., and Nelson, R. S. 2006. Characterization of a *Brome mosaic virus* strain and its use as a vector for gene silencing in monocotyledonous hosts. Mol. Plant-Microbe Interact. 19:1229-1239.
8. De Zoeten, G. A., and Reddick, B. B. 1984. Maize white line mosaic virus. Descriptions of Plant Viruses, No. 283. Commonwealth Mycological Institute and Association of Applied Biologists, Kew, England.
9. Gingery, R. E., and Louie, R. 1985. A satellitelike virus particle associated with maize white line mosaic virus. Phytopathology 75:870-874.
10. Gingery, R. E., Nault, L. R., and Yamashita, S. 1983. Relationship between maize stripe virus and rice stripe virus. J. Gen. Virol. 64:1765-1770.
11. Greber, R. S. 1979. Cereal chlorotic mottle virus—A rhabdovirus of Graminae in Australia transmitted by *Nesoclutha pallida* (Evans). Aust. J. Agric. Res. 30:433-443.
12. Harris, K. F. 1981. Arthropod and nematode vectors of plant viruses. Annu. Rev. Phytopathol. 19:391-426.
13. Horn, F., Habekuss, A., and Stich, B. 2014. Genes involved in barley yellow dwarf virus resistance of maize. Theor. Appl. Genet. 127:2575-2584.
14. Lockhart, B. E. L. 1986. Occurrence of cereal chlorotic mottle virus in northern Africa. Plant Dis. 70:912-915.
15. Louie, R. 1995. Vascular puncture of maize kernels for the mechanical transmission of maize white line mosaic virus and other viruses of maize. Phytopathology 85:139-143.
16. Louie, R., Gordon, D. T., Knoke, J. K., Gingery, R. E., Bradfute, O. E., and Lipps, P. E. 1982. Maize white line mosaic virus in Ohio. Plant Dis. 66:167-170.

Vector	Virus Resistance	Transmission[b]	Remarks	References[c]
Rhopalosiphum padi, *Rhopalosiphum maidis*, *Schizaphis graminum*, *Sitobion avenae*, *Sitobion fragariae*, *Sitobion miscanthi*, other aphid species	Sources of quantitative resistance are available, including Mo17. Transgenic plants expressing BYDV-PAV coat protein show resistance.	Aphids (+); rub inoculation (–)	Luteoviruses are ubiquitous in multiple grass crops worldwide, most studied in wheat	2, 13, 19, 29
Reports of transmission by *Diabrotica* spp., *Longidorus* spp., *Xiphenema* spp.	Disease progression in many corn varieties is limited to local lesions. Variability in susceptibility and systemic movement in corn has been reported.	Beetle and nematode transmission reported; readily mechanically transmitted; VPI (+)	Not currently identified as a major disease problem in corn	1, 3, 7, 12, 20
Cicadellid leafhoppers; *Nesoclutha pallida*	Corn varieties vary in susceptibility to CCMV strains A and M, and strongly resistant cultivars are reported.	Leafhoppers (+); rub inoculation (–)	Limited distribution reported	11, 14
Aphis gossypii, *Myzus persicae*, many other aphid species (nonpersistent stylet-borne)	CMV-M does not infect maize but CMV-Fny does. Resistance is reported in some corn germplasm.	Aphids (+); rub inoculation (+)	Considered a minor disease problem in maize	4, 21, 24, 28
Persistent transmission by *Graminella nigrifrons* nymphs and adults after a latent period of approximately 4 weeks	Some evidence for genetic resistance has been reported in Caribbean germplasm.	Leafhopper (+); rub inoculation (–); VPI (+)	Not currently identified as a disease problem; no control measures have been tested or recommended	22, 26
Low level of soilborne transmission (0.4%) reported; no known vector	No virus resistance has been reported; limited incidence.	Rub inoculation (–); VPI (+)	Not currently identified as a major disease problem in corn	18
Symptomless infections also occur; virus is soilborne; a vector has not been identified	Dent corn is tolerant; sweet corn is very susceptible. There are no studies on resistance because of limited occurrence.	Infested soil and root debris (+); embryo wounding (+); VPI (+)	Often in mixed infections with maize subtle and maize necrotic lesion viruses, which are also soilborne	5, 6, 8, 9, 15–17, 23, 25, 30
Laodelphax striatellus, *Unkanodes sapporonus*, *Ribautodelphax albifascia*; persistent, propagative transmission	Not reported.	Planthopper (+); mechanical (difficult); seed (–)	Primarily studied in rice	10, 27

17. Louie, R., Gordon, D. T., Madden, L. V., and Knoke, J. K. 1983. Symptomless infection and incidence of maize white line mosaic. Plant Dis. 67: 371-373.
18. Louie, R., Redinbaugh, M. G., Gordon, D. T., Abt, J. J., and Anderson R. J. 2000. *Maize necrotic streak virus*, a new maize virus with similarity to species of the family *Tombusviridae*. Plant Dis. 84:1133-1139.
19. Miller, W. A., and Rasochova, L. 1997. Barley yellow dwarf viruses. Annu. Rev. Phytopathol. 35:167-190.
20. Moline, H. E., and Ford, R. E. 1974. Bromegrass mosaic virus infection of seedling roots of *Zea mays*, *Triticum aestivum*, *Avena sativa* and *Hordeum vulgare*. Physiol. Plant Pathol. 4:209-217.
21. Rao, A. L. N., and Francki, R. I. B. 1982. Distribution of determinants for symptom production and host range on the three RNA components of cucumber mosaic virus. J. Gen. Virol. 61:197-205.
22. Redinbaugh, M. G., Seifers, D. L., Meulia, T., Abt, J. J., Anderson, R. J., Styer, W. E., Ackerman, J., Salomon, R., Houghton, W., Creamer, R., Gordon, D. T., and Hogenhout, S. A. 2002. Maize fine streak virus, a new leafhopper-transmitted rhabdovirus. Phytopathology 92:1167-1174.
23. Russo, M., De Stradis, A., Boscia, D., Rubino, L., Redinbaugh, M. G., Abt, J. J., and Martelli, G. P. 2008. Molecular and ultrastructural properties of maize white line mosaic virus. J. Plant Pathol. 90:363-369.
24. Ryu, K. H., Kim, C.-H., and Palukaitis, P. 1998. The coat protein of cucumber mosaic virus is a host range determinant for infection of maize. Mol. Plant-Microbe Interact. 11:351-357.
25. Signoret, P. A. 1983. Maize virus diseases in France. Pages 113-116 in: Proc. Int. Maize Virus Dis. Colloq. Workshop. D. T. Gordon, J. K. Knoke, L. R. Nault, and R. M. Ritter, eds. Ohio State University, Ohio Agricultural Research and Development Center, Wooster, OH.
26. Todd, J. C., Ammar, E.-D., Redinbaugh, M. G., Hoy, C., and Hogenhout, S. A. 2010. Plant host range and leafhopper transmission of *Maize fine streak virus*. Phytopathology 100:1138-1145.
27. Toriyama, S. 1982. Characterization of rice stripe virus: A heavy component carrying infectivity. J. Gen. Virol. 61:187-195.
28. Wahyuni, W. S., Dietzgen, R. G., Hanada, K., and Fracki, R. I. B. 1992. Serological and biological variation between and within subgroup I and II strains of cucumber mosaic virus. Plant Pathol. 41:282-297.
29. Wang, M. B., Abbott, D. C., and Waterhouse, P. M. 2008. A single copy of a virus-derived transgene encoding hairpin RNA gives immunity to barley yellow dwarf virus. Mol. Plant Pathol. 1:347-356.
30. Zhang, L., Zitter, T. A., and Lulkin, E. J. 1991. Artificial inoculation of maize white line mosaic virus into corn and wheat. Phytopathology 81:397-400.

has been identified for many viruses, including *Maize dwarf mosaic virus*, *Sugarcane mosaic virus*, *Wheat streak mosaic virus*, and *Maize streak virus*. However, genetic resistance is not always available or strong and may compromise other important agronomic qualities. Identifying and developing resistant germplasm can also be challenging. An important factor in developing virus-resistant corn lines is the development of effective screening and selection for virus resistance genotypes. Vector-transmitted virus studies have generally relied on natural infection in field evaluations for the selection of virus and vector resistance (e.g., *Maize streak virus* and *Maize rough dwarf virus*). Controlled inoculations of vector-transmitted viruses (e.g., *Maize chlorotic dwarf virus* and *Maize rayado fino virus*) and for mechanically transmissible viruses (e.g., *Maize dwarf mosaic virus*, *Sugarcane mosaic virus*, and *Wheat streak mosaic virus*) may be more effective in identifying resistant germplasm.

When genetic resistance is not available, cultural practices to disrupt the life cycle of the virus or vector may be effective. Planting time, crop rotation patterns, control of weeds harboring the virus, or management of vectors can diminish the impact of disease. However, the effectiveness and economic feasibility of such practices is specific to each disease.

Selected References

Bos, L. 1963. Symptoms of Virus Diseases in Plants. Pudoc, Wageningen, Netherlands.

Gordon, D. T., Bradfute, O. E., Gingery, R. E., Knoke, J. K., Louie, R., Nault, L. R., and Scott, G. E. 1981. Introduction: History, geographical distribution, pathogen characteristics, and economic importance. Pages 1-12 in: Virus and Viruslike Diseases of Maize in the United States. D. T. Gordon, J. K. Knoke, and G. E. Scott, eds. South. Coop. Ser. Bull. 247.

Holmes, F. O. 1964. Symptomatology of viral diseases in plants. Pages 17-38 in: Plant Virology. M. K. Corbett and H. D. Sisler, eds. University of Florida Press, Gainesville.

Jensen, S. G., Lane, L. C., and Seifers, D. L. 1996. A new disease of maize and wheat in the High Plains. Plant Dis. 80:1387-1390.

Louie, R. 1995. Vascular puncture of maize kernels for the mechanical transmission of Maize white line mosaic virus and other viruses of maize. Phytopathology 85:139-143.

Louie, R., Redinbaugh, M. G., Gordon, D. T., Abt, J. J., and Anderson, R. J. 2000. *Maize necrotic streak virus*, a new maize virus with similarity to species of the family *Tombusviridae*. Plant Dis. 84:1133-1139.

Redinbaugh, M. G., Seifers, D. L., Meulia, T., Abt, J. J., Anderson, R. J., Styer, W. E., Ackerman, J., Salomon, R., Houghton, W., Creamer, R., Gordon, D. T., and Hogenhout, S. A. 2002. Maize fine streak virus, a new leafhopper-transmitted maize rhabdovirus. Phytopathology 92:1167-1174.

Stoner, W. N. 1952. Leaf fleck, an aphid-borne persistent virus disease of maize. Phytopathology 42:683-689.

Todd, J. C., Ammar, E.-D., Redinbaugh, M. G., Hoy, C., and Hogenhout, S. A. 2010. Plant host range and leafhopper transmission of *Maize fine streak virus*. Phytopathology 100:1138-1145.

Wangai, A. W., Redinbaugh, M. G., Kinyua, Z. M., Miano, D. W., Leley, P. K., Kasina, M., Mahuku, G., Scheets, K., and Jeffers, D. 2012. First report of *Maize chlorotic mottle virus* and maize lethal necrosis in Kenya. Plant Dis. 96:1582.

(Prepared by R. Louie; Revised by L. R. Stewart)

Maize Dwarf Mosaic

Maize dwarf mosaic was first reported on corn in the Ohio River bottomlands of southern Ohio in 1965. It was the first virus of corn in the United States that caused widespread devastation. It was subsequently reported throughout the world wherever maize is grown. The disease is incited by any of several potyviruses, and the relative importance of different viruses in causing disease differs regionally. The two most important maize-infecting potyviruses are *Maize dwarf mosaic virus* (MDMV) and *Sugarcane mosaic virus* (SCMV; previously known as MDMV-B). Both viruses cause disease in North America and Europe, but SCMV is more prevalent in Asia and South America.

Symptoms

Symptoms of maize dwarf mosaic in corn vary with the stage of host development at the time of infection, genotype, and viral strain (Figs. 219–221). Initial symptoms usually consist of chlorotic spots and streaks against a background of healthy green tissue that then become a mottle or a mosaic. When the virus spreads quickly, there are only small areas of healthy green tissue seen against a background of chlorotic tissue. Stunting occurs in susceptible genotypes infected at an early stage of development. Infection is systemic, with symptoms developing on all tissues that form after infection but no symptoms on tissues formed prior to infection. Ear formation and development are arrested, resulting in losses in grain yield. The appearance of unfertilized kernels at the bottom portion of the ear results from shortened growth of pollen germ tubes in silk of diseased plants. The mosaic or mottle symptoms, but not reddening, are diagnostic of maize dwarf mosaic. Some of the original symptom descriptions attributed to infection by MDMV were undoubtedly actually mixed infections with another unidentified virus, probably *Maize chlorotic dwarf virus*. MDMV and SCMV have very similar symptoms on maize but are distin-

Fig. 219. Maize dwarf mosaic. (Courtesy J. K. Pataky)

Fig. 220. A leaf with maize dwarf mosaic symptoms (top) compared with a healthy leaf (bottom). (Courtesy J. K. Pataky)

Fig. 221. Maize dwarf mosaic. (Courtesy J. K. Pataky)

Fig. 222. Genotype with genetic resistance to maize dwarf mosaic (right row) compared with stunted and chlorotic susceptible genotypes (middle and left rows). (Courtesy J. K. Pataky)

guished by host range (SCMV does not infect Johnsongrass). MDMV is also distinguished from other mechanically transmitted corn viruses in the United States by differential host reactions on corn inbred line Oh28, 'Monon' wheat, and 'Atlas' sorghum. Antiserum is available for serological tests to confirm virus identity. *Potyvirus*-, MDMV- and SCMV-specific molecular assays are available.

Causal Agents

Originally, all potyviruses infecting maize were known as strains of *Maize dwarf mosaic virus* (MDMV), and the virus was thought to be distributed worldwide. However, investigations into the serological and genome sequence diversity of these strains indicated that maize is infected by at least six *Potyvirus* species, including MDMV, *Sugarcane mosaic virus* (SCMV), *Johnsongrass mosaic virus* (JGMV), *Sorghum mosaic virus* (SrMV), *Zea mosaic virus* (ZeMV), and *Pennisetum mosaic virus* (PenMV), that can be distinguished by host range, serology, and genome sequence. JGMV, SrMV, ZeMV, and PenMV have more limited distributions than MDMV and SCMV. ZeMV is found in maize in Israel; JGMV is prevalent in Australia and reported in Texas (United States) and Nigeria; SrMV is found in Asia, India, and the United States but primarily in sugarcane and sorghum; and PenMV is found in northern China.

Potyviruses are members of the family *Potyviridae*. Virions are flexuous rods approximately 12 × 750 nm. The genome is a monopartite, single-stranded RNA of 9–10 kilobases. These viruses are fairly stable and have been well characterized. *Wheat streak mosaic virus* (WSMV), which is also in the family *Potyviridae* but in the genus *Tritimovirus*, also can infect corn but is generally a minor disease since most corn is resistant (see the Wheat Streak Mosaic section).

Vectors

More than 15 aphid species, both nymphs and adults, transmit MDMV in a nonpersistent manner. Aphid vectors acquire the virus in seconds or minutes and transmit it without a latent period. The aphid does not retain the virus when it molts. Alate (winged) adults are primarily responsible for virus spread in the field. Aphids that are the most efficient vectors under laboratory conditions (e.g., *Schizaphis graminum* and *Aphis maidiradicis*) may not be the most important vectors in the field because of their low numbers, population dynamics in relation to crop development, or feeding preferences.

Disease Cycle and Epidemiology

In addition to aphid transmission, MDMV is seed transmitted at a low level. It can also be mechanically transmitted by leaf rubbing, airbrush, solid stream, wounding, and vascular puncture inoculation of seeds.

Corn and sorghum are economically important hosts of MDMV. Although Johnsongrass (*Sorghum halepense*) is considered to be an important virus reservoir, MDMV has been identified in areas some distance from where Johnsongrass overwinters. Its occurrence in these areas has been attributed to viruliferous aphids from the southern United States that have been blown northward by low-level jet stream winds. SCMV infects corn, sugarcane, sorghum, and Johnsongrass, as well as *Panicum*, *Eleusine*, and *Setaria* spp.

Management

Many commercial corn hybrids have high levels of resistance to maize dwarf mosaic; hence, disease control is readily achieved (Fig. 222). A major gene for resistance to MDMV is *Mdm1*, a dominant locus on the short arm of chromosome 6. Jones et al. (2011) also found resistance was modulated by loci on chromosomes 3 and 10. Resistance to SCMV requires both *Scmv1* on chromosome 6 and *Scmv2* on chromosome 3. The *Scmv1* gene is closely linked to or pleiotrophic with the *Mdm1* gene. Chromosomes 3, 6, and 10 loci also confer resistance to JGMV and SrMV. Responses of corn to maize dwarf mosaic also may be modified by other genetic factors. In addition to genetic resistance, use of herbicides to control Johnsongrass, a major overwintering host of the virus, can substantially control disease by eliminating the virus reservoir. Herbicides used in a soybean–corn crop rotation, for example, are effective for the control of Johnsongrass. Systemic insecticides can reduce vector populations and limit disease spread under some conditions but may not be economically justified. Enzyme-linked immunosorbent assay (ELISA)-based seed health testing methods are available and can be used to identify infected seedlots and prevent their use.

Selected References

Ellett, C. W., Janson, B. F., and Williams, L. E. 1965. Maize dwarf mosaic in Ohio in 1964. Plant Dis. Rep. 49:615-616.

Ford, R. E., Tosic, M., and Shukla, D. D. 1989. Maize dwarf mosaic virus. Descriptions of Plant Viruses, No. 341. Commonwealth Mycological Institute and Association of Applied Biologists, Kew, England.

Jones, M. W., Redinbaugh, M. G., and Louie, R. 2011. Responses of maize (*Zea mays* L.) near isogenic lines carrying *Wsm1*, *Wsm2*, and *Wsm3* to three viruses in the Potyviridae. Theor. Appl. Genet. 123:729-740.

Louie, R., and Knoke, J. K. 1981. Symptoms and disease diagnosis. Pages 13-18 in: Virus and Viruslike Diseases of Maize in the United

States. D. T. Gordon, J. K. Knoke, and G. E. Scott, eds. South. Coop. Ser. Bull. 247.
Lubberstedt, T., Ingvardsen, C., Melchinger, A. E., Xing, Y., Salomon, R., and Redinbaugh, M. G. 2006. Two chromosome segments confer multiple potyvirus resistance in maize. Plant Breed. 125:352-356.
MacKenzie, D. R., Wernham, C. C., and Ford, R. E. 1966. Differences in maize dwarf mosaic virus isolates of the northeastern United States. Plant Dis. Rep. 50:814-818.
McDaniel, L. L., and Gordon, D. T. 1985. Identification of a new strain of maize dwarf mosaic virus. Plant Dis. 69:602-607.
McMullen, M. D., Jones, M. W., Simcox, K. D., and Louie, R. 1994. Three genetic loci control resistance to wheat streak mosaic virus in the maize inbred Pa405. Mol. Plant-Microbe Interact. 7:708-712.
Redinbaugh, M. G., Jones, M. W., and Gingery, R. E. 2004. The genetics of virus resistance in maize (*Zea mays* L.). Maydica 49:183-190.
Seifers, D. L., Haber, S., Ens, W., She, Y.-M., Standing, K. G., and Salomon, R. 2005. Characterization of a distinct *Johnsongrass mosaic virus* strain isolated from sorghum in Nigeria. Arch. Virol. 150:557-576.
Shukla, D. D., and Teakle, D. S. 1989. Johnsongrass mosaic virus. Descriptions of Plant Viruses, No. 340. Commonwealth Mycological Institute and Association of Applied Biologists, Kew, England.
Shukla, D. D., Frenkel, M. J., McKern, N. M., Ward, C. W., Jilka, J., Tosic, M., and Ford, R. E. 1992. Present status of the sugarcane mosaic subgroup of potyviruses. Arch. Virol. 5(Suppl.):363-373.
Shukla, D. D., Ward, C. W., and Brunt, A. A. 1994. The Potyviridae. CAB International, Kew, England.
Stewart, L. R., Haque, M. A., Jones, M. W., and Redinbaugh, M. G. 2013. Response of maize (*Zea mays* L.) lines carrying *Wsm1*, *Wsm2*, and *Wsm3* to the potyviruses *Johnsongrass mosaic virus* and *Sorghum mosaic virus*. Mol. Breed. 31:289-297.
Teakle, D. S., Shukla, D. D., and Ford, R. E. 1989. Sugarcane mosaic virus. Descriptions of Plant Viruses, No. 342. Commonwealth Mycological Institute and Association of Applied Biologists, Kew, England.
Williams, L. E., and Alexander, L. J. 1965. Maize dwarf mosaic, a new corn disease. Phytopathology 55:802-804.

(Prepared by L. R. Stewart, R. Louie, and M. G. Redinbaugh)

Maize Chlorotic Dwarf

Rosenkranz and Bradfute et al. independently discovered *Maize chlorotic dwarf virus* (MCDV) during the late 1960s and early 1970s. Maize chlorotic dwarf is found in the southeastern United States where the distribution of Johnsongrass, the overwintering host for MCDV, overlaps the distribution of the leafhopper vector, *Graminella nigrifrons*. This coincidence includes the states from the Gulf of Mexico northward to states along the Ohio River and Pennsylvania and the states from the Atlantic coast to eastern Texas. Maize chlorotic dwarf was also reported from Arizona. The extent of the distribution of maize chlorotic dwarf in the United States is second only to that of maize dwarf mosaic.

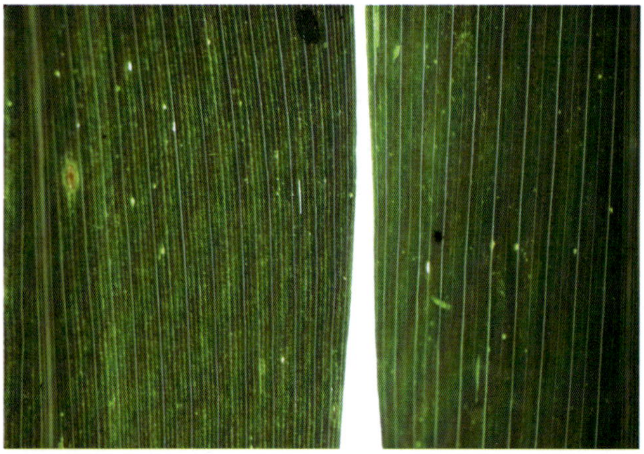

Fig. 223. Leaf striping caused by *Maize chlorotic dwarf virus* (left); healthy leaf on right. (Courtesy R. Louie)

Symptoms

The symptoms caused by MCDV in corn vary greatly (Figs. 223 and 224). Striking symptoms such as reddening, yellowing, twisting, and tearing of the leaves; shortening of the upper internodes or proportionate stunting of the internodes; and veinbanding (also referred to as chlorotic striping) of the tertiary veins and veinclearing are associated with MCDV infection. The veins of leaves of infected plants are swollen, giving the leaves a dull appearance and a rough upper surface. Infected leaves are also more difficult to tear off a plant. Chlorotic mottling occasionally appears at the base of the whorl in very young seedlings. However, the diagnostic symptom of MCDV infection is the chlorotic striping over the tertiary veins that later coalesces to form chlorotic bands alternating with green (i.e., veinbanding). Other leaf colorations associated with MCDV infection are often influenced by environment, genotype, and time of infection. Infection in the field often causes "miniature plants" and plants with shortened upper internodes. Gingery and Nault reported that coinfection with two distinct MCDV strains causes more severe stunting of plants under experimental conditions. The biological properties of these isolates of MCDV are similar, although some sequenced strains are quite divergent from each other (less than 60% nucleotide sequence identity). Symptom severity for isolates ranges from very severe to very mild.

MCDV can be diagnosed by veinbanding or veinclearing symptoms and by commercially available enzyme-linked immunosorbent assays (ELISA), molecular biological assays, or both.

Causal Agent

Maize chlorotic dwarf virus (MCDV) virions are isometric particles approximately 30 nm in diameter containing the single-stranded RNA monopartite genome. MCDV is a member of the genus *Waikavirus* in the recently reclassified virus family *Secoviridae* (formerly *Sequiviridae* and *Comoviridae*). MCDV often is found in mixed infections with *Maize dwarf mosaic virus*. Multiple strains of MCDV have been reported, including the mild strain (MCDV-M1) and the more severe type strain (MCDV-T).

Vectors

In nature, the leafhopper *Graminella nigrifrons* is the principal vector of the virus; experimentally, *Graminella sonora*, *Exitianus exitiosus*, *Amblysellus grex*, *Endria inimica*, *Macrosteles severini*, *Planicephalus flavicostatus*, and *Stirellus bicolor* also transmit MCDV. *Graminella nigrifrons* adults and nymphs, males and females, all transmit MCDV. Females are better transmitters than males under experimental conditions, but because females are more sedentary, males may be more important in disseminating the virus in the field. The virus is acquired in as little as 15 min and transmitted with a minimal inoculation access period of 15 min. Because the leafhopper retains transmission capability for 2–4 days, the relationship is designated semipersistent. No latent period is required for transmission and the ability to transmit the virus is lost after a molt. There is no transovarial transmission. Evidence suggests a viral-encoded helper component is required for leafhopper transmission of MCDV.

Fig. 224. Maize chlorotic dwarf. (Courtesy L. Stewart, C. Nacci, and K. Chamberlain)

Disease Cycle and Epidemiology

The major overwintering host of MCDV is Johnsongrass. In addition to corn, other economically important hosts include proso millet, pearl millet, sorghum, Sudan grass, and wheat. However, only corn appears to be an economically significant host under field conditions. The virus can be mechanically transmitted by vascular puncture inoculation of seeds. It is not transmitted via seed.

Management

Commercial hybrids, with various degrees of resistance or tolerance, in combination with control of Johnsongrass are the most effective management practices. Highly resistant germplasm was identified from the Caribbean. Major quantitative trait loci for MCDV resistance were identified on chromosomes 3 and 10. Whether resistance from this germplasm can be deployed in genotypes appropriate for the other corn-growing regions remains to be determined. Lower levels of resistance to MCDV were reported previously. Corn lines and hybrids are not generally labeled for MCDV resistance currently. A recent survey of corn and Johnsongrass viruses in Ohio found MCDV only rarely. Most MCDV located was the mild strain and in Johnsongrass rather than in corn. Frequent coinfections with potyviruses in the field are not strongly synergistic but may prolong or intensify symptoms. Systemic insecticides can control the vectors of the virus under experimental conditions, but they are not economical in most situations.

Selected References

Bradfute, O. E., Louie, R., and Knoke, J. K. 1972. Isometric viruslike particles in maize with stunt symptoms. (Abstr.) Phytopathology 62:748.

Gingery, R. E. 1976. Properties of maize chlorotic dwarf virus and its ribonucleic acid. Virology 73:311-318.

Gingery, R. E., and Nault, L. R. 1990. Severe maize chlorotic dwarf disease caused by double infection of mild virus strains. Phytopathology 80:687-691.

Gingery, R. E., Bradfute, O. E., Gordon, D. T., and Nault, L. R. 1978. Maize chlorotic dwarf virus. Descriptions of Plant Viruses, No. 194. Commonwealth Mycological Institute and Association of Applied Biologists, Kew, England.

Gordon, D. T., and Nault, L. R. 1977. Involvement of maize chlorotic dwarf virus and other agents in stunting diseases of *Zea mays* in the United States. Phytopathology 67:27-36.

Hunt, R. E., Nault, L. R., and Gingery, R. E. 1988. Evidence for infectivity of maize chlorotic dwarf virus and for a helper component in its leafhopper transmission. Phytopathology 78:499-504.

Jones, M. W., Redinbaugh, M. G., Anderson, R. J., and Louie, R. 2004. Identification of quantitative trait loci controlling resistance to maize chlorotic dwarf virus. Theor. Appl. Genet. 11:48-57.

Louie, R., and Anderson, R. J. 1993. Evaluation of maize chlorotic dwarf virus resistance in maize with multiple inoculations by *Graminella nigrifrons* (Homoptera: Cicadellidae). J. Econ. Entomol. 86:1579-1583.

Louie, R., Knoke, J. K., and Gordon, D. T. 1974. Epiphytotics of maize dwarf mosaic and maize chlorotic dwarf diseases in Ohio. Phytopathology 64:1455-1459.

Louie, R., Knoke, J. K., and Findley, W. R. 1990. Elite maize germplasm: Reactions to maize dwarf mosaic and maize chlorotic dwarf viruses. Crop Sci. 30:1210-1215.

Louie, R., Jones, M. W., Anderson, R., and Redinbaugh, M. G. 2002. Registration of maize germplasm Oh1VI. Crop Sci. 42:991.

Madden, L. V., Knoke, J. K., and Louie, R. 1990. Spread of maize chlorotic dwarf virus in maize fields by its leafhopper vector, *Graminella nigrifrons*. Phytopathology 80:291-298.

Morales, K., Zambrano, J. L., and Stewart, L. R. 2014. Co-infection and disease severity of Ohio *Maize dwarf mosaic virus* and *Maize chlorotic dwarf virus* strains. Plant Dis. 98:1661-1665.

Nault, L. R., and Madden, L. V. 1988. Phylogenetic relatedness of maize chlorotic dwarf virus leafhopper vectors. Phytopathology 78:1683-1687.

Nault, L. R., Styer, W. E., Knoke, J. K., and Pitre, H. N. 1973. Semipersistent transmission of leafhopper-borne maize chlorotic dwarf virus. J. Econ. Entomol. 66:1271-1273.

Pratt, R. C., Anderson, R. J., Louie, R., and McMullen, M. D. 1994. Symptom expression of maize inbreds and hybrids infected by severe and type strains of maize chlorotic dwarf virus. Crop Sci. 34:635-640.

Rosenkranz, E. 1969. A new leafhopper-transmissible corn stunt disease agent in Ohio. Phytopathology 59:1344-1346.

Rosenkranz, E., and Scott, G. E. 1987. Type of gene action in the resistance to maize chlorotic dwarf virus in corn. Phytopathology 77:1293-1296.

Rufener, G. K., II, Mowers, R. P., Pratt, R. C., Louie, R., McMullen, M. D., and Knoke, J. K. 1996. Maize chlorotic dwarf virus resistant maize and the production thereof. U.S. Patent 5563316.

Sanfacon, H., Wellink, J., Le Gall, O., Karasev, A., Van der Vlugt, R., and Wetzel, T. 2009. *Secoviridae*: A proposed family of plant viruses within the order *Picornavirales* that combines the families *Sequiviridae* and *Comoviridae*, the unassigned genera *Cheravirus* and *Sadwavirus*, and the proposed genus *Torradovirus*. Arch. Virol. 154:899-907.

Scott, G. E., and Rosenkranz, E. 1981. Effectiveness of resistance to Maize dwarf mosaic and Maize chlorotic dwarf viruses in maize. Phytopathology 71:937-941.

Stewart, L. R., Teplier, R., Todd, J. C., Jones, M. W., Cassone, B. J., Wijeratne, S., Wijeratne, A., and Redinbaugh, M. G. 2014. Viruses in maize and Johnsongrass in southern Ohio. Phytopathology 104:1360-1369.

(Prepared by L. R. Stewart and R. Louie)

Maize Chlorotic Mottle and Maize Lethal Necrosis

Maize chlorotic mottle virus (MCMV) was first described in 1974 in Peru and was first recognized in the United States in 1976, initially in Kansas and Nebraska. It also has been reported in Colorado, Hawaii, and Texas (United States). Internationally, it has been found in Argentina, China, Democratic Republic of the Congo, Kenya, Mexico, Rwanda, Tanzania, Thailand, and Uganda. Symptoms are typically mild in dent and flint types of corn but more severe in sweet corn and floury types. Yield losses in the field are reported to average 10–15%, and reductions up to 59% have been reported following inoculation.

Mixed infections with any of the members of the family *Potyviridae* (most commonly *Maize dwarf mosaic virus* or *Sugarcane mosaic virus*, but also *Wheat streak mosaic virus*) result in a very severe disease known as maize lethal necrosis or corn lethal necrosis. The effect of coinfection is synergistic, and the symptoms of maize lethal necrosis are much more severe than the additive symptoms of *Maize chlorotic mottle virus* (MCMV) and *Potyviridae* infection. MCMV titer can be five times higher in plants coinfected with a member of the family *Potyviridae* versus those infected with MCMV alone. While severe outbreaks of maize lethal necrosis are sporadic in the continental United States, when they occur, they can cause complete crop loss. In East Africa, maize lethal necrosis has emerged as a devastating problem since 2011.

Symptoms

Fine, chlorotic streaks develop in the youngest leaf 7–10 days after inoculation with MCMV. The streaks coalesce to form a chlorotic mottling, followed by leaf necrosis and epinasty and death of the leaves (Fig. 225). Growth is stunted; male inflorescences are distorted, with hard panicles, a short rachis, and few spikelets; fewer ears are formed; and ears may be shortened, malformed, and partially filled. Severe infection can lead to plant death. Plants infected at earlier growth stages have more severe symptoms.

Symptoms of maize lethal necrosis usually begin with chlorotic mottling of the leaves, starting from the base of the young leaves in the whorl, extending upward and becoming severe (Figs. 226 and 227). Eventually, necrosis develops, starting at the leaf margins and progressing inward to involve the entire leaf. Death of young leaves in the whorl results in the "dead heart" symptom. Severely affected plants die prematurely and form small cobs with little or no grain. Whole fields may be killed before tasseling.

Fig. 225. Maize chlorotic mottle. (Courtesy C. De Leon)

Fig. 226. Maize lethal necrosis. (Courtesy R. Reeder; Reproduced, by permission, from Maize Lethal Necrosis Disease Datasheet. 2014. Crop Protection Compendium. CAB International, Wallingford, U.K.)

Fig. 227. Maize lethal necrosis. (Courtesy R. Reeder; Reproduced, by permission, from Maize Lethal Necrosis Disease Datasheet. 2014. Crop Protection Compendium. CAB International, Wallingford, U.K.)

Causal Agent

Maize chlorotic mottle virus (MCMV) virions are small, isometric particles about 30 nm in diameter, containing monopartite single-stranded RNA genome. MCMV is the sole member of the genus *Machlomovirus* in the family *Tombusviridae*. MCMV often is found in mixed infections with other viruses, particularly *Maize dwarf mosaic virus*, *Sugarcane mosaic virus*, and *Wheat streak mosaic virus*.

Vectors

Chrysomelid beetles are the primary vectors involved in MCMV transmission in the continental United States, but corn thrips (*Frankliniella williamsi*) is the major vector in Hawaii, Africa, and probably Taiwan. Beetles that transmitted MCMV experimentally include *Diabrotica virgifera virgifera*, *D. barberi* (syns. *Diabrotica longicornis*, *Diabrotica longicornis barberi*), *D. undecimpunctata howardi*, *Chaetocnema pulicaria*, *Systena frontalis*, and *Oulema melanopus*. *Diabrotica* spp. are the only natural vectors among the beetle species. Beetle and thrips transmission is by both adults and larvae. Acquisition access and inoculation access periods are about 24 h for beetles and 3 h for thrips. There is no latent period for beetles or thrips. Vectors do not retain the virus after molting but the ability to transmit is retained up to 6 days. The virus is not persistent or propagative in the vectors.

Potyviruses *Maize dwarf mosaic virus* and *Sugarcane mosaic virus* are transmitted by aphids in several genera, and *Wheat streak mosaic virus* is transmitted by the wheat-curl mite (see the Maize Dwarf Mosaic section).

Disease Cycle and Epidemiology

MCMV is spread primarily by insects in the field, but it also is mechanically transmissible, soil transmissible, and can be seed transmitted at frequencies of 0–0.33%. The importance of mechanical transmission in locations where MCMV is widespread requires controlled study to ascertain. MCMV levels are very elevated in plants coinfected with *Maize dwarf mosaic virus*, *Sugarcane mosaic virus*, or *Wheat streak mosaic virus*. Corn, sugarcane, and millets are reported natural hosts, but other (Poaceae) hosts, including small grain cereals, sorghum, and grasses in several genera, can be experimentally infected.

Management

Genetic resistance is a highly desirable management tool, but strong resistance to MCMV remains to be identified. Tolerant corn lines have been identified and are among Hawaiian-grown corn, but they are still systemically infected and can be a virus reservoir for infection of more susceptible lines. Crop rotation and a fallow season are probably important in reducing soil transmission, vector populations, and virus reservoirs in the continental United States. The length and efficacy of fallow periods or crop rotation to manage MCMV in locations where there is currently continuous corn remain to be experimentally determined. Several sources of resistance/tolerance are available; resistance is more common in tropical germplasm. Transgenic plants expressing *Sugarcane mosaic virus* coat protein are resistant to maize lethal necrosis but not directly to MCMV. Control of insect vectors in tropical or subtropical areas may require repeated insecticide applications. Weed control in and around fields can help prevent the buildup of vector populations. Seed health testing can prevent spread of the disease through seed.

In the U.S. mainland, resistance to potyviruses combined with MCMV tolerance, crop rotation, and weed control reduce disease severity significantly. In Hawaii, recommendations include using *Maize dwarf mosaic virus*-resistant/tolerant hybrids, using up to weekly insecticide sprays to control thrips populations, controlling weed hosts of MCMV, removing infected plants, and minimizing movement of personnel and machines through fields.

Selected References

Adams, I. P., Harju, V. A., Hodges, T., Hany, U., Skelton, A., Rai, S., Deka, M. K., Smith, J., Fox, A., Uzayisenga, B., Ngaboyisonga, C., Uwumukiza, B., Rutikanga, A., Rutherford, M., Ricthis, B., Phiri, N., and Boonham, N. 2014. First report of maize lethal necrosis disease in Rwanda. New Dis. Rep. 29:22.

Bockelman, D. L., Claflin, L. E., and Uyemoto, J. K. 1982. Host range and seed-transmission studies of maize chlorotic mottle virus in grasses and corn. Plant Dis. 66:216-218.

Cabanas, D., Watanabe, S., Higashi, C. H., and Bressan, A. 2013. Dissecting the mode of maize chlorotic mottle virus transmission (*Tombusviridae: Machlomovirus*) by *Frankliniella williamsi* (*Thysanoptera*: *Thripidae*). J. Econ. Entomol. 106:16-24.

Carrera-Martinez, H., Lozoya-Saldana, H., Mendoza-Zamora, C., and Alvizo-Villasana, H. 1989. Immunoabsorcion enzimatica (ELISA) en Ia identificacion y distribucion del virus moteado clorotico del maize (VMCM) en el estado de Mexico. Rev. Mex. Fitolpatol. 7:20-25.

Castillo, J., and Herbert, T. T. 1974. Nueva enfermedad virosa afectando al maiz en el Peru. Fitopatologia 9:79-84.

Deng, T.-C., Chou, C.-M., Chen, C.-T., Tsai, C.-H., and Lin, F.-C. 2014. First report of *Maize chlorotic mottle virus* on sweet corn in Taiwan. Plant Dis. 98:1748.

Jensen, S. G. 1985. Laboratory transmission of maize chlorotic mottle virus by three species of corn rootworms. Plant Dis. 69:864-868.

Jensen, S. J., Wysong, D. S., Ball, E. M., and Higley, P. M. 1991. Seed transmission of maize chlorotic mottle virus. Plant Dis. 75:497-498.

Jiang, X. Q., Meinke, L. J., Wright, R. J., Wilkinson, D. R., and Campbell, J. E. 1992. Maize chlorotic mottle virus in Hawaiian-grown maize: Vector relations, host range and associated viruses. Crop Prot. 11:248-254.

Klingkong, T., and Sutabutra, T. 1982. A new virus disease of maize in Thailand. Pages 191-193 in: Proc. Int. Maize Virus Dis. Colloquium and Workshop. D. T. Gordon, J. K. Knoke, L. R. Nault, and R. M. Ritter, eds. OARDC, Wooster, OH.

Lapierre, H., and Signoret, P.-A., eds. 2004. Viruses and Virus Diseases of *Poaceae* (*Gramineae*). INRA, Paris.

Lukanda, M., Owati, A., Ogunsanya, P., Valimunzigha, K., Katsongo, K., Ndemere, H., and Kumar, P. L. 2014. First report of *Maize chlorotic mottle virus* infecting maize in the Democratic Republic of the Congo. Plant Dis. 98:1448-1449.

Murry, L. W., Elliott, L. G., Capitant, S. A., West, J. A., Hanson, K. K., Scarafia, L., Johnston, S., DeLuca-Flaherty, C., Nichols, S., Cunanan, D., Dietrich, P. S., Mettler, I. J., Dewald, S., Warnick, D. A., Rhodes, C., Sinibaldi, R. M., and Brunke, K. J. 1993. Transgenic corn plants expressing MDMV strain B coat protein are resistant to mixed infections of maize dwarf mosaic virus and maize chlorotic mottle virus. Nat. Biotechnol. 11:1559-1564.

Nault, L. R., Styer, W. E., Coffey, M. E., Gordon, D. T., Negi, L. S., and Niblett, C. L. 1978. Transmission of maize chlorotic mottle virus by Chrysomelid beetles. Phytopathology 68:1071-1074.

Niblett, C. L., and Claflin, L. E. 1978. Corn lethal necrosis: A new virus disease of corn in Kansas. Plant Dis. Rep. 62:15-19.

Scheets, K. 1998. Maize chlorotic mottle machlomovirus and wheat streak mosaic rymovirus concentration increase in the synergistic disease corn lethal necrosis. Virology 242:28-38.

Stenger, D. C., and French, R. 2008. Complete nucleotide sequence of a maize chlorotic mottle virus isolate from Nebraska. Arch. Virol. 153:995-997.

Teyssandier, E. E., Dal Bo, E., and Nome, S. F. 1983. Maize virus diseases in Argentina. Pages 93-99 in: Proc. Int. Maize Virus Dis. Colloquium and Workshop. Ohio Agricultural Research and Development Center, Wooster, OH.

Uyemoto, J. K. 1983. Biology and control of maize chlorotic mottle virus. Plant Dis. 67:7-10.

Wang, Q., Xhou, X. P., and Wu, J. X. 2014. First report of *Maize chlorotic mottle virus* infecting sugarcane (*Saccharum officinarum*). Plant Dis. 98:572.

Wangai, A. W., Redinbaugh, M. G., Kinyua, Z. M., Miano, D. W., Leley, P. K., Kasina, M., Mahuku, G., Scheets, K., and Jeffers, D. 2012. First report of *Maize chlorotic mottle virus* and maize lethal necrosis in Kenya. Plant Dis. 96:1582.

Xie, L., Zhang, J., Wang, Q., Meng, C., Hong, J., and Zhou, X. 2010. Characterization of maize chlorotic mottle virus associated with maize lethal necrosis disease in China. J. Phytopathol. 159:191-193.

(Prepared by R. Louie, G. P. Munkvold,
L. R. Stewart, and M. G. Redinbaugh)

Maize Mosaic

Maize mosaic is a damaging disease in tropical and subtropical areas. It was first recognized in Hawaii in 1914, and now its distribution is widespread, including Australia, islands of the Caribbean and West Indies (Cuba, French Guyana, Guadeloupe, Trinidad, Puerto Rico), Mexico, Central America (Belize, Guatemala, Nicaragua), South America (Peru, Surinam, Venezuela), Africa (Tanzania), western Indian Ocean islands (Mauritius, Reunion), the United States (Florida, Hawaii), Africa, and parts of the Middle East. In Venezuela, disease incidence can reach 60% and it can greatly reduce yield; in fact, maize mosaic has been postulated as a factor in the collapse of Mayan civilization. The causal agent, *Maize mosaic virus* (MMV), has often been confused with *Maize stripe virus* because of its similar symptoms, host range, and vector relationship.

Symptoms

Maize mosaic also is known as *enanismo rayado*, corn stripe, or sweet corn mosaic. Symptoms first appear as chlorotic spots and short lines. Fine, yellow stripes develop along the veins, and leaves eventually display long, broad yellow stripes that ultimately become necrotic. Stripes or chlorotic spots rarely appear in the interveinal areas. Chlorotic stripes also can be found on leaf sheaths, husks, and stalks. Stunting of plants is common, with shortened internodes and a rosette appearance. Despite the disease name, mosaic patterns do not appear on leaves (Fig. 228).

Causal Agent

The causal agent is *Maize mosaic virus* (MMV), a member of the genus *Nucleorhabdovirus* of the family *Rhabdoviridae*. MMV is characterized by 45–70 × 225–325-nm, bullet-shaped enveloped virions containing negative-sense single-stranded RNA.

Vector

MMV is transmitted in a persistent propagative manner (it infects both the plant host and insect vector) by the planthopper

Fig. 228. Maize mosaic. (Courtesy R. Louie)

Peregrinus maidis, which is the only known vector. Transmission can occur with both adults and nymphs. Acquisition rates of 5–42% have been reported. Acquisition time and incubation period vary with MMV strain and *P. maidis* biotype; the acquisition time can range from 15 min to 24 h and incubation period reports vary from 9–12 days up to 11–58 days. Transovarial transmission has not been reported. Transmission patterns can be erratic.

Disease Cycle and Epidemiology

MMV infects only members of the family Poaceae. Natural hosts include corn, sorghum, and several grass species, including itchgrass and *Setaria vulpiseta*. The virus has been transmitted experimentally to several other grasses, barley, and *Sorghum verticilliflorum*. MMV is not known to be seedborne and is not readily sap transmissible. The vector, *P. maidis*, is limited to tropical and subtropical areas, and the virus likely persists in the vectors or uncultivated hosts in endemic areas.

Management

MMV-resistant inbred lines have been developed and resistance is the most effective means of disease control. Major genes for resistance to MMV in maize are located on chromosomes 2 and 3 (*mv1*). Resistance associated with *mv1* has been introduced into more than 100 inbred lines and cultivars and 150 genetic stocks by backcrossing. Tropical adapted inbred lines of maize resistant to MMV have been registered and released. More recently, a maize line (Hi40), which is highly resistant to MMV, has been developed. Integrated management programs for *P. maidis* can reduce spread of the disease, but the insect is difficult to manage with insecticide applications. Destruction of weed hosts near fields reduces inoculum and vector populations.

Selected References

Ammar, E. D., Aboul-Ata, A. E., and Hussain, A. M. 2008. Viruses affecting maize and sorghum. Pages 399-431 in: Viral Diseases on Economic Crops in the Arab World. K. Makkouk and S. Kumari, eds. Macmillan, Beirut.

Bradfute, O. E., and Tsai, J. H. 1983. Identification of maize mosaic virus in Florida. Plant Dis. 67:1339-1342.

Dintinger, J., Boissot, N., Chiroleu, F., Hamon, P., and Reynaud, B. 2005. Evaluation of maize inbreds for *Maize stripe virus* and *Maize mosaic virus* resistance: Disease progress in relation to time and the cumulative number of planthoppers. Phytopathology 95:600-607.

Falk, B. W., and Tsai, J. H. 1985. Serological detection and evidence for multiplication of maize mosaic virus in the planthopper, *Peregrinus maidis*. Phytopathology 75:852-855.

Lapierre, H., and Signoret, P.-A., eds. 2004. Viruses and Virus Diseases of *Poaceae* (*Gramineae*). INRA, Paris.

(Prepared by L. R. Stewart, G. Munkvold, and R. Louie)

Maize Rayado Fino

Maize rayado fino virus (MRFV) was originally described in Costa Rica in 1969. Its range extends from the southern United States to northern Argentina and Uruguay (approximately 30°N latitude in Texas to 32°S latitude in Uruguay). It is also likely that MRFV is found in the Caribbean, since the vector, *Dalbulus maidis*, is known to exist in these islands. It has been reported only from the Americas: United States (Texas, Florida), Mexico, Central America (Costa Rica, El Salvador, Guatemala, Honduras, Nicaragua, Panama), and South America (Colombia, Ecuador, Peru, Paraguay, Venezuela, Uruguay, Brazil). Yield reductions of 40–50% have been reported in individual plants, and greater losses can occur with nonadapted cultivars introduced into Central America. MRFV is sometimes confused with Brazilian maize streak and maize rayado Colombiano viruses, which are serologically distinct; and it often is found in combination with the corn stunt spiroplasma and the maize bushy stunt phytoplasma, which are transmitted by the same vector.

Symptoms

MRFV initially causes fine, chlorotic stipples that coalesce into chlorotic stripes over the veins (Fig. 229). Generalized yellowing appears when large areas of the leaf are affected. Stunting and chlorosis may occur. Symptoms can vary because of the frequent coinfection with corn stunt and maize bushy stunt phytoplasmas. Symptoms are more evident during the seedling stages. Infected plants may have small ears with few grains. Plants infected at tasseling or later may not show symptoms.

Causal Agent

Maize rayado fino virus (MRFV) is the type species of the genus *Marafivirus* of the family *Tymoviridae*. Particles are isometric, filled capsids 33 nm in diameter with monopartite single-stranded RNA genomes.

Vectors

The maize leafhopper, *Dalbulus maidis*, is the only natural vector. Experimental vectors include *Graminella nigrifrons*, *D. elimatus*, *Stirellus bicolor*, and *Baldulus tripsaci*. *D. maidis* acquires the virus in 6–48 h and has a long incubation period of 8–37 days, with an average of 12–17 days at 25°C. The insect transmits MRFV with an inoculation access period of 8 h to 4 days. The virus is persistent and multiplies in the vector and is likely propagative. Both adults and nymphs are vectors, but nymphs are more efficient. There are no significant differences between males and females as vectors. Transmission is intermittent and the insects retain the ability to transmit for 1–20 days; however, infectivity decreases with time. The virus is retained after molting, but MRFV is not transmitted transovarially.

Disease Cycle and Epidemiology

Hosts of MRFV are species and subspecies of teosinte (*Zea* spp.) and maize, including *Rottboellia exaltata*, *Tripsacum australe*, *Z. diploperennis*, *Z. mays* subsp. *mexicana*, *Z. mays*

Fig. 229. Maize rayado fino. (Courtesy R. Louie)

subsp. *parviglumis* var. *parviglumis*, *Z. mays* subsp. *parviglumis* var. *huehuetenangensis*, and *Z. mays* subsp. *mays*. However, maize and teosinte are the only known natural hosts. MRFV is not known to be seed transmitted, nor is it easily transmitted mechanically. MRFV causes more damage at higher altitudes in Mexico and Colombia, while the phytoplasmas are more predominant at lower altitudes.

Management

Partially resistant cultivars are available. Recent identification of highly resistant germplasm and a major quantitative trait locus for resistance on chromosome 10 should aid in development of more resistant cultivars. Crop rotation, mixed planting, insect vector control, and adjusting planting dates can reduce the impact of the disease.

Selected References

Bradfute, O. E., Nault, L. R., Gordon, D. T., Robertson, D. C., Toler, R. W., and Boothroyd, C. W. 1980. Identification of maize rayado fino virus in the United States. Plant Dis. 64:50-53.

Bustamante, P. I., Hammond, R., and Ramirez, P. 1998. Evaluation of maize germ plasm for resistance to maize rayado fino virus. Plant Dis. 82:50-56.

Gamez, R. 1969. A new leafhopper-borne virus of corn in Central America. Plant Dis. Rep. 53:929-932.

Gamez, R. 1980. Maize rayado fino virus. Descriptions of Plant Viruses, No. 220. Commonwealth Mycological Institute and Association of Applied Biologists, Kew, England.

Gingery, R. E., Gordon, D. T., and Nault, L. R. 1982. Purification and properties of an isolate of maize rayado fino virus from the United States. Phytopathology 72:1313-1318.

Lapierre, H., and Signoret, P.-A., eds. 2004. Viruses and Virus Diseases of *Poaceae* (*Gramineae*). INRA, Paris.

Nault, L. R., Gingery, R. E., and Gordon, D. T. 1980. Leafhopper transmission and host range of maize rayado fino virus. Phytopathology 70:709-712.

Rivera, C., and Gamez, R. 1986. Multiplication of maize rayado fino virus in the leafhopper vector *Dalbulus maidis*. Intervirology 25:76-82.

Zambrano, J. L., Francis, D. M., and Redinbaugh, M. G. 2013. Identification of resistance to *Maize rayado fino virus* in maize inbred lines. Plant Dis. 97:1418-1423.

Zambrano, J. L., Jones, M. W., Francis, D. M., Tomas, A., and Redinbaugh, M. G. 2014. Quantitative trait loci for resistance to *Maize rayado fino virus*. Mol. Breed. 34:989-996.

(Prepared by G. P. Munkvold)

Maize Streak

In 1901, C. Fuller described a disease in corn that he called mealie variegation. It was renamed maize streak by H. H. Storey in 1925. Maize streak is still an economically important virus disease of corn. It is found throughout sub-Saharan Africa, where it is a major problem, in Madagascar and Mauritius, and in southeastern Asia.

Maize streak virus (MSV) causes economically important diseases of corn, sugarcane, and millet. Wheat, sorghum, barley, oats, rye, finger millet, bulrush millet, rice, and pearl millet also are infected.

Symptoms

The severity of disease symptoms depends greatly on maize and virus genotypes (Figs. 230 and 231). Initially, symptoms consist of pale, chlorotic, circular spots on the basal portions of leaves that emerge after infection. The spots vary from 0.1 to 2 mm in diameter, contrast intensely against the healthy green tissue, and have diffuse margins. These spots coalesce and, in susceptible genotypes, form extensive, elongated, chlorotic streaks. Plants infected at an early stage are stunted and suffer significant yield losses because of unfilled cobs and immature seeds.

Causal Agent

Maize streak virus (MSV) is a single-stranded DNA, nonenveloped virus classified in the family *Geminiviridae*, genus *Mastrevirus*. Geminate isometric particles that can be viewed with an electron microscope are a distinguishing characteristic of the virus.

MSV isolates are classified into eleven strains: MSV-A to MSV-K, but most maize streak is caused by MSV-A. MSV-A strains are further subdivided into groups MSV-A_1 to MSV-A_6, based on small sequence differences, but have similar biological characteristics.

Vectors

Several species of *Cicadulina* leafhoppers transmit MSV, but *C. mbila* is responsible for most transmission under field conditions. Other vectors include *C. storeyi* (syn. *C. triangula*), *C. arachidis*, *C. similis*, and *C. ghaurii*. *C. niger* is also a vector but only under experimental conditions. *C. mbila* acquires MSV in as little as 15 sec, and after a latent period of 6–12 h,

Fig. 230. Maize streak. (Courtesy B. M. Anderson)

Fig. 231. Maize streak. (Courtesy R. Louie)

transmits it to corn after a 5-min inoculation access period. Although the virus is circulative, it does not propagate in the vector. The leafhopper transmits the virus during all developmental stages, even after molting. Furthermore, once the leafhoppers acquire MSV, they transmit it for life. Mechanical transmission is possible by vascular puncture inoculation of corn seeds. MSV is not seed transmitted.

Management

Epidemics of maize streak are sporadic but can be devastating. Incidence and severity from year to year depend mostly on vector–environment interactions. Cultural practices, choosing a site away from irrigated fields, leaving bare ground barriers between early- and late-planted fields, and use of insecticides all contribute to disease management but can be impractical or unjustified economically. Insecticidal seed treatments are effective at reducing early-season transmission by vectors. Host resistance remains the most effective means of control. A study in 2006 by Mawere et al. indicated that MSV resistance in four inbred lines developed by CIMMYT (i.e., CML312, CML206, CML216, and MSR) was consistent for 20 different isolates of MSV.

A major quantitative trait locus, designated *Msv1*, on chromosome 1 provides resistance in all currently developed lines. Several contributing minor quantitative trait loci are also described. Tolerant and resistant genotypes are available. Pathogen-derived transgenic resistance for MSV was recently developed using the viral replicase sequence. Transgenic plants have not yet been deployed in the field but may be an option in the future.

Selected References

Barrow, M. R. 1993. Increasing maize yields in Africa through the use of maize streak virus resistant hybrids. Afr. Crop Sci. J. 1:139-144.

Bock, K. R. 1974. Maize streak virus. Descriptions of Plant Viruses, No. 133. Commonwealth Mycological Institute and Association of Applied Biologists, Kew, England.

Efron, Y., Kim, S. K., Fajemisin, J. M., Mareck, J. M., Tang, C. Y., Dabrowski, Z. T., Rossel, H. W., Thottappilly, G., and Buddenhagen, I. 1989. Breeding for resistance to maize streak virus: A multidisciplinary approach. Plant Breed. 103:1-36.

Kim, S. K., Efron, Y., Fajemisin, J. M., and Buddenhagen, I. W. 1989. Mode of gene action for resistance in maize to maize streak virus. Crop Sci. 29:890-894.

Kyetere, D. T., Ming, R., McMullen, M. D., Pratt, R. C., Brewbaker, J., and Musket, T. 1999. Genetic analysis of tolerance to maize streak virus in maize. Genome 42:20-26.

Martin, D. P., Willment, J. A., Billharz, R., Velders, R., Odhiambo, B., Njuguna, J., James, D., and Rybicki, E. P. 2001. Sequence diversity and virulence in *Zea mays* of maize streak virus isolates. Virology 288:247-255.

Mawere, S., Vincent, V., De Meyer, J., and Pixley, K. V. 2006. Resistance of four inbred maize lines to inoculation with 20 isolates of *Maize streak virus* from Zimbabwe. Plant Dis. 90:1485-1489.

Rose, D. J. W. 1978. Epidemiology of maize streak disease. Annu. Rev. Entomol. 23:259-282.

Shepherd, D. N., Mangwende, T., Martin, D. P., Bezuidenhoust, M., Kloppers, F. J., Carolissen, C. H., Monjane, A. L., Rybicki, E. P., and Thomson, J. A. 2007. Maize streak virus-resistant transgenic maize: A first for Africa. Plant Biotechnol. J. 5:759-767.

Shepherd, D. N., Martin, D. P., Van Der Walt, E., Dent, K., Varsani, A., and Rybicki, E. P. 2010. Maize streak virus: An old and complex 'emerging' pathogen. Mol. Plant Pathol. 11:1-12.

Soto, P. E., Buddenhagen, I. W., and Asnani, V. L. 1982. Development of streak virus resistant maize populations through improved challenge and selection methods. Ann. Appl. Biol. 100:539-546.

Storey, H. H. 1928. Transmission studies of maize streak disease. Ann. Appl. Biol. 15:1-25.

Tang, C. Y., and Bjarnason, M. S. 1993. Two approaches for the development of maize germplasm resistant to maize streak virus. Maydica 38:301-307.

Varsani, A., Shepherd, D. N., Monjane, A. L., Owor, B. E., Erdmann, J. B., Rybicki, E. P., Peterschmitt, M., Briddon, R. W., Markham, P. G., Oluwafemi, S., Windram, O. P., Lefeuvre, P., Lett, J.-M., and Martin, D. P. 2008. Recombination, decreased host specificity and increased mobility may have driven the emergence of maize streak virus as an agricultural pathogen. J. Gen. Virol. 89:2063-2074.

(Prepared by R. Louie, L. R. Stewart, and M. G. Redinbaugh)

Maize Stripe

Maize stripe was first described from East Africa and has been reported from several African countries (Egypt, Nigeria, Sao Tome, Kenya, and Botswana), Costa Rica, the United States (Florida), South America (Venezuela and Peru), Guadeloupe, western Indian Ocean islands (Mauritius and Reunion), and Australia. The disease is known as *hoja blanca* or Tsai's disease in some locations. Maize stripe is caused by *Maize stripe virus* (MStV) and is likely present in all tropical regions throughout the range of the vector, *Peregrinus maidis*. Maize stripe is sometimes confused with maize mosaic because of their similar symptoms, vector associations, and host range. Coinfection of MStV and *Maize mosaic virus* is common.

Symptoms

Initial symptoms are numerous spots and narrow streaks that appear at the leaf base and extend outward on the youngest leaves. Chlorosis of secondary veins is similar to that of maize mosaic. Later, severe chlorotic striping and streaking coalesce and become wide chlorotic bands, and there is apical bending (Fig. 232). Chlorotic patterns involve the entire leaf blade of later-developing leaves. Severely infected plants are stunted and may die prematurely; yields are greatly reduced.

Causal Agent

Maize stripe virus (MStV) is a member of the plant virus genus *Tenuivirus*, unassigned family, with long, fine filamentous "panhandle" particles ranging from about 3–12 nm wide and up to 900 nm long, consisting of a single capsid protein; the genome contains five single-stranded RNA molecules. There can be large amounts of noncapsid protein in infected tissues.

Vectors

MStV is transmitted in a persistent propagative manner by *Peregrinus maidis*, the corn planthopper. Both nymphs and adults transmit the virus. Nymphs can transmit the virus with high efficiency after an acquisition access period of 1–8 days, and the minimum acquisition period is 4 h. The minimum incubation period is 4–5 days, and the retention period is up to 16 days for nymphs and up to 7 days for adults. Transovarial transmission occurs.

Fig. 232. Leaves with maize stripe (top and bottom); healthy leaf in the center. (Courtesy R. Louie)

Disease Cycle and Epidemiology

MStV naturally infects corn, sorghum, and itchgrass (*Rottboellia exaltata*). It can be experimentally transmitted to barley, oats, rye, triticale, and teosinte. The virus is not seedborne and generally not sap transmissible.

Management

Resistant germplasm has been reported. Destruction of vectors and weed hosts reduces incidence of the disease. Integrated pest management for *P. maidis* can reduce disease spread. Roguing infected plants can be effective in nurseries.

Selected References

Dintinger, J., Boissot, N., Chiroleu, F., Hamon, P., and Reynaud, B. 2005. Evaluation of maize inbreds for *Maize stripe virus* and *Maize mosaic virus* resistance: Disease progress in relation to time and the cumulative number of planthoppers. Phytopathology 95:600-607.

Gingery, R. E. 1985. Maize stripe virus. Descriptions of Plant Viruses, No. 300. Association of Applied Biologists, National Vegetable Research Station, Wellesbourne, England.

Lapierre, H., and Signoret, P.-A., eds. 2004. Viruses and Virus Diseases of *Poaceae* (*Gramineae*). INRA, Paris.

Nault, L. R., and Gordon, D. T. 1988. Multiplication of maize stripe virus in *Peregrinus maidis*. Phytopathology 78:991-995.

(Prepared by L. R. Stewart and R. Louie)

High Plains Disease

High Plains disease was first reported in the High Plains region of the United States (Texas, Utah, Colorado, Idaho, Nebraska, Kansas) in sweet corn, dent corn, blue corn, and wheat in 1993 to 1994 by Jensen et al. (1996), and the viral etiology of High Plains disease was described by 2006. Economically important hosts are corn and wheat.

Disease incidence varies from location to location and from year to year. The disease is broadly distributed in the High Plains of the United States, suggesting that this disease has been present much longer than it has been described.

Symptoms

High Plains disease in corn often appears in conjunction with *Wheat streak mosaic virus* infection. A new agent was suspected of causing the disease, because symptoms were much more severe than those of *Wheat streak mosaic virus* alone. As for most virus diseases, symptom severity is variable depending on the time of inoculation and host genotype. Susceptible field and sweet corn exhibit stunting (considered by the first descriptors to be more severe than *Maize dwarf mosaic virus* or *Maize chlorotic dwarf virus* stunting), severe chlorosis, and a flecking or streaking mosaic (Figs. 233 and 234). The losses are greatest for sweet corn but significant for dent corn. Additional symptoms of High Plains disease in corn include pronounced chlorosis at the basal parts of older leaves and reddening to necrosis of leaf tips and margins progressing downward. Symptoms also include discrete, small white spots in linear clusters or rows associated with the veins. Symptoms in wheat are small chlorotic spots that expand to a general yellowing that is not readily distinguished from other viral diseases in this host.

Symptoms are variable and include mosaics, chlorosis and necrosis, and red striping on maize. It is also suggested that mite feeding exacerbates symptoms and might be responsible for pinpoint chlorotic spots on wheat and maize leaves and that mixed infections with *Wheat streak mosaic virus* (which has the same vector and overlapping distribution) is more severe than with *Wheat streak mosaic virus* alone.

Diagnostic methods include enzyme-linked immunosorbent assays (ELISA) using antisera and analyses by sodium dodecyl sulfate-polyacrylamide gel electrophoresis (SDS-PAGE) for the unique 32-kDa protein associated with infection and reverse-transcription polymerase chain reaction (RT-PCR).

Causal Agent

The causal agent of High Plains disease is *Wheat mosaic virus* (WMoV) (also referred to as *Maize red stripe virus* and *High Plains virus* in early literature) in the genus *Emaravirus*. This genus is similar to viruses in the family *Bunyaviridae* but is distinguished by a four-segment structure. Like other bunya-like plant viruses, such as *Fig mosaic virus* and *European mountain ash ringspot-associated virus*, the High Plains disease agent is transmitted by eriophyid mites and is associated with double-membrane bound bodies in infected cells. Initially, the inability to separate the effects of mite feedings from the necessity to have mites as vectors for the transmission of High Plains disease precluded the demonstration of Koch's postulates. Mechanical transmission of High Plains disease in the absence of mites is now possible by vascular puncture inoculation and has resolved the uncertainty of the viral etiology of the causal agent.

The characterized Nebraska isolate of WMoV has eight negative-sense single-stranded RNAs in the genome, found in double-membraned bodies.

Fig. 233. Stunting caused by High Plains disease. (Courtesy R. Hammon)

Fig. 234. High Plains disease. (Courtesy R. Louie)

Vectors

WMoV is transmitted by the wheat curl eriophyid mite, *Aceria tosichella*. Epiphytotics in corn are highly dependent on migrations of viruliferous mites from infected wheat to corn. Wheat curl mites also transmit *Wheat streak mosaic virus* (WSMV), an unrelated virus in the family *Potyviridae*, and is frequently found in coinfections (see the Wheat Streak Mosaic section).

Management

Genetic variability in maize for resistance to High Plains disease has been reported. Marçon et al. (1997) tested 30 dent corn inbred lines and found resistance to WMoV–WSMV coinfection. Resistance to WMoV–WSMV coinfection is associated with chromosomes 3 and 6 in B73. Near-isogenic lines carrying *Scmv1* and *Scmv2* derived from FAP1360A, or *Wsm1* derived from Pa405 in a susceptible background, were resistant to WMoV. Results indicate that resistance to WMoV is found at the same loci as potyvirus resistance.

Disease management is most important where winter wheat and corn are grown in adjacent fields. The mite vector is suspected to move from mature diseased wheat in the spring to corn seedlings. Systemic insecticide treatment of wheat has limited effectiveness and is not generally economical. Planting time relative to wheat maturity in adjacent plots is an important factor in High Plains disease incidence. Corn is most susceptible when planted within 10–30 days after winter wheat heading. Older corn plants are less susceptible.

Selected References

Fritts, D. A., Michels, G. J., Jr., and Rush, C. M. 1999. The effects of planting date and insecticide treatments on the incidence of High Plains disease in corn. Plant Dis. 83:1125-1128.

Jensen, S. G., Lane, L. C., and Seifers, D. L. 1996. A new disease of maize and wheat in the high plains. Plant Dis. 80:1387-1390.

Lebas, B. S. M., Ochoa-Corona, F. M., Elliott, D. R., Tang, Z., and Alexander, B. J. R. 2005. Development of an RT-PCR for High Plains virus indexing scheme in New Zealand post-entry quarantine. Plant Dis. 89:1103-1108.

Louie, R., Seifers, D. L., and Bradfute, O. E. 2006. Isolation, transmission and purification of the High Plains virus. J. Virol. Methods 135:214-222.

Lubberstedt, T., Ingvardsen, C., Melchinger, A. E., Xing, Y., Salomon, R., and Redinbaugh, M. G. 2006. Two chromosome segments confer multiple potyvirus resistance in maize. Plant Breed. 125:352-356.

Marçon, A., Kaeppler, S. M., and Jensen, S. G. 1997. Genetic variability among maize inbred lines for resistance to the High Plains virus–Wheat streak mosaic virus complex. Plant Dis. 81:195-198.

Marçon, A., Kaeppler, S. M., Jensen, S. G., Senior, L., and Stuber, C. 1999. Loci controlling resistance to High Plains virus and wheat streak mosaic virus in a B73 × Mo17 population of maize. Crop Sci. 39:1171-1177.

Mielke-Ehret, N., and Muhlback, H. P. 2012. *Emaravirus*: A novel genus of multipartite, negative strand RNA plant viruses. Viruses 4:1515-1536.

Seifers, D. L., Harvey, T. L., Martin, T. J., and Jensen, S. G. 1997. Identification of the wheat curl mite as the vector of the high plains virus of corn and wheat. Plant Dis. 81:1161-1166.

Seifers, D. L., Harvey, T. L., Martin, T. J., and Jensen, S. G. 1998. A partial host range of the High Plains virus of corn and wheat. Plant Dis. 82:875-879.

Seifers, D. L., She, Y.-M., Harvey, T. L., Martin, T. J., Haber, S., Ens, W., Standing, K. G., Louie, R., and Gordon, D. T. 2004. Biological and molecular variability among *High Plains virus* isolates. Plant Dis. 88:824-829.

Skare, J. M., Wijkamp, I., Denham, I., Rezende, J. A. M., Kitajima, E. W., Park, J.-W., Desvoyes, B., Rush, C. M., Michels, G., Scholthof, K,-B. G., and Scholthof, H. B. 2006. A new eriophyid mite-borne membrane-enveloped virus-like complex isolated from plants. Virology 347:343-353.

Tatineni, S., McMechan, A. J., Wosula, E. N., Wegulo, S. N., Graybosch, R. A., French, R., and Hein, G. L. 2014. An eriophyid mite-transmitted plant virus contains eight genomic RNA segments with unusual heterogeneity in the nucleocapsid protein. J. Virol. 88:11834-11845. doi: 10.1128/JVI.09101-14

(Prepared by L. R. Stewart and R. Louie)

Maize Rough Dwarf and Other *Fijivirus* Diseases

The discovery of maize rough dwarf in corn in Italy coincided with the introduction of corn from the United States into Italian agriculture as part of the postwar reconstruction effort in the 1940s. Originally named *nanismo del mais*, it was later renamed maize rough dwarf when the viral nature of the disease was demonstrated. Maize rough dwarf and similar or related diseases have since been reported in Europe (France, Italy, Spain, Yugoslavia, the former Czechoslovakia, Italy, and Switzerland) and in the Middle East (Iran). In China and Korea, maize rough dwarf is caused by the serologically related and molecularly similar *Rice black streak dwarf virus* (RBSDV). In South America (Argentina and Brazil), *mal de Rio Cuarto* incited by a related fijivirus (*Mal de Rio Cuarto virus* [MRCV]) causes losses in maize.

The most economically significant host of *Maize rough dwarf virus* (MRDV) is corn, but barley, oats, wheat, sorghum, and rice have been shown to be susceptible to the virus in experimental planthopper inoculations. RBSDV causes extensive problems in rice as well as corn, and MRCV causes losses in wheat and corn. There are recent news reports of maize rough dwarf emergence in Africa.

Symptoms

Maize rough dwarf symptoms in young, field-grown corn at about 1–2 weeks after emergence consist of irregularly shaped swellings (rarely exceeding 2 mm in length) protruding along the lower surfaces of the leaf blades, leaf sheaths, ligules, and husks (Figs. 235 and 236). Abnormal proliferation of the phloem elements resulting in veinal hyperplasia causes these enations on the undersides of leaves. Short, chlorotic streaks up to 5 mm long develop in the mesophyll and veins of mature leaves and coalesce into yellowish green stripes parallel to the veins. The internodes are shortened and fewer in number, resulting in stunted plants with increased girth (Figs. 237 and 238); stunted plants may die prematurely. Plants often develop a darkened green color, which is most pronounced for RBSDV. Tassels are sterile (Fig. 239), and many of the ovules at the

Fig. 235. Maize rough dwarf. (Courtesy R. Louie)

Fig. 236. Galls on leaves caused by *Maize rough dwarf virus*. (Courtesy G. P. Munkvold)

apical ends of the ears atrophy and are unfertile. Leaf margins may have a tattered appearance, and leaves may be deformed or truncated, especially with MRCV (Fig. 239). The root system is reduced to less than half the normal volume (Fig. 240), and roots are grayish because of superficial lesions that form longitudinal fissures. The first symptoms of infection in greenhouse-grown plants (10–12 days after vascular puncture inoculation of corn seeds) are very short, chlorotic streaks on the tertiary veins. Enations develop 2–3 weeks after inoculation.

Causal Agents

Maize rough dwarf virus (MRDV), *Rice black streak dwarf virus* (RBSDV), and *Mal de Rio Cuarto virus* (MRCV) are members of the genus *Fijivirus* in the family *Reoviridae* and have double-stranded RNA genomes. The 70-nm icosahedral virions are nonenveloped and consist of a double-shelled protein capsid with external spikes (A spikes) containing 10 double-stranded RNA viral genome segments. The outer shell is easily degraded in vitro, leaving a B-spiked subviral particle.

Fig. 237. Stunting caused by *Mal de Rio Cuarto virus*. (Courtesy G. P. Munkvold)

Fig. 238. *Mal de Rio Cuarto virus*. (Courtesy G. P. Munkvold)

Fig. 239. Sterile tassel and leaf deformation caused by *Mal de Rio Cuarto virus*. (Courtesy G. P. Munkvold)

Fig. 240. Root system of a healthy plant (left) compared with the reduced root system of a *Mal de Rio Cuarto virus*-infected plant (right). (Courtesy G. P. Munkvold)

Vectors

Under field conditions, MRDV and RBSDV are transmitted by the delphacid planthopper, *Laodelphax striatellus*. Experimental vectors of MRDV include *Javesella pellucida* and *Sogatella vibix*. RBSDV is also transmitted by *Unkanodes* spp. MRCV is transmitted by *Delphacodes kuscheli* and *D. haywardii*. Under optimal conditions, the planthopper acquires MRDV in 1 day. After a latent period of 10–15 days, it transmits the virus after an inoculation access period of 5 h. MRDV propagates in the vector and is also pathogenic to it. The virus can also be transmitted by planthoppers injected with purified virus preparations or with sap from infected plants. MRDV can be mechanically transmitted by syringe injection into stalks of young corn seedlings and by vascular puncture inoculation of corn seeds.

Management

Early-maturing cultivars have higher levels of disease incidence and severity. Under field conditions, there is a higher incidence of disease among U.S. corn hybrids than among local types, although both are very susceptible in experimental inoculations. Thus, there may be some effect on the probing or feeding behavior of the planthopper by different genotypes. Maize inbred lines with resistance to MRDV, RBSDV, and MRCV have been identified, and quantitative trait loci for fijivirus resistance in maize have recently been identified.

Selected References

Boccardo, G., and Milne, R. G. 1975. The maize rough dwarf virion. I. Protein composition and distribution of RNA in different viral fractions. Virology 68:79-85.

Di Renzo, M. A., Bonamico, N. C., Diaz, D. G., Ibanez, M. A., Faricelli, M. E., Balzarini, M. G., and Salerno, J. C. 2004. Microsatellite markers linked to QTL for resistance to Mal de Rio Cuarto disease in *Zea mays* L. J. Agric. Sci. 142:289-295.

Distefano, A. J., Conci, L. R., Hidalgo, M. M., Guzman, F. A., Hopp, H. E., and del Vas, M. 2002. Sequence analysis of genome segments S4 and S8 of Mal de Rio Cuarto virus (MRCV): Evidence that the virus should be a separate *Fijivirus* species. Arch. Virol. 147:1699-1702.

Harpaz, I. 1972. Maize Rough Dwarf, A Planthopper Virus Disease Affecting Maize, Rice, Small Grains and Grasses. Israel Universities Press, Jerusalem.

Lenardon, S. L., March, G. J., Nome, S. F., and Ornaghi, J. A. 1998. Recent outbreak of "Mal de Rio Cuarto" virus on corn in Argentina. Plant Dis. 82:448.

Li, A. H., Pan, C. H., Wu, L. B., Dai, Z. Y., Zuo, S. M., Xiao, N., Yu, L., Li, Y. H., Zhang, X. X., Xue, W. X., Zhang, H. X., and Pan, X. B. 2013. Identification and fine mapping of *qRBSDV-6*^{MH}, a major QTL for resistance to rice black-streaked dwarf virus disease. Mol. Breed. 32:1-13.

Louie, R., and Abt, J. J. 2004. Mechanical transmission of maize rough dwarf virus. Maydica 49:231-240.

Lovisolo, O. 1971. Maize rough dwarf virus. Descriptions of Plant Viruses, No. 72. Commonwealth Mycological Institute and Association of Applied Biologists, Kew, England.

Luan, J. W., Wang, F., Li, Y. J., Zhang, B., and Zhang, J. R. 2012. Mapping quantitative trait loci conferring resistance to rice black-streaked dwarf virus in maize (*Zea mays* L.). Theor. Appl. Genet. 125:781-791.

Martin, T., Franchino, J. A., Kreff, E. D., Procopiuk, A. M., Tomas, A., Luck, S., and Shu, G. G. 2010. Major QTLs conferring resistance of corn to Fijivirus. U.S. Patent US2010/0325750.

Milne, R. G., Conti, M., and Lisa, V. 1973. Partial purification, structure, and infectivity of complete maize rough dwarf virus particles. Virology 53:130-141.

Nakkazi, E. 2010. Maize disease threatens to devastate East Africa. Science and Development Network. August 31, 2010. www.scidev.net/en/news/maize-disease-threatens-to-devastate-east-africa.html

Rodriguez Pardina, P. E., Giménez Pecci, M. P., Laguna, I. G., Dagoberto, E., and Truol, G. 1998. Wheat: A new natural host for the Mal de Río Cuarto virus in the endemic disease area, Río Cuarto, Córdoba Province, Argentina. Plant Dis. 82:149-152.

(Prepared by R. Louie, L. R. Stewart, and M. G. Redinbaugh)

Wheat Streak Mosaic

Wheat streak mosaic was first recognized on wheat in 1922 in Nebraska, and the virus was first described in 1937. It is now widely distributed throughout North America and has been reported from eastern and southern Europe, the Middle East (Jordan, Syria, Turkey, and Iran), other parts of Asia (Russia, China, Kazakhstan, and Uzbekistan), Australia, and New Zealand. It is not a major disease of corn, but infection is common in fields adjacent to maturing wheat fields, when vectors spread from wheat into the corn.

Symptoms

Initial symptoms are chlorotic rings, spots, and streaks at the tips of new leaves. Symptoms develop on leaves that emerge after infection. Depending on the germplasm, the spots and streaks later coalesce to form mosaics or mottle patterns that are parallel, delimited by the larger veins (Fig. 241), and often brilliant yellow. Very susceptible lines may have poorly developed ears with few kernels, and plants may be stunted.

Causal Agent

The causal agent is *Wheat streak mosaic virus* (WSMV), genus *Tritimovirus*, family *Potyviridae*. WSMV was previously classified as a *Rymovirus*. Virions are nonenveloped flexuous rods (15 × 700 nm) containing a monopartite single-stranded RNA genome. Pinwheels and aggregates of virions can be observed in infected cells.

Vector

The wheat curl mite, *Aceria tosichella*, is the vector of WSMV. Nymphs, but not adults, acquire the virus in 15 min. Adults are vectors only when the virus is acquired as nymphs. Virus is retained in the mid- and hind-gut of all larval and adult stages, and mites can be infective for as long as 9 days.

Disease Cycle and Epidemiology

WSMV infection of corn is mainly a concern in late-planted corn adjacent to maturing wheat when wind carries the vectors between crops. Hosts include corn, wheat, oats, barley, rye, and numerous grass species. Mite populations build up on young leaves of wheat and grasses and persist on the living plants.

Fig. 241. Wheat streak mosaic. (Courtesy D. G. White)

Mites cannot survive on ripened grain or grasses, but they can only move short distances on their own. Most movement of the vector is by wind. WSMV is seed transmitted in wheat, but this has not been reported in corn.

Management

Resistance is available in commercial corn hybrids. Because any of three resistance genes on chromosomes 3, 6, and 10 provide resistance, many inbred lines are resistant. Late planting of susceptible corn adjacent to maturing wheat should be avoided.

Selected References

Brakke, M. K. 1971. Wheat streak mosaic virus. Descriptions of Plant Viruses, No. 48. Commonwealth Mycological Institute and Association of Applied Biologists, Kew, England.

Sill, W. H., Jr., and del Rosario, M. S. 1959. Transmission of wheat streak mosaic virus to corn by the eriophyid mite, *Aceria tulipae*. Phytopathology 49:396.

Slykhuis, J. T. 1953. Wheat streak mosaic in Alberta and factors related to its spread. Can. J. Agric. Sci. 33:195-197.

(Prepared by L. R. Stewart, R. Louie, and M. G. Redinbaugh)

Diseases Caused by Nematodes

Nematodes, or roundworms, compose a phylum of nonsegmented, pseudocoelomate invertebrates that are among the most numerous and diverse animal groups found in freshwater, marine, and terrestrial environments. Numbers can exceed 10 billion per acre of soil; nonetheless, the importance of nematodes often is overlooked because of their small size (typically less than 2 mm long). Many terrestrial species feed on soil microorganisms or are parasites or predators of other invertebrates, including insect pests. Approximately 10% of species, however, are obligate parasites of plants, causing billions of dollars in losses annually. Plant parasitism has converged from among several diverse nematode clades, with all species sharing a superficially similar feeding structure called a stylet, which usually is hollow. The morphology of the stylet (e.g., length and robustness) and labial region (e.g., degree of sclerotization), along with other digestive and reproductive structures, form the basis of nematode identification. While nematode specimens usually can be resolved to the genus level by characters visible under low magnification, high magnification and specialized training are required for species-level diagnostics.

Many genera of plant-parasitic nematodes, including those that parasitize corn, are globally distributed; species distributions, in contrast, typically exhibit zonal patterns related to specific temperature and soil-type preferences. Active migration occurs through water films around soil particles and, while limited horizontally, can be marked vertically (greater than 1 m within a growing season). Passive movement of nematodes can occur over greater distances and includes both local (e.g., within or between fields) and long-distance (e.g., between geographic regions) dissemination due to wind, water, migratory animals, and human activity (e.g., soil attached to agricultural equipment). Many species exhibit adaptive survival strategies, such as anhydrobiosis, that enhance survival during unfavorable environmental conditions (or in the absence of a host) and promote dissemination.

The life history of most plant-parasitic nematodes is similar, beginning with an egg and passing through four juvenile stages of increasing size and reproductive maturity to the final adult stage, with each stage separated by a molt, in which the old cuticle is shed and replaced by a new cuticle. Juveniles resemble adults in appearance, with the exception of secondary sex organs that are clearly differentiated only in adults. Reproduction is either sexual, with similar numbers of males and females present, or parthenogenetic, with males rare or absent entirely. Reproductive strategy is species specific, and both types of reproduction can be found within a genus and, occasionally, within the same species. Eggs are deposited in the soil or within root tissue by adult females and hatch as either first- or second-stage juveniles. Most species complete their life cycle in 3–4 weeks under optimum conditions, although members of the family Longidoridae (*Longidorus* and *Xiphinema* spp.) may undergo only a single generation per year.

Nematode feeding on plant cells consists of stylet perforation of the cell wall, followed by injection of esophageal gland secretions and ingestion of host cytoplasm. Injury results from mechanical wounding and the enzymatic activity of secretions and often involves interactions with fungal and bacterial pathogens in disease complexes. Nematode parasites of corn exhibit a range of feeding habits that largely can be discerned from the morphology of the stylet and cephalic framework. Ectoparasitic species are characterized by a relatively weak, transparent cephalic framework and inhabit the rhizosphere throughout their life cycle, inserting only their stylets into root tissue. Some of these species have comparatively short stylets and browse on epidermal cells and root hairs (e.g., *Tylenchorhynchus* spp., Fig. 242A), while others possess elongate stylets (e.g., *Belonolaimus* spp., Fig. 242B) and feed for prolonged periods on cortical and vascular tissues, often preferentially at or near root tips. Endoparasitic species generally exhibit a heavily sclerotized cephalic framework, presumably to facilitate root penetration, and either remain migratory while feeding on cortical parenchyma (e.g., *Pratylenchus* spp., Fig. 242C) or establish a specialized vascular feeding cell and adopt a swollen, sedentary lifestyle (e.g., *Meloidogyne* spp., Fig. 242D). A few species exhibit intermediate feeding behaviors (i.e., becoming partially embedded in root tissue) and are referred to as semiendoparasites (e.g., *Helicotylenchus* spp., Fig. 242E; *Hoplolaimus* spp., Fig. 242F). In many endoparasitic and semiendoparasitic species, feeding actually encompasses a range of behaviors depending on the nematode life stage, host plant, and size of the root being fed upon. Nematode feeding behavior also dictates sampling strategy, since ectoparasites can be recovered using standard passive soil extraction techniques such as sieving and centrifugation; whereas root sampling and active extraction using various root incubation methods are required to accurately diagnose nematode problems involving endoparasites. Nematode extraction methods exhibit variable efficiencies and, therefore, knowledge of both the accuracy and precision of population estimates is essential. It is generally unreliable to compare the results from different laboratories and different extraction methods unless extraction efficiencies are well documented.

Symptoms of nematode injury to corn primarily consist of reduced plant growth and vigor, with conspicuous root abnormalities present for some species. Mechanical and enzymatic disruption of root function caused by nematode feeding, endoparasitic migration, or both results in reduced water and nutrient availability and nonspecific aboveground symptoms, such as stunting, chlorosis, and wilting. The most characteristic manifestation of a nematode problem in corn often is the irregular, patchy distribution of symptoms in the field (Fig. 243). Yield losses, however, can take place in the absence of visible aboveground symptoms.

Belowground symptoms can be more diagnostic, but many indicators of nematode feeding, such as swollen or discolored

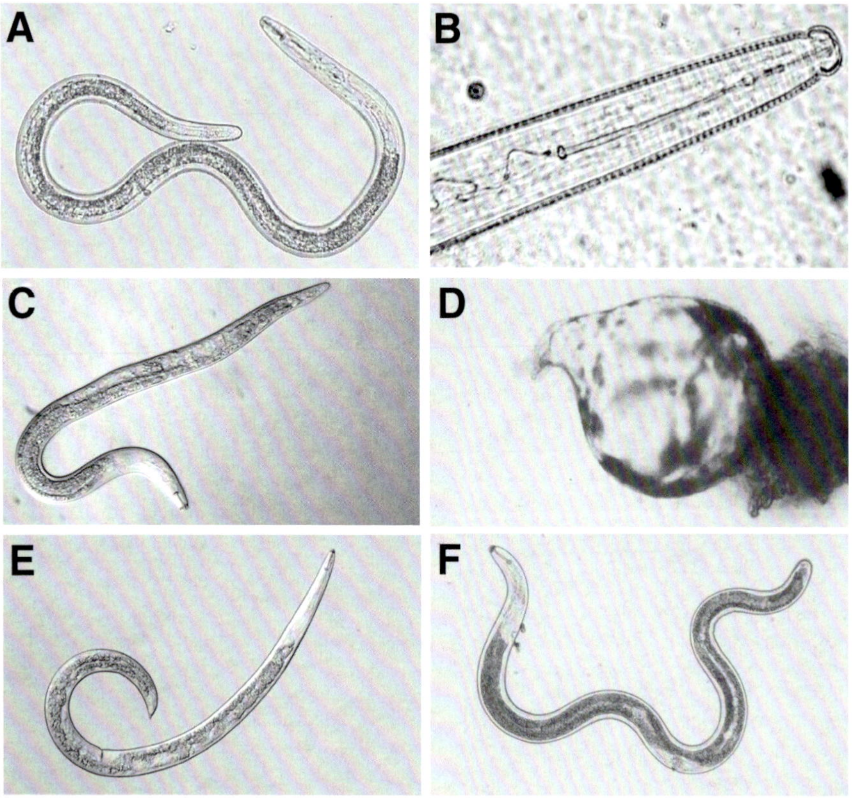

Fig. 242. Nematodes that feed on corn roots. **A,** Stunt nematode (*Tylenchorhynchus* sp.); **B,** sting nematode (*Belonolaimus* sp.); **C,** root lesion nematode (*Pratylenchus* sp.); **D,** root-knot nematode (*Meloidogyne* sp.); **E,** spiral nematode (*Helicotylenchus* sp.); and **F,** lance nematode (*Hoplolaimus* sp.). (Courtesy T. C. Todd)

Fig. 243. Nematode damage in the field (sting nematode). (Courtesy T. Jackson-Ziems)

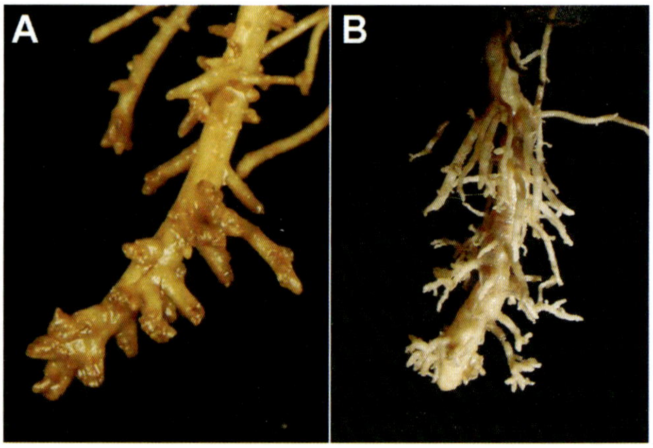

Fig. 244. Root tip damage in corn associated with nematode feeding: **A,** root tip necrosis caused by a *Belonolaimus* sp.; and **B,** stubby root caused by *Paratrichodorus minor.* (Courtesy T. C. Todd)

root tips and the proliferation of lateral roots (Fig. 244), also are characteristic of damage caused by dinitroaniline herbicide carryover, anhydrous ammonia excesses, and aluminum toxicity. Accurate diagnosis requires proper sampling and extraction of nematode populations from both soil and root tissues, followed by a determination of both the kinds (genus or species level of resolution) and numbers of parasitic nematodes present. The number of nematodes for which an economic loss can be attributed is not absolute, however, and varies with climatic and edaphic conditions, as well as sample timing. Under favorable conditions, the rapid growth and development of an extensive fibrous root system allows corn plants to tolerate moderate levels of parasitism by most nematode species without noticeable effects on growth or yield. The complexity of nematode damage relationships for corn has resulted in a wide range of threshold estimates for some nematode species, while threshold values for numerous other species remain to be critically evaluated.

The most recent crop loss estimates for corn-parasitic nematodes are nearly two decades old, but they suggest a total loss of approximately 1% across the United States, with estimates from individual states ranging from less than 1% to 20%. More than 60 species of plant-parasitic nematodes encompassing more than a dozen genera are associated with corn in North America, but only eight genera were specified in a 1994 survey of crop losses (Table 8). The most commonly reported species of corn-parasitic nematodes were members of the genus *Pratylenchus*, the root lesion nematodes. These nematodes were specified by 54% of the reporting states and were associated

TABLE 8. Plant-Parasitic Nematodes Associated with Corn Yield Loss in the United States[a]

Common Name (Genus)	States Reporting Loss (%)	Average Loss (%)[b]	Total Loss (%)[c]
Root lesion (*Pratylenchus*)	54	1.2	0.28
Stubby-root (*Paratrichodorus*)	25	2.2	0.04
Root-knot (*Meloidogyne*)	25	1.1	0.03
Needle (*Longidorus*)	18	0.9	0.27
Lance (*Hoplolaimus*)	18	0.6	0.15
Sting (*Belonolaimus*)	14	3.3	0.07
Dagger (*Xiphinema*)	7	0.3	0.13
Spiral (*Helicotylenchus*)	7	1.1	0.01
Unspecified	21	0.5	0.11
Total			1.12

[a] Based on responses to a 1994 survey of extension personnel in 28 states.
[b] Average loss attributed to each nematode; if multiple nematodes were listed for a given state, then the loss estimate was divided equally among them.
[c] Total yield loss as a percent of U.S. crop value.

with the highest estimated total yield loss. Surveys conducted in the midwestern Corn Belt during 2007 confirm the importance of *Pratylenchus* spp., with a prevalence of 81% observed across nearly 2,200 sampled corn production fields in 16 states. Other nematode genera, such as *Belonolaimus* and *Longidorus*, are capable of causing far greater damage to corn roots than *Pratylenchus* spp., but these nematodes are much more limited in their distribution. Some corn-parasitic nematodes display regional patterns of importance, with *Xiphinema* spp. more often associated with losses in the midwestern Corn Belt, while *Meloidogyne* and *Paratrichodorus* spp. are more often associated with losses in the southeastern United States. Although damage relationships have been described for several of these genera, the risk of damage from other common nematode genera or from nematode complexes remains poorly documented for corn. Additionally, established damage thresholds for individual nematodes often vary widely among states and regions. Nonetheless, there is general consensus regarding the relative pathogenic potential of the various corn-parasitic nematodes, and consequently, recent focus has shifted to developing a relative risk assessment model based on (i) the combined damage potential of all plant-parasitic nematodes present and (ii) site-specific risk factors such as soil sand content (increases risk) and irrigation capacity (lowers risk).

Nematode management strategies for corn traditionally have relied on cultural practices such as crop rotation to limit nematode population increases and on soil-applied carbamate or organophosphate insecticide-nematicides to induce transient reductions in nematode population densities or activity and, therefore, in damage risk. Host resistance has been identified for several nematode species but generally is quantitative in nature and remains to be exploited commercially. The migratory behavior of the most important corn-parasitic nematodes (e.g., root lesion, sting, and needle nematodes) severely compromises the primary basis for commercial resistance to nematode pests in other crops, i.e., resistance based on hypersensitivity. Removal of effective nematicides from the market because of environmental concerns, limitations of remaining management options (e.g., crop rotation versus the polyphagous nature of corn-parasitic nematodes), and reduced funding opportunities resulted in a significant decline in research activity and product development for nematode management in corn. Recently, however, there has been renewed interest in nematode parasites of corn because of concerns about unforeseen impacts on nematode pest populations resulting from changing corn production management practices, including increased utilization of conservation tillage and a reduced reliance on conventional soil insecticide-nematicides. The development of novel seed treatments for nematode control in corn and other crops also has contributed to this resurgence. For example, an abamectin-based seed treatment for early-season control of nematodes became widely available for use on corn in the United States in 2010, followed by the introduction of a bionematicide corn seed treatment for the 2011 growing season. Early efficacy reports for these products are encouraging when damaging levels of one or more nematode species are present, but extensive field testing is needed before a reliable assessment of their potential for managing corn-parasitic nematodes can be made.

Selected References

Jones, K., Long, D., Shetty, K., Watrin, C., Xing, L., and Ulmer, B. 2009. Avicta® brand nematicide for early season nematode management in United States cropping systems. (Abstr.) J. Nematol. 41:342.

Koenning, S. R., Overstreet, C., Noling, J. W., Donald, P. A., Becker, J. O., and Fortnum, B. A. 1999. Survey of crop losses in response to phytoparasitic nematodes in the United States for 1994. J. Nematol. 31(Suppl.):587-618.

Nickle, W. R., ed. 1991. Manual of Agricultural Nematology. Marcel Dekker, New York.

Perry, R. N., and Moens, M. 2006. Plant Nematology. CAB International, Wallingford, United Kingdom.

Tylka, G. L., Todd, T. C., Niblack, T. L., MacGuidwin, A. E., and Jackson, T. 2011. Sampling for plant-parasitic nematodes in corn strip trials comparing nematode management products. Online. Plant Health Progress doi: 10.1094/PHP-2011-0901-01-DG

Windham, G. L. 1998. Corn. Pages 335-357 in: Plant and Nematode Interactions, Agronomy Monograph No. 36. K. R. Barker, G. A. Pederson, and G. L. Windham, eds. American Society of Agronomy, Crop Science Society of America, Soil Science Society of America, Madison, WI.

(Prepared by T. C. Todd)

Root Lesion Nematodes

Root lesion nematodes are more prevalent in corn fields than any other type of plant-parasitic nematode and likely account for the majority of the crop injury. At least six species are associated widely with corn in North America. *Pratylenchus hexincisus*, *P. neglectus*, *P. penetrans*, and *P. scribneri* are reported commonly from the midwestern Corn Belt, while *P. brachyurus* and *P. zeae* are prevalent in the southeastern United States.

Description

The genus *Pratylenchus* is characterized by a strongly sclerotized labial framework, an adaptation that reflects the group's endoparasitic behavior, and a short (approximately 16 µm long) but robust stylet (stomatostyle) with well-developed knobs (Fig. 242C). The glands of the esophagus form a lobe overlapping the intestine ventrally, and the reproductive system of the adult female consists of a single anteriorly directed gonad with a posterior vulva.

Species identification requires high magnification and is extremely difficult because of the limited number of diagnostic characters, many of which exhibit considerable intraspecific variability. Morphological characters used to distinguish *Pratylenchus* spp. include number of labial annules, stylet length, vulva ratio, length of esophageal gland overlap, and the number of lateral field lines. Reproductive strategy as indicated by the presence or absence of males is an additional diagnostic trait. Although species identification in this genus continues to be based on morphology, molecular diagnostic procedures, such as the comparative analysis of ribosomal DNA sequences, are becoming increasingly important.

Symptoms

Aboveground symptoms of root lesion nematode damage are nonspecific and may include reduced stalk height and diameter, nutrient deficiencies, chlorosis, and drought stress. These symptoms often remain inconspicuous, however, and can easily be overlooked. The most frequently ascribed symptom of *Pratylenchus* damage is the presence of root lesions, which may be discolored or necrotic (Fig. 245). Intracellular migration by the nematode destroys cortical cells, and large infestations may lead to sloughing of the cortical tissue of infected roots. Predisposition of damaged root tissues to secondary bacterial and fungal infections is commonly reported, and synergistic disease interactions between *P. penetrans* and several seedling pathogens, including *Fusarium verticillioides* and *Pythium ultimum*, have been documented for corn.

Biology

The life cycle of *Pratylenchus* spp. is typical for migratory plant-parasitic nematodes, with all juvenile stages outside of the egg (second through fourth) and the adults functioning as infective stages. Little information is available about life cycle duration under field conditions, but based on laboratory studies, the time to complete the life cycle on corn varies from 3 to 5 weeks, depending on temperature and nematode species. Optimum temperature for reproduction on corn ranges from 24°C for *P. penetrans* to 30°C for *P. hexincisus*. Reproductive strategy is species specific, with roughly equal numbers of examples of sexual reproduction (with males common) and parthenogenesis (with males rare).

Root lesion nematodes are classified as root endoparasites, although ectoparasitic feeding also occurs. Invasion can take place along the entire length of the root. Both invasion and migration through the cortex are thought to involve a combination of mechanical (stylet thrusting) and enzymatic (secretion of cell wall-degrading enzymes) processes. Nematode population densities increase rapidly in the seminal roots during the first month after seedling emergence, with subsequent increases primarily restricted to the nodal roots. Populations remain concentrated in the roots until host senescence, when the relative proportion of nematodes in the soil increases again. Research has demonstrated the need for assessment of root lesion nematode population densities from both soil and root habitats, regardless of the time of year. More than half of the total viable nematode population may be present in dead root fragments prior to corn planting, for example, leading to a nontrivial underestimation of population density if only the soil fraction is considered.

Fig. 245. Root damage caused by the root lesion nematode. (Courtesy University of Illinois)

Pratylenchus spp. distributions exhibit patterns related to both temperature and soil-type preferences. Most *Pratylenchus* spp. commonly associated with corn are favored by sandy soils, but some species, such as *P. hexincisus*, are equally abundant in fine-textured soils. *Pratylenchus* populations in corn fields generally are concentrated in the upper 15–30 cm of the soil profile. Vertical distributions are determined primarily by nematode reproduction, as influenced by soil environmental conditions and root distributions. There is little evidence to indicate that there is substantial vertical migration.

Economic Importance

Damage relationships vary significantly among the *Pratylenchus* spp. found on corn and often with environment. Yield loss is more consistently related to root population densities during the growing season than to soil population densities at planting, especially if dead root fragments are not included in the assay. Significant losses (greater than 5%) in seed yield are associated with early-season root population densities of *P. scribneri* and *P. hexincisus* exceeding 5,000 per g dry root weight, and populations at this level are relatively common across the midwestern Corn Belt. Yield reductions up to 20% have been observed for *P. zeae* on corn and as high as 49% for *P. penetrans* on sweet corn.

Management

The potential for managing root lesion nematodes with genetic resistance is poorly developed, despite the fact that large differences in susceptibility to *P. hexincisus* and *P. scribneri* were documented decades ago for both corn inbred lines and commercial hybrids. Resistance to *P. hexincisus* and *P. scribneri* also has been identified in the perennial teosintes, which support up to 98% lower nematode populations than standard corn lines. No commercial hybrids have been specifically developed for resistance to root lesion nematodes; however, and the degree to which modern cultivars vary in susceptibility is unknown. The stability of resistant reactions across *Pratylenchus* spp. and populations also remains to be determined.

The usefulness of crop rotation as a management tool for root lesion nematodes is severely compromised by the broad host ranges generally exhibited by *Pratylenchus* spp., combined with conflicting or incomplete information on the host status of suitable rotational crops. It is clear that the popular corn–soybean rotation of the midwestern Corn Belt guarantees an uninterrupted sequence of host crops for several of the most common *Pratylenchus* spp. found on corn in that region, although there is evidence that even this rotation may reduce nematode numbers compared with continuous corn production. Effective crop rotations that have been identified are species specific. Alfalfa, for example, has proven to be an effective rotational crop for managing midwestern populations of *P. scribneri*, but its status as a good host for *P. penetrans* limits its usefulness across the region. In the southeastern United States, multiyear rotations with cotton and peanut have been reported to suppress *P. zeae* populations, but both of these crops are good hosts for at least some populations of *P. brachyurus*. Pearl millet exhibits relatively broad promise as a rotational crop for management of several corn-parasitic *Pratylenchus* spp., including *P. brachyurus* in Georgia (United States) and *P. penetrans* in Canada.

Numerous studies have demonstrated the efficacy of preplant applications of carbamate and organophosphate nematicides for controlling root lesion nematodes in corn. Yield responses (and economic returns) are often erratic, however. Reductions of *P. neglectus* and *P. scribneri* population densities by approximately 50% in nematicide trials in western Kansas (United States) have produced yield increases ranging from nearly 20%

across two environments in one study to 0.2% across nine environments in another study. The expense of treatment combined with the moderate damage potential of most *Pratylenchus* spp. generally results in marginal economic return. Additional disadvantages of nematicides include their transient effectiveness, which often allows recovery of nematode populations by the end of the growing season, and the fact that the availability of soil-applied insecticide-nematicides labeled for use on corn has declined in recent years because of environmental concerns. Meanwhile, there have been new developments in protectant seed treatments, and several commercial products are currently available for nematode management in corn. These compounds have fewer environmental issues than conventional nematicides, but information on their efficacy and economics is currently limited. In one preliminary report, abamectin seed treatment was associated with reductions of 25–72% in root lesion nematode population densities in corn roots and increases of 17–69% in corn yields across two locations and 3 years. The effective concentration of abamectin seed treatment for a 50 and 80% reduction in penetration of corn roots by *P. zeae* was determined to be 0.16 and 1.0 mg of a.i. per seed, respectively, under greenhouse conditions.

Selected References

Belair, G., Fournier, Y., Dauphinais, N., and Dangi, O. P. 2002. Reproduction of *Pratylenchus penetrans* on various rotation crops in Quebec. Phytoprotection 83:111-114.

Cabrera, J. A., Kiewnick, S., Grimm, C., Dababat, A. A., and Sikora, R. A. 2009. Efficacy of abamectin seed treatment on *Pratylenchus zeae*, *Meloidogyne incognita* and *Heterodera schachtii*. J. Plant Dis. Prot. 116:124-128.

Castillo, P., and Vovlas, N. 2007. *Pratylenchus* (Nematoda: Pratylenchidae): Diagnosis, Biology, Pathogenicity and Management. Nematology Monographs and Perspectives Volume 6. Brill, Lieden, Netherlands.

Cochran, A., Watrin, C., and Ulmer, B. 2007. Review of nematode protection benefits from abamectin seed treatment in corn. (Abstr.) J. Nematol. 39:78.

da Silva, M. P. 2010. Interactions between lesion nematodes and corn pathogens. M.S. thesis. Iowa State University, Ames, IA.

Lengkeek, V. H., Sanden, G. E., and Lash, L. D. 1981. Efficacy of nematicides in corn in western Kansas. Trans. Kans. Acad. Sci. 84:209-215.

MacGuidwin, A. E. 1989. Distribution of *Pratylenchus scribneri* between root and soil habitats. J. Nematol. 21:409-415.

MacGuidwin, A. E., and Bender, B. E. 2012. Estimating population densities of root lesion nematodes, *Pratylenchus* spp., from soil samples using dual active and passive assays. Online. Plant Health Progress doi: 10.1094/PHP-2012-1120-01-RS

MacGuidwin, A. E., and Stanger, B. A. 1991. Changes in the vertical distribution of *Pratylenchus scribneri* under potato and corn. J. Nematol. 23:73-81.

McSorley, R., and Dickson, D. W. 1989. Effects and dynamics of a nematode community on maize. J. Nematol. 21:462-471.

McSorley, R., and Dickson, D. W. 1990. Vertical distribution of plant-parasitic nematodes in sandy soil under maize. Plant Soil 123:95-100.

Norton, D. C. 1983. Maize nematode problems. Plant Dis. 67:253-256.

Norton, D. C. 1989. Host efficiencies of *Zea diploperennis* and *Z. perennis* for *Pratylenchus* spp. J. Nematol. 21:547-548.

Smolik, J. D., and Evenson, P. D. 1987. Relationship of yields and *Pratylenchus* spp. population densities in dryland and irrigated corn. Ann. Appl. Nematol. 1:71-73.

Smolik, J. D., and Wicks, Z. W. 1987. Reproduction of *Pratylenchus hexincisus* and *P. scribneri* in corn inbreds. Ann. Appl. Nematol. 1:29-31.

Timper, P., and Hanna, W. W. 2005. Reproduction of *Belonolaimus longicaudatus*, *Meloidogyne javanica*, *Paratrichodorus minor*, and *Pratylenchus brachyurus* on pearl millet (*Pennisetum glaucum*). J. Nematol. 37:214-219.

Todd, T. C. 1991. Effect of cropping regime on populations of *Belonolaimus* sp. and *Pratylenchus scribneri* in sandy soil. J. Nematol. 23(Suppl.):646-651.

Todd, T. C., and Oakley, T. O. 1996. Seasonal dynamics and yield relationships of *Pratylenchus* spp. in corn roots. J. Nematol. 28(Suppl.):676-681.

Zirakparvar, M. E., Norton, D. C., and Cox, C. P. 1980. Population increase of *Pratylenchus hexincisus* on corn as related to soil temperature and type. J. Nematol. 12:313-318.

(Prepared by T. C. Todd)

Sting Nematodes

Sting nematodes, belonging to the genus *Belonolaimus*, are among the most economically important nematode parasites of corn and one of the few examples of a plant-parasitic nematode with the capacity to cause total crop yield loss in localized areas. The genus consists of a small group of related species, all of which are found exclusively in North America. The geographical distribution of *Belonolaimus* spp. is severely restricted, with populations confined to soils containing at least 80% sand content. There are only five described species, but morphological as well as molecular evidence suggests that the genus is far more complex, with several unique populations inhabiting the midwestern Corn Belt. *B. longicaudatus* is most commonly associated with damage to corn in the southeastern United States, but recent evidence suggests that *B. longicaudatus*, and perhaps other species, should be considered as species complexes.

Description

Sting nematodes are large, slender nematodes exceeding 2 mm in length. The most diagnostic character for the genus is a slender elongate stylet, which exceeds 100 µm in most species (Fig. 242B). The esophageal glands overlap the intestine and the reproductive system of adult females consists of a median (midbody) vulva with two oppositely directed gonads. Important species diagnostic characters include stylet, tail, and total body lengths of adult females.

Symptoms

Damage caused by sting nematode feeding typically is manifested in the seedling stage, usually within 2–3 weeks after seedling emergence. Aboveground symptoms consist of conspicuous, irregular patches of severely stunted and chlorotic plants that often display symptoms of nutrient deficiencies and drought stress (Fig. 246). As the season progresses, the boundary between stunted and healthy plants becomes sharply

Fig. 246. Severe sting nematode damage in the field. (Courtesy T. Jackson-Ziems)

delineated, with plants in severely infested areas tasseling at heights of a foot or less. Belowground symptoms consist of similarly stunted root systems with necrotic root tips (Figs. 244A and 247). The destruction of the root tips results in a proliferation of lateral roots, which in turn are attacked and suffer the same fate. In less severely infested areas, symptoms of root feeding may be concentrated deeper (e.g., below 20 cm) in the soil profile.

Biology

The life cycle of *B. longicaudatus* is completed in 24 days on excised corn roots at 28°C, and field observations for other *Belonolaimus* spp. suggest that this is a representative generation time for the genus. Developmental rates vary among isolates of *B. longicaudatus*, however, with a range of 18 to 25 days observed for completion of the life cycle at 28°C. Reproduction is sexual, with a sex ratio of approximately 3:2 in favor of females. Eggs are deposited in the soil soon after feeding, followed by hatching of second-stage juveniles 5 days later. Cultured females were observed to produce an average of 124 eggs over a 90-day period.

Fig. 247. Healthy roots (left) and roots damaged by sting nematodes (center and right). (Courtesy M. L. Carson)

Fig. 248. Plants treated (left) and untreated (right) for sting nematodes. (Courtesy M. L. Carson)

Sting nematodes feed as ectoparasites on primary and lateral root tips. The elongate flexible stylet is inserted deeply into the root tissue and ingestion of plant nutrients can proceed for several hours. This feeding behavior produces marked seasonal fluctuations in the vertical distribution of field populations, with nematodes concentrated in the upper 15–30 cm of soil during the corn seedling stage but migrating progressively deeper in the soil profile during the growing season. The majority of late-season populations often are concentrated below normal sampling depths (0–15 cm) and this, in turn, can lead to the erratic diagnosis of sting nematode problems. Increasing sampling depth to at least 30 cm improves recovery reliability and precision from the soil. Both active (e.g., Baermann funnel) and passive (e.g., centrifugation) soil extraction methods are suitable for sting nematodes.

Economic Importance

The economic threshold level for *Belonolaimus* spp. typically is set at the detection limit (i.e., one nematode). The relationship between corn yield and sting nematode population density is linear, with complete yield loss predicted for populations exceeding 30 nematodes per 100 cm^3 of soil at planting or 100 nematodes per 100 cm^3 of soil 1 month after seedling emergence. The appearance of sting nematodes is localized in most corn-producing states, and severe damage usually is localized within individual fields. Nonetheless, economic production in infested fields nearly always requires management of the nematode.

Management

Management of sting nematodes in corn traditionally has relied heavily on applications of granular carbamate and organophosphate nematicides (Fig. 248). Yield increases exceeding 100% have been documented for corn following nematicide applications in infested fields, but yields may remain below acceptable levels in the presence of severe nematode pressure. The availability of effective compounds has declined in recent years because of environmental concerns, and the efficacy of newer protectant seed treatments remains to be adequately demonstrated.

Sting nematodes have broad host ranges, and very few crop species have been confirmed as nonhosts. Tobacco and watermelon are poor hosts for *Belonolaimus* populations in the southeastern United States, and alfalfa has been demonstrated to be comparable to fallow in reducing populations in the midwestern Corn Belt. Soybean is a good to excellent host for populations in both regions. Contradictory results have been reported for peanut and pearl millet. Host ranges vary considerably among populations of *B. longicaudatus*, so caution must be exercised in selecting crop rotation sequences. Effective weed control, especially for grass species, is a critical component of any management strategy for sting nematodes. Host resistance to sting nematodes has not been identified.

Selected References

Cherry, T., Szalanski, A. L., Todd, T. C., and Powers, T. O. 1997. The internal transcribed spacer region of *Belonolaimus* (Nemata: Belonolaimidae). J. Nematol. 29:23-29.

Gozel, U., Adams, B. J., Nguyen, K. B., Inserra, R. N., Giblin-Davis, R. M., and Duncan, L. W. 2006. A phylogeny of *Belonolaimus* populations in Florida inferred from DNA sequences. Nematropica 36:155-171.

Han, H.-R., Dickson, D. W., and Weingartner, D. P. 2006. Biological characterization of five isolates of *Belonolaimus longicaudatus*. Nematropica 26:25-35.

Huang, X., and Becker, J. O. 1997. In vitro culture and feeding behavior of *Belonolaimus longicaudatus* on excised *Zea mays* roots. J. Nematol. 29:411-415.

Huang, X., and Becker, J. O. 1999. Life cycle and mating behavior of *Belonolaimus longicaudatus* in gnotobiotic culture. J. Nematol. 31:70-74.

McSorley, R., and Dickson, D. W. 1989. Effects and dynamics of a nematode community on maize. J. Nematol. 21:462-471.
McSorley, R., and Dickson, D. W. 1990. Vertical distribution of plant-parasitic nematodes in sandy soil under maize. Plant Soil 123:95-100.
Smart, G. C., and Nguyen, K. B. 1991. Sting and awl nematodes: *Belonolaimus* spp. and *Dolichodorus* spp. Pages 627-667 in: Manual of Agricultural Nematology. W. R. Nickle, ed. Marcel Dekker, New York.
Timper, P., and Hanna, W. W. 2005. Reproduction of *Belonolaimus longicaudatus*, *Meloidogyne javanica*, *Paratrichodorus minor*, and *Pratylenchus brachyurus* on pearl millet (*Pennisetum glaucum*). J. Nematol. 37:214-219.
Todd, T. C. 1989. Population dynamics and damage potential of *Belonolaimus* sp. on corn. J. Nematol. 21(Suppl.):697-702.
Todd, T. C. 1991. Effect of cropping regime on populations of *Belonolaimus* sp. and *Pratylenchus scribneri* in sandy soil. J. Nematol. 23(Suppl.):646-651.

(Prepared by T. C. Todd)

Needle Nematode

The corn needle nematode, *Longidorus breviannulatus*, is widely distributed in sandy soils throughout the midwestern Corn Belt. It is found only in soils with greater than 50% sand content, with the highest populations and greatest damage restricted to soils composed of more than 90% sand. At high populations (more than 100 nematodes per 100 cm^3 of soil), *L. breviannulatus* is capable of killing seedling corn plants.

Description

Adult females of *L. breviannulatus* average nearly 5 mm in length, making this the largest species of corn-parasitic nematode. The stylet, called an odontostyle, is long and slender, averaging 85 μm in length, and lacks basal swellings or knobs. The reproductive system consists of a median (midbody) vulva with two oppositely directed, reflexed ovaries.

Symptoms

Aboveground symptoms include severe stunting, chlorosis, and occasional mortality of seedling corn plants. Damage characteristically is in irregular patches that are evident within 2 weeks after seeding emergence and remain pronounced during the first 6–8 weeks after planting (Figs. 249 and 250). Affected plants display symptoms of nutrient deficiencies and drought stress. Plants may recover in height by the end of the growing season but stalks remain slender and ears are absent or greatly reduced in size.

Belowground, the root tips of the primary seminal and nodal roots appear swollen and discolored, although necrosis usually is absent (Fig. 251). Root tip injury often results in a proliferation of enlarged stubby lateral roots with few feeder roots, giving the root system a coarse appearance. Symptoms on nodal roots produced at later growth stages generally appear at progressively greater depths.

Biology

Needle nematodes feed as ectoparasites on primary and lateral root tips. The details of the life cycle of the corn needle nematode remain to be described, but males are extremely rare, so reproduction is assumed to be parthenogenetic. Seasonal

Fig. 249. Severe needle nematode damage in the field. (Courtesy T. Jackson-Ziems)

Fig. 250. Stunting caused by needle nematodes. (Courtesy L. R. Krusberg)

Fig. 251. Root damage caused by needle nematodes. (Courtesy D. G. White)

vertical migration results in the concentration of nematode populations in the top 15 cm of soil after seedling emergence, followed by concentrations at progressively greater depths throughout the season. Detection at normal sampling depths (15–20 cm) often becomes unreliable by the end of the growing season. In addition to sand content, irrigation is a major determinant for the distribution and population dynamics of *L. breviannulatus*, with the highest population densities found in irrigated fields. Soil moisture also appears to influence the vertical distribution of this nematode during the growing season. As for other root tip-feeding nematodes, increasing sampling depth to at least 30 cm improves the reliability and precision of recovery from the soil. Both active (e.g., Baermann funnel) and passive (e.g. centrifugation) soil extraction methods are suitable for needle nematodes, although individuals often are collected on sieves with larger openings (e.g., 250 µm) than those traditionally used for nematodes (38–44 µm). The use of 250-µm-opening sieves for collecting sample debris can, if discarded, result in erratic detection of needle nematodes.

Economic Importance

Stunting of seedling corn has been observed with as few as 10 needle nematodes per 100 cm^3 of soil, but the action threshold probably should be set at the detection limit (i.e., one nematode). Yield reductions up to 62% have been reported for localized areas.

Management

Information on the efficacy of available nematicidal compounds against *L. breviannulatus* is limited and inconsistent. Rotation with leguminous crops, such as soybean and alfalfa, is the most frequently recommended management strategy. Effective weed control, especially for grass species, is essential. No resistant corn hybrids are available.

Selected References

MacGuidwin, A. E. 1989. Abundance and vertical distribution of *Longidorus breviannulatus* associated with corn and potato. J. Nematol. 21:404-408.

Malek, R. B., Norton, D. C., Jacobsen, B. J., and Acosta, N. 1980. A new corn disease caused by *Longidorus breviannulatus* in the Midwest. Plant Dis. 64:1110-1113.

Norton, D. C., and Hoffman, J. K. 1975. *Longidorus breviannulatus* n. sp. (Nematoda: Longidoridae) associated with stunted corn in Iowa. J. Nematol. 7:168-171.

(Prepared by T. C. Todd)

Stubby-Root Nematodes

Paratrichodorus minor is widespread in sandy soils throughout the United States, although it is more commonly reported as a serious pest of corn in the southeastern coastal plain region. The highest population densities are found in soils with greater than 80% sand content. *P. porosus* also has been associated with damage to corn in the southeastern United States.

Description

Stubby-root nematodes are cigar shaped, with a distinctively curved stylet, called an onchiostyle, which lacks basal knobs. The reproductive system of adult females consists of a median (midbody) vulva with two oppositely directed, reflexed ovaries. Species identification is based largely on reproductive morphology, including that of males, which are rare in corn-parasitic species.

Symptoms

Aboveground symptoms are characteristic of seedling root damage and include stunting and drought sensitivity. Belowground symptoms are more distinctive, consisting of swollen root tips and a proliferation of lateral roots, resulting in a condition known as "stubby root" (Figs. 244B and 252). Nematode feeding at or near root tips causes a cessation of mitosis and root elongation, with little evidence of additional pathogenic effects, such as mechanical injury and necrosis.

Biology

Stubby-root nematodes feed as ectoparasites at or near root tips. The life cycle of *P. minor* is completed in 21–22 days on corn at 22°C but requires only 16–17 days for completion at 30°C. Males are rare in field populations and reproduction is assumed to be parthenogenetic. Populations can exhibit marked seasonal variation, with the highest population densities typically observed in midseason. Vertical distribution is similarly erratic, although populations often are concentrated at soil depths below 30 cm. Increasing sampling depth to at least 30 cm improves the reliability and precision of recovery of root tip-feeding nematodes from the soil. Both active (e.g., Baermann funnel) and passive (e.g., centrifugation) soil extraction methods are commonly used for stubby-root nematodes.

Economic Importance

Stubby-root nematodes are capable of causing severe injury to corn but the population densities required are rarely attained. Moderate injury to corn is more common. Damage threshold estimates for these nematodes vary widely, but population densities of 100–200 nematodes per 100 cm^3 of soil generally are reported to represent some risk of economic damage.

Management

More than 100 species of plants are known hosts of *P. minor*, but effective crop rotations do exist. In the southeastern United States, peanut generally is considered to be a poor host of *P. minor* and has been observed to suppress population densities of this nematode in the field, and while increases in *P. minor* population densities have been observed on soybean, this crop generally supports lower population densities of the nematode than corn. Caution must be exercised, however, because considerable morphological and physiological variation exists among *P. minor* populations. Nematode reproduction also varies among crop genotypes. For example, the pearl mil-

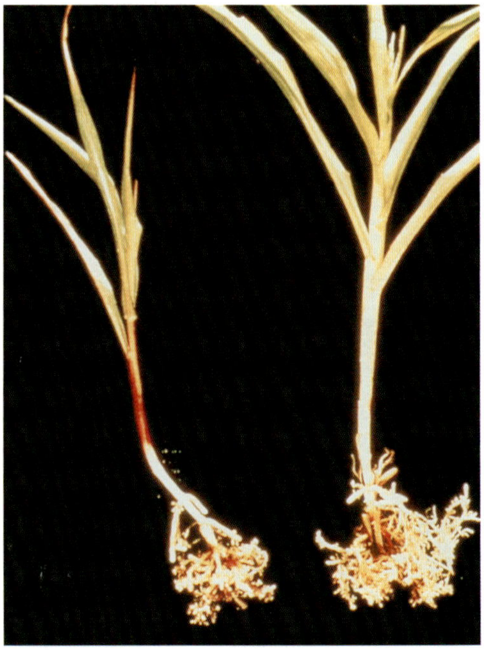

Fig. 252. Root damage caused by the stubby-root nematode. (Courtesy R. S. Hussey)

let hybrid Tifgrain 102 was reported to be equivalent to peanut in host status, while reproduction on the hybrid HGM-100 was equivalent to that on corn. Moderate resistance to *P. minor* also was observed for 4 of 33 commercial corn hybrids evaluated in Georgia (United States), but none of the 22 public inbred lines tested displayed any resistance.

Stubby-root nematodes are reported to be sensitive to soil disturbance and desiccation; tillage, therefore, is a frequently recommended management strategy. Effective control of stubby-root nematodes in corn also has been demonstrated using carbamate and organophosphate nematicides.

Selected References

Decraemer, W. 1991. Stubby root and virus vector nematodes: *Trichodorus*, *Paratrichodorus*, *Allotrichodorus*, and *Monotrichodorus*. Pages 587-625 in: Manual of Agricultural Nematology. W. R. Nickle, ed. Marcel Dekker, New York.

Johnson, A. W., Dowler, C. C., and Hauser, E. W. 1975. Crop rotation and herbicide effects on population densities of plant-parasitic nematodes. J. Nematol. 7:159-168.

McSorley, R., and Dickson, D. W. 1990. Vertical distribution of plant-parasitic nematodes in sandy soil under maize. Plant Soil 123:95-100.

Timper, P., and Hanna, W. W. 2005. Reproduction of *Belonolaimus longicaudatus*, *Meloidogyne javanica*, *Paratrichodorus minor*, and *Pratylenchus brachyurus* on pearl millet (*Pennisetum glaucum*). J. Nematol. 37:214-219.

Timper, P., Krakowsky, M. D., and Snook, M. E. 2007. Resistance in maize to *Paratrichodorus minor*. Nematropica 37:9-20.

(Prepared by T. C. Todd)

Lance Nematodes

Hoplolaimus galeatus is widely distributed in the United States and has been implicated in corn yield losses in the midwestern Corn Belt. This relationship has not been documented for the southeastern United States, however. This species generally is associated with coarse-textured soils, with sandy loam soils providing the optimum environment for nematode reproduction. *H. columbus* is a serious pest of cotton and soybean in the southeastern coastal plain, but corn appears to be relatively tolerant to this nematode.

Description

Lance nematodes are large (1–2 mm long), robust nematodes with a massive labial framework and stylet, 40–50 µm long (Fig. 242F). Esophageal glands overlap the intestine dorsally and laterally, and the female reproductive system consists of a median (midbody) vulva with two oppositely directed outstretched gonads. Males exhibit secondary sexual dimorphism, with a less-developed labial region. Species identification is based on stylet and total body lengths, as well as the fine structure of the labial region and cuticle (e.g., number of longitudinal lines in the basal lip annulus and in the lateral field). Molecular analysis of ribosomal DNA sequences supports morphological schemes, although considerable intraspecific variation has been documented for the internal transcribed spacer regions.

Symptoms

Little is known about the effects of lance nematode parasitism on corn. *Hoplolaimus* spp. are known to penetrate the cortex of host roots, causing significant mechanical damage and resulting in necrotic lesions and even cortical sloughing. Stunting and chlorosis of cotton and soybean have been observed in fields infested with *H. columbus*.

Biology

Lance nematodes generally are classified as semiendoparasites but often feed both ectoparasitically and endoparasitically. Reproduction is sexual in *H. galeatus*, with a sex ratio of 2:1 in favor of females. The mode of reproduction in *H. columbus* is parthenogenesis, and males are rare. The feeding behavior of lance nematodes necessitates enumeration of population densities from both soil and root samples. Underestimation of lance nematode prevalence and population density may occur if either fraction is ignored. Both active (e.g., Baermann funnel) and passive (e.g., centrifugation) soil extraction methods are suitable for lance nematodes.

Economic Importance

Damage potential is moderate for *H. galeatus*, with risk of significant (greater than 5%) yield loss associated with populations above 100 nematodes per 100 cm^3 of soil or 500 nematodes per g dry root weight. Lance nematodes were recovered from 23% of samples submitted for diagnosis from Iowa corn fields during 2000 to 2010, but established damage thresholds (300–400 per g dry root weight) were never exceeded.

Management

A large number of legumes and graminaceous plants are hosts for both *H. galeatus* and *H. columbus*, limiting crop rotation as a management strategy. Applications of carbamate and organophosphate nematicides have produced corn yield increases as great as 26% in fields infested with *H. galeatus* in the Midwest. No resistant corn hybrids are available.

Selected References

Ahmad, M., and Chen, T. A. 1980. Effect of certain environmental factors and host plants on reproduction of *Hoplolaimus galeatus*. Plant Dis. 64:479-480.

Bae, C. H., Szalanski, A. L., and Robbins, R. T. 2008. Molecular analysis of the lance nematode *Hoplolaimus* spp., using the first internal transcribed spacer and the D1-D3 expansion segments of 28S ribosomal DNA. J. Nematol. 40:201-209.

Krusberg, L. R., and Sasser, J. N. 1956. Host-parasite relationships of the lance nematode in cotton roots. Phytopathology 46:505-510.

Lewis, S. A., and Smith, F. H. 1976. Host plants, distribution, and ecological associations of *Hoplolaimus columbus*. J. Nematol. 8:264-270.

Lewis, S. A., Smith, F. H., and Powell, W. M. 1976. Host-parasite relationships of *Hoplolaimus columbus* on cotton and soybeans. J. Nematol. 8:141-145.

Norton, D. C., and Hinz, P. 1976. Relationship of *Hoplolaimus galeatus* and *Pratylenchus hexincisus* to reduction of corn yields in sandy soils in Iowa. Plant Dis. Rep. 60:197-200.

Tylka, G. L., Sisson, A. J., Jesse, L. C., Kennicker, J., and Marett, C. C. 2011. Testing for plant-parasitic nematodes that feed on corn in Iowa 2000-2010. Online. Plant Health Progress doi: 10.1094/PHP-2011-1205-01-RS

(Prepared by T. C. Todd)

Root-Knot Nematodes

Meloidogyne incognita (southern root-knot nematode), *M. arenaria* (peanut root-knot nematode), and *M. javanica* (javanese root-knot nematode) are important parasites of corn in the southeastern United States and other warm regions of the world. Corn is a host to the four host races of *M. incognita* and both races of *M. arenaria*. *M. chitwoodi* (Columbia root-knot nematode) is a pest of corn in the Pacific Northwest. Corn is not considered a host for *M. hapla* (northern root-knot nematode), but two Pacific Northwest biotypes have been reported to reproduce on corn.

Description

Adult females of the genus *Meloidogyne* are found embedded in root tissue and have pearly white, rounded to pear-shaped bodies with a protruding neck (Fig. 242D). Females are 0.4–1.3 mm long and 0.3–0.7 mm wide. Eggs are deposited in protective gelatinous masses, which, for most species, can be observed on the surface of the root. Adult males are vermiform, motile, and nonfeeding. Morphological identification of females is based on the perineal pattern, a fingerprint-like pattern in the cuticle of the perineal region. Isozyme patterns and ribosomal DNA sequences also are used to differentiate species. Species-specific ribosomal DNA primers have been developed for some *Meloidogyne* spp.

Symptoms

Damaged plants generally are localized in aggregated patterns in the field. Aboveground symptoms may include stunting, unthrifty growth, and signs of moisture or nutrient deficiencies. Belowground symptoms may include galled and stunted roots; however, on corn, the galls typically are very small. Root systems supporting a large root-knot nematode population may have numerous small galls but display only mild symptoms of stunting or necrosis caused by secondary infections (Fig. 253).

Biology

Infective second-stage juveniles hatch from eggs and penetrate regions of cell elongation in apical and lateral root tips, eventually becoming sedentary near the vascular cylinder. Feeding initiates the formation of hypertrophied, multinucleate giant cells in the phloem and adjacent parenchyma, which act as metabolic sinks. Juveniles molt three times without feeding, becoming saccate adult females or vermiform males. Reproduction in most economically important species is parthenogenetic, with males absent, rare, or abundant depending on environmental conditions.

Swollen, saccate preadults and adult females are found in susceptible plants as early as 15–18 days postinoculation at 26°C. Egg-producing females have been detected at 25 days after inoculation. Nematode development is slower with resistant hybrids than with susceptible hybrids, and the nematode population density in roots is generally smaller. Root-knot nematodes are commonly found in corn fields at depths of 0–15 cm, but during the summer months, greater numbers may be found at 15–45 cm. As for other corn-parasitic nematodes, increasing sampling depth to at least 30 cm improves the reliability and precision of recovery from the soil. Both active (e.g., Baermann funnel) and passive (e.g., centrifugation) soil extraction methods are commonly used for infective juveniles.

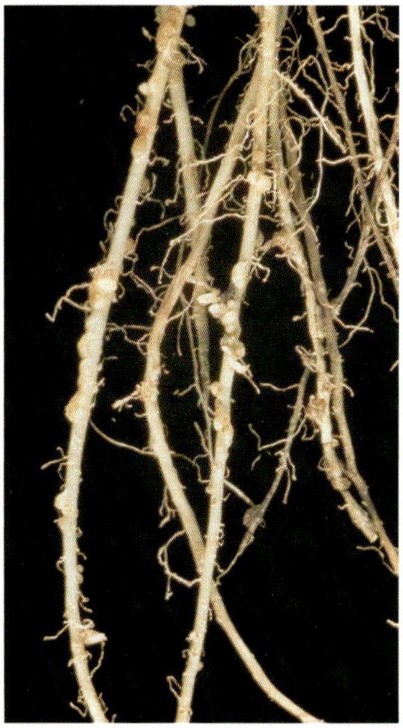

Fig. 253. Root damage caused by root-knot nematodes. (Courtesy E. A. Brown)

Economic Importance

Because of its extensive root system, corn may exhibit tolerance to moderate population densities of root-knot nematodes, but yields in heavily infested fields may be reduced by 30% or more. The economic impact of *Meloidogyne* spp. may be greater on root-knot nematode-susceptible crops planted in rotation with corn. For advisory purposes in North Carolina (United States), the probability of nematode damage to corn by *M. incognita* and *M. arenaria* is rated as "slight" for autumn population densities less than 1,000 infective juveniles per 100 cm^3 of soil and "moderate" for autumn population densities greater than 1,000 infective juveniles per 100 cm^3 of soil. Root-knot nematodes are rarely recovered from midwestern corn fields and do not represent a significant threat to corn production in that region.

Management

Evaluations of corn genotypes for resistance to root-knot nematodes were initiated during the mid-1980s and continue to the present. Initial studies did not find resistance to *M. incognita* in commercial hybrids, but several commercial hybrids with resistance to *M. arenaria* and *M. javanica* were identified. High levels of resistance to *M. incognita* subsequently were identified in several corn inbred lines and open-pollinated cultivars. Reproduction on corn inbred lines and hybrids was observed to be positively correlated for *M. arenaria* and *M. javanica*, but reproduction of *M. incognita* was unrelated to that of either *M. arenaria* or *M. javanica*. Sources of resistance were used to make diallel crosses that effectively limited reproduction of *Meloidogyne* spp. in field tests, and hybrids made from these root-knot nematode-resistant materials have been used successfully in Florida (United States) to suppress populations of *M. incognita*. Nonetheless, the potential for root-knot nematode resistance in commercial corn hybrids remains to be realized.

Meloidogyne spp. associated with corn have extremely broad host ranges but effective rotations have been identified. Sorghum grown in rotation with corn has proven to be one effective strategy for suppressing populations of *M. incognita*. Root-knot nematode-resistant soybean cultivars provide an additional rotation option. Caution must be exercised, however, because plants that are nonhosts of *M. incognita* can serve as good hosts for *M. arenaria* or *M. javanica*. Fields containing multiple *Meloidogyne* spp. are believed to be fairly common.

Use of conventional, soil-applied nematicides to manage root-knot nematodes in corn rarely is economical, and the efficacy of newer protectant seed treatments remains to be adequately demonstrated in the field. Variable suppression of *M. incognita* has been reported for abamectin seed treatments on cotton. Culture filtrates of *Bacillus firmus*, an antagonist bacterium and bionematicide seed treatment for corn, has been shown to induce paralysis and mortality of *M. incognita* in vitro.

Selected References

Aung, T., Windham, G. L., and Williams, W. P. 1990. Reproduction of *Meloidogyne incognita* on open-pollinated maize varieties. J. Nematol. 22(Suppl.):651-653.

Davis, R. F., and Timper, P. 2000. Resistance in selected corn hybrids to *Meloidogyne arenaria* and *M. incognita*. J. Nematol. 32(Suppl.):633-640.

Faske, T. R., and Starr, J. L. 2007. Cotton root protection from plant-parasitic nematodes by abamectin-treated seed. J. Nematol. 39:27-30.

Ibrahim, I. K., Lewis, S. A., and Harshman, D. C. 1993. Host suitability of graminaceous crop cultivars for isolates of *Meloidogyne arenaria* and *M. incognita*. J. Nematol. 25(Suppl.):858-862.
Koenning, S. R., Duncan, H. E., Bailey, J. E., Barker, K. R., and Imbriani, J. L. 1990. Nematode thresholds for soybeans, corn, cotton, and peanuts. N. C. Agric. Ext. Serv. Publ. AG-394.
McSorley, R., and Gallaher, R. N. 1993. Effect of crop rotation and tillage on nematode densities in tropical corn. J. Nematol. 25(Suppl.):814-819.
Mendoza, A. R., Kiewnick, S., and Sikora, R. A. 2008. In vitro activity of *Bacillus firmus* against burrowing nematode *Radopholus similis*, the root-knot nematode *Meloidogyne incognita* and the stem nematode *Ditylenchus dipsaci*. Biocontrol Sci. Technol. 18:377-389.
Santo, G. S., and O'Bannon, J. H. 1981. Pathogenicity of the Columbia root-knot nematode (*Meloidogyne chitwoodi*) on wheat, corn, oat, and barley. J. Nematol. 13:548-550.
Williams, W. P., and Windham, G. L. 1992. Reaction of a diallel cross of maize to *Meloidogyne incognita* under field conditions. Field Crops Res. 30:167-171.
Windham, G. L., and Williams, W. P. 1987. Host suitability of commercial corn hybrids to *Meloidogyne arenaria* and *M. incognita*. Ann. Appl. Nematol. 1:13-16.
Windham, G. L., and Williams, W. P. 1988. Resistance of maize inbreds to *Meloidogyne incognita* and *M. arenaria*. Plant Dis. 72:67-69.
Windham, G. L., and Williams, W. P. 1988. Reproduction of *Meloidogyne javanica* on corn hybrids and inbreds. Ann. Appl. Nematol. 2:25-28.
Windham, G. L., and Williams, W. P. 1994. Penetration and development of *Meloidogyne incognita* in roots of resistant and susceptible corn genotypes. J. Nematol. 26:80-85.

(Prepared by T. C. Todd, G. L. Windham, and D. I. Edwards)

Other Nematodes

Awl Nematode

The awl nematode, *Dolichodorus heterocephalus*, is similar to the sting nematode in size, shape, and general appearance. A female awl nematode has a bluntly pointed tail and the male has a short, wide bursa, whereas the sting female has a blunt, rounded tail and the male has a long, narrow bursa. Awl and sting nematodes have similar feeding habits and inflict the same type of damage on corn roots. Awl nematodes feed mainly at root tips but may feed along the sides of roots and also at the base of the hypocotyl. Seeds may not germinate if awl nematodes feed on the seed embryo. Plants may be severely stunted and may have greatly depleted root systems and numerous necrotic root lesions. The awl nematode can cause damage as severe as that caused by sting nematodes, but the awl nematode is more localized and has a less extensive distribution. Because the sting nematode is more aggressive, it is usually dominant in fields where the two nematodes are found together. The awl nematode has been a problem primarily on corn in Florida (United States) in areas that have high soil moisture levels.

Selected Reference

Rhoades, H. L. 1985. Effects of separate and concomitant populations of *Belonolaimus longicaudatus* and *Dolichodorus heterocephalus* on *Zea mays*. Nematropica 15:171-174.

(Prepared by G. L. Windham and D. I. Edwards)

Cyst Nematodes

The corn cyst nematode, *Heterodera zeae* (Fig. 254), is an important parasite of corn in Egypt, India, and Pakistan. It was first reported in the United States in 1981 in Kent County, Maryland, and subsequently was found in a field with severely stunted corn plants in Cumberland County, Virginia, in 1992. The known distribution of the nematode in the United States has not expanded beyond these two states. The optimum temperature for reproduction of *H. zeae* on corn is 36°C, with minimal reproduction occurring at 24°C. The life cycle is completed in 15–18 days at 33°C but requires 42 days at 25°C. Evidence of damage potential for *H. zeae* in Maryland is inconsistent, with no yield responses observed for nematicide trials in naturally infested fields, while artificial infestation of field microplots resulted in the suppression of corn grain yields in 4 out of 5 years (21–73% loss) in sandy soil and 1 out of 5 years (15% loss) in silty soil. The warm climate preference of *H. zeae* suggests that it could become a more serious problem for corn production if introduced into the southeastern United States. The host range of *H. zeae* is limited to plants in the family Gramineae, including barley, corn, oats, rice, sorghum, sugarcane, teosinte, and wheat.

Two additional cyst-forming species are associated with corn in North America. The Mexican corn cyst nematode, *Punctodera chalcoensis*, is a serious pest of corn in Mexico. It is widely distributed at high elevations and may represent a potential threat to corn production in the more temperate areas of the United States. *Vittatidera zeaphila* is a new genus and species of cyst nematode recently described from corn in Tennessee

Fig. 254. Corn cyst nematode. (Courtesy L. R. Krusberg)

(United States). This nematode currently is known only from Tennessee and Kentucky and appears to have a restricted host range (only corn and goosegrass have been confirmed as hosts to date). Resistance has been identified in several corn inbred lines but does not appear to be a dominant trait and may be conferred through the cytoplasm. Assessment of the broader potential risk for corn production in the United States from this species awaits further information on its host range and environmental requirements.

Selected References

Bernard, E. C., Handoo, Z. A., Powers, T. O., Donald, P. A., and Heinz, R. D. 2010. *Vittatidera zeaphila* (Nematoda: Heteroderidae), a new genus and species of cyst nematode parasitic on corn (*Zea mays*). J. Nematol. 42:139-150.

Donald, P. A., Heinz, R., Bernard, E., Hershman, D., Hensley, D., Flint-Garcia, S., and Joost, R. 2012. Distribution, host status and potential sources of resistance to *Vittatidera zeaphila*. Nematropica 42:91-95.

Eisenback, J. D., Reaver, D. M., and Stromberg, E. L. 1993. First report of corn cyst nematode (*Heterodera zeae*) in Virginia. Plant Dis. 77:647.

Hashmi, S., Krusberg, L. R., and Sardanelli, S. 1993. Reproduction of *Heterodera zeae* and its suppression of corn plant growth as affected by temperature. J. Nematol. 25:55-58.

Hutzell, P. A., and Krusberg, L. R. 1990. Temperature and the life cycle of *Heterodera zeae*. J. Nematol. 22:414-417.

Koshy, P. K., Swarup, G., and Sethi, C. L. 1970. *Heterodera zeae* n. sp. (Nematoda: Heteroderidae), a cyst-forming nematode on *Zea mays*. Nematologica 16:511-516.

Krusberg, L. R., Sardanelli, S., and Grybauskas, A. P. 1997. Damage potential of *Heterodera zeae* to *Zea mays* as affected by edaphic factors. Fundam. Appl. Nematol. 20:593-599.

Ringer, C. E., Sardanelli, S., and Krusberg, L. R. 1987. Investigations of the host range of the corn cyst nematode, *Heterodera zeae*, from Maryland. Ann. Appl. Nematol. 1:97-106.

Sardanelli, S., Krusberg, L. R., and Golden, A. M. 1981. Corn cyst nematode, *Heterodera zeae*, in the United States. Plant Dis. 65:622.

Sosa Moss, C. 1987. Cyst nematodes in Mexico, Central America, and South America. Nematol. Mediterr. 15:1-12.

(Prepared by T. C. Todd, G. L. Windham, and D. I. Edwards)

Dagger Nematodes

Although dagger nematodes are important pests of many agriculturally important plants, very limited information has been collected on their relationship with corn. Large population densities are rarely encountered on corn, but when they occur, particularly if early in the growing season, significant damage is possible. The dagger nematode is slender and up to 2 mm long with a dorylaimoid esophagus and a flanged odontostyle that may be 130 μm long. *Xiphinema americanum* is the most commonly found dagger species associated with corn in the United States. Others include *X. chambersi* and *X. rivesi*. Maximum numbers of dagger nematodes generally are found on corn at depths of 0–15 cm during August and September. Numbers of *X. americanum* have been reported to be negatively correlated with corn yields, and population densities of this nematode exceeded established damage thresholds (30–40 individuals per 100 cm^3 of soil) in 6% of corn samples submitted for diagnosis in Iowa during 2000 to 2010.

Selected References

Griffin, G. D. 1964. Association of nematodes with corn in Wisconsin. Plant Dis. Rep. 48:458-459.

Norton, D. C. 1984. Nematode parasites of corn. Pages 61-94 in: Plant and Insect Nematodes. W. R. Nickle, ed. Marcel Dekker, New York.

Norton, D. C., Tollefson, J., Hinz, P., and Thomas, S. H. 1978. Corn yield increases relative to nonfumigant chemical control of nematodes. J. Nematol. 10:160-166.

Tylka, G. L., Sisson, A. J., Jesse, L. C., Kennicker, J., and Marett, C. C. 2011. Testing for plant-parasitic nematodes that feed on corn in Iowa 2000-2010. Online. Plant Health Progress doi: 10.1094/PHP-2011-1205-01-RS

(Prepared by T. C. Todd, G. L. Windham, and D. I. Edwards)

Ring Nematodes

Criconemella ornata and *C. sphaerocephala* are commonly associated with corn in the southeastern United States. Ring nematodes are named for the prominent annulations found on their cuticles. Although ring nematodes are not very large (0.2–1.0 mm long), they have massive stylets. Population densities usually peak on corn during July. Studies of the vertical distribution of *C. sphaerocephala* in corn fields have shown that the majority are found in the soil at depths of 15–45 cm. Although huge increases in nematode numbers on corn have been observed over a growing season, it has been difficult to establish their pathogenicity on corn. In field studies, nematicide treatments were inconsistent in demonstrating yield increases.

Selected References

Ayers, A. R., Duncan, H. E., Barker, K. R., and Beute, M. K. 1989. Effects of crop rotation and nonfumigant nematicides on peanut and corn yields in fields infested with *Criconemella* species. J. Nematol. 21:268-275.

Johnson, A. W., and Chalfant, R. B. 1973. Influence of organic pesticides on nematode and corn earworm damage and on yield of sweet corn. J. Nematol. 5:177-180.

McSorley, R., and Gallaher, R. N. 1992. Comparison of nematode population densities on six summer crops at seven sites in north Florida. J. Nematol. 24(Suppl.):699-706.

(Prepared by G. L. Windham and D. I. Edwards)

Spiral Nematodes

Spiral nematodes (*Helicotylenchus dihystera*, *H. digonicus*, *H. pseudorobustus*, and *Scutellonema brachyurum*) are commonly found in association with corn (Fig. 242E). Generally, spiral nematodes are considered semiendoparasites, although they may be found completely embedded in corn roots. Plants attacked by spiral nematodes may have numerous small, light to dark brown lesions. Infected areas in the roots are characterized by damage to cortical cell walls and often by the presence of a ruptured epidermis. In most cases, these nematodes are considered weak to moderate pathogens, although high numbers have been associated with declining corn yields. The damage threshold density for *H. dihystera* on corn in field microplots is greater than 200 per 100 cm^3 of soil. Population densities of *H. pseudorobustus* can increase rapidly on corn, and there is some evidence of pathogenicity for this species. Fields infested with *H. pseudorobustus* and treated with granular nematicides have shown yield increases of 15–20%. Spiral nematodes were recovered from Iowa diagnostic samples submitted during 2000 to 2010 more frequently (77% of total samples) than any other plant-parasitic nematode, but only 1% of samples exceeded established thresholds of 500–1,000 nematodes per 100 cm^3 of soil.

Selected References

Barker, K. R., and Olthof, T. H. 1976. Relationships between nematode population densities and crop responses. Annu. Rev. Phytopathol. 14:327-353.

Norton, D. C. 1977. *Helicotylenchus pseudorobustus* as a pathogen on corn, and its densities on corn and soybean. Iowa State J. Res. 51:279-285.

Norton, D. C., Tollefson, J., Hinz, P., and Thomas, S. H. 1978. Corn yield increases relative to nonfumigant chemical control of nematodes. J. Nematol. 10:160-166.

Tylka, G. L., Sisson, A. J., Jesse, L. C., Kennicker, J., and Marett, C. C. 2011. Testing for plant-parasitic nematodes that feed on corn in Iowa 2000-2010. Online. Plant Health Progress doi: 10.1094/PHP-2011-1205-01-RS

(Prepared by T. C. Todd, G. L. Windham, and D. I. Edwards)

Stunt Nematodes

Stunt nematodes (*Tylenchorhynchus* spp.) are very common ectoparasites of corn. *T. claytoni* and *T. maximus* are important in the southeastern United States and the Midwest, respectively. The stunt nematode is easily identified by its tylenchoid esophagus, which has a well-developed medium bulb and a very prominent, nonoverlapping, glandular basal bulb (Fig. 242A). More information about the pathogenicity of *T. claytoni* on corn is available than for any other *Tylenchorhynchus* sp. These nematodes feed almost exclusively at the root surface, although they may be found partially embedded or completely inside the root. *T. maximus* also has been shown to be pathogenic to corn.

Selected References

Nelson, R. R. 1956. Resistance to the stunt nematode in corn. Plant Dis. Rep. 40:635-639.

Norton, D. C. 1984. Nematode parasites of corn. Pages 61-94 in: Plant and Insect Nematodes. W. R. Nickle, ed. Marcel Dekker, New York.

(Prepared by G. L. Windham and D. I. Edwards)

Disease Caused by a Parasitic Seed Plant

Witchweed

Witchweed is a parasitic seed plant native to the grasslands of Africa and Asia. It is also reported in Australia and, in 1956, was discovered in the United States in North Carolina and South Carolina. Losses vary with degree of infestation and cultural conditions. The most severe losses are associated with high levels of infestation, low levels of soil fertility, and low plant populations.

Symptoms

Corn plants infested by witchweed exhibit symptoms of drought stress and nutrient deficiencies, even though moisture and fertility are adequate. Extremely high levels of infestation may result in death of plants. One to several witchweed plants usually grow aboveground next to a corn plant (Fig. 255). Below ground, fleshy stems of the witchweed plant are attached to the corn root system (Fig. 256).

Causal Organisms

There may be as many as 60 species of witchweed; however, only two are recognized as important pathogens of corn. *Striga hermonthica* is the most widely distributed species, causing the most severe damage in Africa and Asia. *S. asiatica* (syn. *S. lutea*) is also found in many areas of the world, including the United States. *S. asiatica* is an obligately parasitic flowering plant with square, aerial stems and small, bright green, elongated leaves. The plant is erect and slender, rarely grows more than 15–30 cm tall, and has opposite lower leaves and alternate upper leaves. Flowers are tubular with well-developed two-lipped corollas, are 5–11 mm in diameter, and usually have 10 calyx ribs. They may be brick red, yellowish red, lemon yellow, or almost white with straw yellow centers and yellow undersides. Seeds are

Fig. 255. *Striga* sp. (witchweed). (Courtesy P. M. LeRoux and D. R. Smith)

Fig. 256. *Striga* sp. parasitizing corn roots. (Courtesy M. Bolarin)

minute (less than 0.25 mm long), striated, brown, and borne in pods or capsules. A single plant can produce up to a half million seeds. *S. asiatica* is highly self pollinated.

S. hermonthica is generally larger than *S. asiatica* and has bigger leaves and flowers with five calyx ribs. Its flowers are pink to pink white. This species is believed to be an obligate outcrosser.

Disease Cycle

Witchweed survives as seeds, most of which require 90 days to 18 months of dormancy before germination. Seeds are numerous and can remain viable for as long as 20 years. They germinate in the presence of the host or certain nonhost plants that produce a chemical stimulant necessary to trigger germination. When the witchweed rootlets contact corn roots, they develop haustoria in the corn roots that serve as morphological and physiological connections between the host and the parasite. The haustoria dissolve and penetrate host roots within 8 to 24 h, and eventually, tracheids of the witchweed form and connect to xylem vessels of the host. The witchweed grows aboveground, produces large numbers of minute seeds in pods or capsules, and then dies. The time from seed germination to seed production is 60–90 days. The plants emerge about 30 days after germination and may produce flowers 14–20 days after emergence and the first mature seed pods approximately 3 weeks after flowering. The witchweed plant depends on the host roots for the uptake of water and minerals. Although it has green leaves with functional chlorophyll, manufactured food stuffs still move from the host to the witchweed plant.

Epidemiology

Seeds are disseminated in contaminated soil on machinery, equipment, and tools, in water, or by the wind. Corn plants are usually infested in a circular pattern; the center is the most seriously affected. The size of the infested area increases as the seeds are spread to surrounding areas. Low soil fertility, drought, and low plant populations all favor the development of witchweed and are unfavorable to corn. It is under these conditions that damage is greatest.

Management

Research is being conducted to determine the mechanisms and inheritance of resistance to develop cultivars that are suitable for use in areas where witchweed is a serious problem. Generally, the application of fertilizer or manure reduces both the infestation by witchweed and losses in yield. Dense corn plant populations may shade the witchweed and provide some control. Adequate control can be achieved by the use of herbicides, particularly in combination with good fertility and adequate corn plant populations. Herbicidal seed treatments, coupled with herbicide-tolerant (non-GMO) corn hybrids, have been successful in Africa. Trap crops, which stimulate the germination of the witchweed seeds but are not parasitized by it, have been used for management of witchweed. However, this method is not widely employed because of the cost. In many countries, phytosanitary regulations specify that international seed shipments must be free of *Striga* seeds.

Selected References

Kanampiu, F. K., Friesen, D., and Gressel, J. 2002. CIMMYT collaborative efforts unveils herbicide-coated maize seed technology to curb problematic striga. Integr. Pest Manage. Rev. 7:63-64.

Sand, P. F., Eplee, R. E., and Westbrooks, R. G., eds. 1990. Witchweed Research and Control in the United States. Monogr. 5. Weed Science Society of America, Champaign, IL.

(Prepared by D. G. White)

Part II. Noninfectious or Abiotic Disorders

Nutrient Toxicities

Fertilizer Burn

Excessive amounts of soluble nitrogen or potassium fertilizer placed close to the seed may prevent germination. Seedlings often emerge and then die or are severely stunted if they survive. Leaves turn yellow and sometimes die. Leaf tip necrosis is evidence of salt accumulation in the plant. Boron in starter fertilizer is especially toxic to corn seedlings. Fertilizer burn development is favored by sandy soil, dry weather, and use of fertilizer with a high salt index.

Fig. 257. Anhydrous ammonia injury to corn seedling roots. (Courtesy J. Sawyer)

Fig. 258. Anhydrous ammonia "burn" injury to corn seedlings. (Courtesy R. G. Hoeft)

Anhydrous Ammonia or Aqua-Ammonia Damage

Short or long gaps or patterns of dead plants in rows scattered throughout the field may indicate anhydrous ammonia damage. In plants injured after emergence, roots appear to be sheared off or dried up at varying distances from the seed (Fig. 257). Leaves may be chlorotic or dried and not fully expanded (Fig. 258). The latter injury almost always occurs at the end of the field when ammonia (NH_3) is released while the knives are out of the soil or the fertilizer slice fails to close. Fertilizers containing free NH_3 (e.g., diammonium phosphate) or those that readily convert to NH_3 (e.g., urea) may be toxic if placed near the seed. This type of damage is promoted by soil that is sandy, dry, compact, and low in organic matter and has an uneven surface; by deep planting; and by shallow placement of NH_3.

Other Nutrient Excesses

In highly acid or alkaline soils, aluminum and iron oxides may accumulate and cause internal necrosis near shoot apexes or in vascular bundles in the nodes. Nickel and chromium excesses may cause stunting, interveinal chlorosis, and puckering of the leaf margins.

Selected Reference

Voss, R. D. 1993. Corn. Pages 11-14 in: Nutrient Deficiencies and Toxicities in Crop Plants. W. F. Bennett, ed. American Phytopathological Society, St. Paul, MN.

(Prepared by E. L. Stromberg, E. S. Hagood, and D. G. White)

Nutrient Deficiencies

Nutrient deficiencies are difficult to diagnose on the basis of visual symptoms alone because adverse weather, diseases, nematodes, insects, and various other injuries or factors may cause similar reactions in plants. These symptoms are most reliable when all other factors favor rapid growth and symptoms generally appear on several plants in a broad area following a specific soil or management pattern. If possible, field diagnosis should be verified by plant analyses (chemical tissue tests) or soil tests. Often, a careful appraisal of cropping history and fertility treatments confirms the reliability of a tentative diagnosis. Nutrient deficiencies are least likely to be present where the crop is grown for maximum yield within the pH range of 5.6 to not more than 7.3. Nutrient deficiency symptoms are most difficult to interpret during very cold or wet weather.

Nitrogen

Nitrogen deficiency is characterized by stunted, spindly growth and pale, yellowish green foliage in young plants. Older plants develop a V-shaped yellowing along the midribs to the tips of the bottom leaves (Fig. 259). Later, the lower leaf tips die (a process called firing); the margins remain green. Stalks are thin and spindly and the entire plant may be pale green (Fig. 260). Ears tend to be small, with poorly filled kernels at the tips (Fig. 261). Mature kernels are glossy, hard, and flinty. Nitrogen deficiency development is favored by cold, wet, or flooded soils; drought (especially after midseason); sandy soils low in organic matter; heavy, leaching rains; standing water when the weather is warm; and intensive row cropping.

Phosphorus

Mild phosphorus deficiencies are characterized by stunted growth. More severe deficiencies cause purpling or reddening of the leaves beginning early in the growing season (Fig. 262). Leaf tips die and turn dark brown. Phosphorus deficiency is most pronounced in young plants growing in cool soils that are too wet or too dry; when insufficient phosphorus is stored in or applied to the soil; when root growth is restricted in compacted soils; and when root damage is caused by insects, cultivation, fertilizer sidedressing knives, or herbicides. When deficiency is caused by cool, wet soils, plants are unable to take up adequate phosphorus because of restricted root systems. Usually, young plants outgrow this condition when warmer weather begins. Ears on affected plants are often small and may appear twisted with irregular kernel rows and imperfectly developed tips.

Potassium

The first symptoms of potassium deficiency are yellowing and death of the leaf margins beginning at the tips of the lower leaves (Fig. 263). Plants often lodge late in the season because of stalk rot, to which they are more susceptible. Ears on potassium-deficient plants may be small, chaffy, and dull with pointed, poorly developed tips. Potassium deficiency development is favored by sandy, organic, wet, or compacted, weathered soil and the removal of large amounts of potassium by preceding crops.

Calcium

Calcium is rarely deficient in corn. Leaf tips tend to stick together, so leaves fail to unfold. Affected plants may have a slight yellowish green tint and be severely stunted. Calcium deficiency development is favored by very low soil pH (below 5.5 in mineral soils and below 4.8 in organic soils) and soils high in potassium and magnesium.

Magnesium

Magnesium deficiency manifests itself in seedlings as a general yellowing of the upper leaves. Eventually, yellow to white

Fig. 259. Nitrogen deficiency symptoms on lower leaves. (Courtesy G. P. Munkvold)

Fig. 261. Symptoms of nitrogen deficiency on an immature corn ear. (Courtesy J. Sawyer and B. Pringnitz)

Fig. 260. Nitrogen deficiency symptoms. (Courtesy J. Sawyer)

Fig. 262. Phosphorus deficiency in plants in cold soil. (Courtesy J. Sawyer)

interveinal striping develops, and older leaves appear reddish purple along their edges and tips. Severe deficiencies cause leaf tips and margins to die, and the entire plant may show interveinal striping. Stunting may occur but is not as severe as with phosphorus or potassium deficiency. Magnesium deficiency development is favored by strongly acidic, sandy soils in regions of moderate to high rainfall; high potassium levels; and soil treated with limestone low in magnesium.

Sulfur

Symptoms of sulfur deficiency are similar to those of nitrogen deficiency: stunting, delayed maturity, a general yellowing of the foliage, and in some cases, interveinal chlorosis. With sulfur deficiency, the yellowing is more pronounced on the younger leaves than with nitrogen deficiency. Sulfur deficiency development is favored by acidic, sandy soils low in organic matter and cold, wet soils.

Zinc

Symptoms of zinc deficiency appear usually within 2 weeks of seedling emergence and include shortening of the internodes, lack of vigor, and light streaking of leaves, followed by a broad stripe of bleached tissue on each side of the midrib (Fig. 264). The midrib, margins, and leaf tip remain green. New leaves are sometimes almost white ("white bud" condition). Occasionally, the leaf edges and interior of the stalks at the nodes appear purplish. Plants generally outgrow the deficiency unless it is severe. Very seldom is an entire field uniformly affected. Zinc deficiency development is favored by high phosphorus levels in soil or fertilizer; high soil pH (above 6.3); cool, wet soil; and soil with low organic matter (exposed subsoil).

Fig. 263. Potassium deficiency. (Courtesy K. Jensen)

Fig. 264. Zinc deficiency. (Courtesy L. E. Claflin)

Iron

Entire upper leaves of iron-deficient plants become pale green to almost white between the veins (Fig. 265). Because corn has a low iron requirement, this deficiency is rare except in alkaline soils. Iron deficiency development is favored by high pH; wet, compacted, poorly aerated soils; and cool weather.

Manganese

Symptoms of manganese deficiency are vague because of the low manganese requirement of corn. Leaves become olive green and may be somewhat streaked. If the deficiency is severe, leaves develop elongated, whitish streaks, the centers of which turn brown, die, and may fall out. Manganese deficiency development is favored by high pH and peat, muck, and sandy soils high in organic matter.

Copper

Copper is rarely deficient in corn. The youngest leaves turn yellow, as with iron deficiency. Severely deficient plants are stunted; younger leaves become chlorotic; and some dieback of older leaves occurs, as with potassium deficiency. Stalks of affected plants are soft and flexible. Copper deficiency development is favored by soils high in organic matter.

Boron

Boron is rarely deficient in corn. Symptoms appear first on the youngest leaves as white, irregularly shaped spots between the veins. These spots may coalesce to form white stripes that appear waxy and raised from the leaf tissue. The internodes do not elongate, and the tassels and ear shoots often fail to emerge, giving plants a bushy appearance. Boron deficiency development is favored by drought, high pH, and sandy soils low in organic matter.

Molybdenum

Molybdenum is rarely, if ever, deficient in corn. Older leaves die at the tips, along the margins, and finally between the veins. Younger leaves wilt and then die along the margins and, in some cases, twist. Molybdenum deficiency development is favored by low pH and heavily weathered soils.

Fig. 265. Iron deficiency. (Courtesy L. E. Claflin)

Selected References

Hoeft, R. G., Nafziger, E. D., Johnson, R. R., and Aldrich, S. R. 2000. Modern Corn and Soybean Production. MCSP Publications, Champaign, IL.

Krantz, B. A., and Melsted, S. W. 1964. Nutrient deficiencies in corn, sorghums, and small grains. Pages 25-58 in: Hunger Signs in Crops, 3rd ed. H. B. Sprague, ed. David McKay, New York.

Voss, R. D. 1993. Corn. Pages 11-14 in: Nutrient Deficiencies and Toxicities in Crop Plants. W. F. Bennett, ed. American Phytopathological Society, St. Paul, MN.

(Prepared by E. L. Stromberg, E. S. Hagood, and D. G. White)

Symptoms Associated with Herbicide Injury

If herbicides are used at excessive rates, herbicides to which corn has a low tolerance are used, or there is unintended carryover or drift from applications to other crops, injury may occur. Herbicides may cause plant death, abnormal color, leaf spots, twisted or malformed shoots or roots, lack of emergence, stunting, loss of chlorophyll, irregularities in ear shape, poor kernel set, and fill problems. Damage caused by herbicides can be difficult to diagnose because symptoms can be confused with those caused by diseases, insects, and nutrient deficiencies. Because herbicide injury can predispose plants to seedling disease and root rot, it can be difficult or impossible to separate injury from disease. Common herbicide injuries and symptoms are summarized here. Precision application and proper procedures for application are the most important first steps in reducing herbicide injury to a minimum. Application rates should be carefully selected and the equipment accurately calibrated. It is important to remember that corn hybrids may differ in their susceptibility to herbicide injuries.

Herbicides are the most common method of weed control in the major agronomic crops grown in the continental United States. Selective herbicides have the ability to remove undesirable plants (weeds) while leaving other plants in the area undamaged. There are numerous selectivity mechanisms that allow a plant to survive herbicides, but the most common mechanism in conventional crops is differential metabolism. Most herbicides are metabolized by plants, but the rate at which they are degraded varies widely among species. Species that are sensitive to the herbicide are inefficient at metabolizing the herbicide, whereas tolerant species can degrade the herbicide before it has a toxic effect. Any factor that reduces a crop's ability to metabolize a herbicide may result in crop injury. Herbicide degradation is most efficient when a plant is actively growing, thus any factor that stresses the plant reduces the speed at which a herbicide is detoxified. In addition, factors that increase exposure of the plant to the herbicide may overwhelm the plant's ability to degrade the compound. Misapplication resulting in a greater than labeled rate of herbicide is a common cause of crop injury.

The margin of safety varies widely among herbicides and crops. Glyphosate-resistant crops utilize an altered target site for selectivity rather than differential metabolism. This provides a high margin of safety since glyphosate is unable to bind to the target site. Corn is very efficient at metabolizing atrazine, and thus, injury from atrazine is rare. With other herbicides, there may be a lower margin of safety, and thus, situations in which the crop's tolerance is exceeded are more common. Many herbicides are now formulated with safeners that increase a crop's ability to metabolize the herbicide, thus increasing the margin of safety. Examples of safened herbicides include acetochlor (Harness; Monsanto Co., St. Louis), dicamba (Status; BASF Ag Products, Research Triangle Park, NC), isoxaflutole (Balance Flexx; Bayer CropScience, Research Triangle Park, NC), and nicosulfuron (Accent Q; E. I. du Pont de Nemours and Co., Wilmington, DE).

The most common scenario for herbicide injury to crops results from herbicides intentionally applied to the crop for weed control. Another cause of crop injury occurs when a crop is unintentionally exposed to a herbicide to which it has little tolerance or selectivity. For example, if a herbicide applied to last year's soybean crop did not degrade sufficiently because of dry soil conditions following its application, enough of the herbicide may persist in the soil to cause damage to the following year's corn crop. Unintentional exposure of a sensitive crop may also occur via contaminated sprayers or from physical or vapor drift from applications made to fields adjacent to the corn.

Diagnosing which herbicide might have caused the observed crop injury or determining how the crop was exposed to a damaging herbicide can often be challenging. Each herbicide has a specific site of action, usually a protein, in the plant where it binds. Binding of the herbicide to the protein interferes with its function, causing sufficient disruption of a plant pathway (mode of action) that results in plant death. Herbicides are classified based on their site of action, and each site of action has been assigned a group number. The group number is prominently displayed on most herbicide labels.

An understanding of how and where herbicides work within a plant is a valuable asset when attempting to diagnose which herbicide or herbicide family caused the observed injury symptoms. Also important to diagnosis is acquiring all pertinent information related to the field in question, including herbicides and other chemicals applied this year and last, planting date and seeding rate, soil pH and nutrient levels, and environmental conditions immediately before and after herbicide application. Looking across the affected field for telltale patterns that might be related to uniformity of herbicide application, tillage, or planting; differences in soil characteristics; and the presence of old building sites can be helpful in diagnosing the cause of symptoms. Injury appearing on a border of a field may be an indicator that the herbicide responsible for the damage was applied to an adjacent area rather than the field with the injury.

Herbicides can be grouped by mode of action (Table 9); the following is a discussion of the herbicide groups that are used in corn production or of herbicides that are commonly used in areas adjacent to where corn is grown.

Group 1 herbicides. These herbicides block the activity of acetyl-CoA carboxylase (ACCase), an enzyme involved in the synthesis of fatty acids. They are used postemergence in soybean and other broadleaf crops to control grasses. Injury may occur because of drift from applications made to soybean or from contaminated sprayers. In addition, corn may be exposed to these herbicides if they are applied to fields at or prior to corn planting, a situation that may happen when corn fields are replanted because of stand failure and corn plants from the previous planting need to be eliminated. Group 1 herbicides are systemic and move to the growing points, therefore initially damaging tissue that develops following exposure. At sublethal doses, leaves that emerge following application develop bands of chlorotic, distorted tissue. Occasionally, these bands of damaged tissue are clear or white. At higher concentrations, the shoot's growing point is killed, resulting in rotted tissue within the whorl. In these situations, the newer leaves can be easily pulled out of the whorl.

Group 2 herbicides. These herbicides inhibit the activity of acetolactate synthase (ALS), an enzyme involved in the synthesis of branched chain amino acids. They are used in a wide variety of crops and are applied both preemergence and postemergence. Injury may occur because of direct application, drift, or carryover from the prior year's application. Symptoms are highly variable depending upon dose and plant development stage at the time of exposure. Symptoms associated with

TABLE 9. Herbicide Groups, Modes of Action, and Common Examples

Herbicide Group	Herbicide Family	Products: Common Name (Tradename)[a]
1	Acetyl-CoA carboxylase (ACCase) inhibitors	Quizalofop (Assure II[b]); fluazifop (Fusilade[c]); sethoxydim (Poast[d]); clethodim (Select[e])
2	Acetolactate synthase (ALS) inhibitors	Imazethapyr (Pursuit[d]); imazaquin (Sceptor[d]); nicosulfuron (Accent[b]); chlorimuron (Classic[b]); thifensulfuron (Harmony[b]); foramsulfuron (Option[f]); halosulfuron (Permit[g])
3	Dinitroanilines	Pendimethalin (Prowl[d]); trifluralin (Treflan[h])
4	Growth regulators	2,4-Dichlorophenoxyacetic acid (2,4-D); dicamba (Banvel[i]/Clarity[d]/Status[d])
5	Photosystem II (PSII) inhibitors (triazine)	Atrazine (AAtrex[c]); simazine (Princep[c])
6	Photosystem II (PSII) inhibitors (unclassified)	Bromoxynil (Buctril[f]); bentazon (Basagran[d])
9	5-Enolpyruvylshikimate-3-phosphate synthase (EPSPS) inhibitor	Glyphosate (Roundup[j]/Touchdown[c]/etc.)
10	Glutamine synthetase inhibitor	Glufosinate (Liberty[f]/Ignite[f])
13	Unclassified	Clomazone (Command[k])
14	Protoporphyrinogen oxidase (PPO) inhibitors	Acifluorfen (Ultra Blazer[l]); lactofen (Cobra[e]/Phoenix[l]); fomesafen (Flexstar[c]/Reflex[c]); carfentrazone (Aim[k]); sulfentrazone (Authority[k]); flumioxazin (Valor[e]); saflufenacil (Sharpen[d]/Kixor[d]); fluthiacet (Cadet[k])
15	Amide/acetamide	Metolachlor (Dual II Magnum[c]/Cinch[b]/etc.); acetochlor (Harness[j]/Surpass[h]); dimethenamid (Outlook[d]); pyroxasulfone (Zidua[d])
22	Bipyridyliums	Paraquat (Gramoxone[c]); diquat
27	4-Hydroxyphenylpyruvate dioxygenase (HPPD) inhibitors	Isoxaflutole (Balance Pro[f]); mesotrione (Callisto[c]); tembotrione (Laudis[f]); topramezone (Armezon[d]/Impact[m])

[a] Many herbicides are sold in premixes that contain two or more active ingredients. Package mixes are not listed in this table. In addition, numerous herbicides are off patent and sold under a variety of tradenames.
[b] E. I. du Pont de Nemours and Co., Wilmington, DE.
[c] Syngenta Crop Protection LLC, Greensboro, NC.
[d] BASF Ag Products, Research Triangle Park, NC.
[e] Valent U.S.A. Corp., Walnut Creek, CA.
[f] Bayer CropScience, Research Triangle Park, NC.
[g] Gowan Co., Yuma, AZ.
[h] Dow AgroSciences, LLC, Indianapolis.
[i] Arysta LifeScience North America LLC, Cary, NC.
[j] Monsanto Co., St. Louis.
[k] FMC Corp., Philadelphia.
[l] United Phosphorus Inc., King of Prussia, PA.
[m] Amvoc Chemical Corp., Newport Beach, CA.

Fig. 266. Injury caused by the postemergence application of an acetolactate synthase-inhibiting herbicide (group 2). (Courtesy A. G. Hager)

Fig. 267. Sulfonylurea injury (group 2 herbicide). (Courtesy D. G. White)

carryover include inhibited development of primary or lateral roots and a proliferation of secondary root development, resulting in a "bottle-brush" appearance of the roots. The roots of injured plants may resemble roots damaged by nematodes. Inhibited root development often leads to overall plant stunting. Imazaquin, imazethapyr, and chlorimuron are products used in soybean that have the potential to persist and damage rotational corn. Injury symptoms after foliar applications of ALS-inhibiting herbicides include stunting because of shortening of internode length, chlorotic leaves emerging from the whorl (Fig. 266), and crinkling of leaf margins. Plants may exhibit interveinal yellowing or purpling (Fig. 267). Applications of group 2 herbicides after the V6 corn stage may not be manifested until ear development, when ears may be greatly reduced in size, be constricted about the middle, or have abrupt changes in kernel row numbers (Figs. 268 and 269).

Group 3 herbicides. Dinitroaniline herbicides (group 3) inhibit cell division in susceptible plants. They are used preemergence in soybean and other crops; pendimethalin can be used in corn. Injury to corn is generally associated with carryover from applications made to soybean the prior year. Direct application of pendimethalin to corn may cause injury if the herbicide comes in direct contact with the roots, usually from shallow planting or improper closure of the seed furrow. Symptoms include inhibited development of roots, resulting in swollen root tips and stunted plants (Fig. 270). The roots of plants injured by these herbicides may look similar to roots damaged by nematodes (Fig. 271).

Fig. 268. Injury caused by the interaction of sulfonylurea (group 2 herbicide) and organophosphate. (Courtesy A. G. Hager)

Fig. 269. Sulfonylurea injury (group 2 herbicide). (Courtesy B. M. Anderson)

Group 4 herbicides. The synthetic auxin herbicides (group 4) bind to the auxin receptor site, mimicking the activity of indole acetic acid (IAA). They are used primarily postemergence to control broadleaf weeds in grass crops, although they are also used preplant in no-till to control emerged weeds. Symptoms are dependent upon dose and growth stage at the time of exposure. Injury symptoms usually develop on tissues that develop shortly after exposure to the herbicide. Symptoms associated with applications to corn include improper unfurling of leaves, brittle stalks, root fasciation (Figs. 272 and 273), and malformed ears. Leaves may have yellow streaks extending down most of the length of the blade. These stripes on the leaves could be confused with immature lesions of Stewart's bacterial wilt. Mesocotyl twisting can occur when emerging plants come in contact with group 4 herbicides applied preemergence. Damage to roots from group 4 herbicides may resemble symptoms of nematode feeding.

Group 5 and 6 herbicides. These herbicides bind to a protein involved in electron transfer in photosystem II (PSII), but the two groups attach at different sites on the protein. Members of these herbicide groups may be foliar applied, soil applied, or both. Injury symptoms after foliar applications first appear on leaf tissue directly contacted by the spray droplets. The injury begins with a water-soaked area where spray droplets contacted the leaf, followed by leaf chlorosis and necrosis (often referred to as leaf burn) (Figs. 274 and 275). The risk and severity of foliar symptoms is dependent upon the specific herbicide as well as environmental conditions. Soil applications may result in chlorosis and necrosis of leaf margins, generally on older leaves (Fig. 276). Atrazine is the most commonly used herbicide in this PSII inhibitor, and corn has a high tolerance to atrazine, resulting in few cases of injury.

Group 9 herbicides. Glyphosate is a potent inhibitor of 5-enolpyruvylshikimate-3-phosphate synthase (EPSPS), an enzyme in the shikimate acid pathway. Glyphosate has little soil activity and is used to control emerged weeds in no-till prior to planting and postemergence in glyphosate-resistant crops (e.g., Roundup Ready; Monsanto Co., St. Louis). Injury symptoms caused by glyphosate are variable and depend upon dose, environment, and corn developmental stage. At very low doses, the most characteristic symptom is a bleached or chlorotic leaf tissue emerging from the whorl after exposure (Fig. 277). Leaves may fail to unfurl properly, resulting in tightly rolled leaves

Fig. 270. Dinitroaniline injury (group 3 herbicide). (Courtesy D. R. Pike)

Fig. 271. Dinitroaniline injury (group 3 herbicide). (Courtesy D. R. Sumner)

Fig. 272. Phenoxyacetic acid injury (group 4 herbicide). (Courtesy D. R. Pike)

emerging from the whorl. At higher doses, growth of corn seasons and the plant becomes chlorotic or tan colored. Often, bacterial soft rot may follow injury to tissue in the whorl. Corn is extremely sensitive to glyphosate, thus injury may occur when exposed to low levels associated with drift or sprayer contamination. The majority of corn planted in the Midwest is now glyphosate resistant, therefore greatly reducing the incidence of glyphosate damage.

Group 10 herbicides. Glufosinate inhibits glutamine synthetase, an enzyme involved in the incorporation of ammonia. Disruption of this process interferes with photosynthesis, which ultimately results in plant death. Glufosinate has little soil activity, and its primary use is postemergence in glufosinate-resistant crops (e.g., LibertyLink). Injury may result because of drift from applications made on adjacent fields. Injury symptoms are similar to those of group 5 herbicides, with chlorotic and necrotic lesions developing on contacted tissue. Many corn hybrids, including but not limited to LibertyLink hybrids, contain a gene that codes for an enzyme that metabolizes glufosinate, therefore providing a high margin of safety. Delayed applications (beyond the V7 growth stage) of glufosinate to LibertyLink hybrids may result in improper unfurling of the leaves emerging from the whorl and abnormal stalk or ear development.

Group 14 herbicides. These herbicides are commonly referred to as protoporphyrinogen oxidase (PPO) inhibitors, blocking the production of a precursor involved in chlorophyll synthesis. Some group 14 herbicides are used solely preemergence, whereas others are primarily postemergence materials. Foliar contact from direct applications of carfentrazone, or drift from group 14 herbicides used in soybean, result in chlorotic and necrotic lesions on foliage wetted by spray droplets. Rain shortly after applications of carfentrazone may wash the herbicide into the whorl, killing the tissue and preventing normal unfurling of leaves. Late-season applications of fomesafen in soybean followed by dry soil conditions may result in carryover injury to corn the next season. Carryover causes necrotic

Fig. 273. Benzoic acid injury (group 4 herbicide). (Courtesy B. M. Anderson)

Fig. 274. Triazine injury (group 5 herbicide). (Courtesy A. G. Hager)

Fig. 275. Damage caused by postemergence application of bromoxynil (group 6 herbicide). (Courtesy A. G. Hager)

Fig. 276. Triazine injury (group 5 herbicide). (Courtesy D. R. Pike)

Fig. 277. Glyphosate injury (group 9 herbicide). (Courtesy A. G. Hager)

veins on corn seedlings, resulting in striping of the leaves. The root system of affected plants usually shows no symptoms.

Group 15 herbicides. Acetamide herbicides and related compounds interfere with the synthesis of long-chain fatty acids. They have little activity on emerged weeds and are used primarily preemergence. Injury is usually manifested when soils are cold and very moist or when other conditions slow the rate of herbicide metabolism within young corn plants. These herbicides may cause the corn plant to leaf out underground, resulting in emergence problems. Additional injury symptoms include leaves that do not emerge correctly from the whorl or that do not unfurl completely, causing the next leaf to be trapped and resulting in a buggy whip appearance (Fig. 278). Most products utilize safeners, therefore reducing the risk of injury.

Fig. 278. Chloroacetamide injury (group 15 herbicide). (Courtesy A. G. Hager)

Group 22 herbicides. Paraquat (and other bipyridylium herbicides) interferes with electron transfer in photosystem I. It has little soil activity and is used primarily in no-till to control emerged weeds at planting. Symptoms appear quickly after spray droplets land on foliage, resulting in necrotic lesions of contacted tissue (Figs. 279 and 280). Symptoms of paraquat drift may be confused with symptoms of holcus spot.

Fig. 279. Bipyridylium injury (group 22 herbicide). (Courtesy E. L. Stromberg)

Fig. 280. Bipyridylium injury (group 22 herbicide). (Courtesy G. C. Bergstrom)

Group 27 herbicides. These herbicides block the activity of an enzyme (4-hydroxyphenylpyruvate dioxygenase [HPPD]) involved in the synthesis of carotene pigments. They are used both preemergence and postemergence in corn. Their use on corn occasionally results in crop injury. Symptoms include chlorosis and bleaching of foliage. Clomazone (e.g., Command; FMC Corp., Philadelphia) is a group 13 herbicide that also blocks the synthesis of carotene pigments, but it affects a different enzyme than the group 27 herbicides. Clomazone may damage corn with drift (both particle and vapor) or from carryover.

Selected Reference

Hoeft, R. G., Nafziger, E. D., Johnson, R. R., and Aldrich, S. R. 2000. Modern Corn and Soybean Production. MCSP Publications, Champaign, IL.

(Prepared by R. G. Hartzler and A. G. Hager)

Environmental Factors

Drought

Lack of moisture causes the upper leaves of corn plants to roll upward and inward and appear dull (Fig. 281). On young plants that have sufficient nitrogen but lack moisture, the leaves roll and become dull in appearance but do not turn yellow. In older

Fig. 281. Drought stress symptoms. (Courtesy G. P. Munkvold)

plants, drought produces nitrogen deficiency, even if enough nitrogen is applied. Severe stunting and irregular brown or tan patches of dead leaf tissue may develop, especially under marginal drought conditions (Figs. 282 and 283). Seedlings may die, particularly in a loose, coarse seedbed and where seed located in moist soil germinates in an area surrounded by dry soil. Dry surface soil, shallow planting, and lack of crown root development result in poor emergence or seedling death.

Excess Moisture

Roots in waterlogged soils suffer oxygen starvation, causing the lower leaves of young plants to turn yellow and die. The crown area may also die and the entire plant may turn yellow, wilt, and die. Warmer weather during flooding hastens the death of corn plants. Plants up to 25 cm tall have a low tolerance to wet or flooded soils; plants with nodal roots are better able to tolerate "wet feet." After flooding, there can be soil crusting, and soil deposits in the whorl of the plant may interfere with plant growth and lead to decay. Excessive moisture also favors seed rot and seedling disease, Pythium root rot, and crazy top; root damage caused by flooding can make plants less tolerant to later season stresses, such as drought, and can predispose them to stalk rot.

Frost

Young corn plants are sometimes killed by a severe frost or freeze. Damage is most severe on organic (muck and peat) soils and is most common in low-lying areas. Small microclimate changes—such as weed growth, cultivation, row orientation—affect plant injury. When the growing point is still underground, only the leaves are usually killed (Fig. 284) and plants normally recover. However, periods of cold, wet weather following a frost event can slow plant growth and result in plant death. Bacteria (*Pseudomonas syringae* and *Erwinia herbicola*) serve as ice nuclei on the surfaces of seedling corn leaves, and the severity of frost damage increases as the density of bacterial cells increases.

High Temperatures

High temperatures (38–52°C) cause scalding of corn leaf tissues, usually in association with low soil moisture and hot, dry winds. Terminal leaves are at first chlorotic and then roll and become bleached. Hybrids differ in their tolerance to high temperatures. During pollination, high temperatures (≥35°C), especially in combination with low relative humidity, can reduce pollen viability and dessicate silks, resulting in poor seed set. Emerging tassels may be sterile and appear "blasted" or bleached. High temperatures in conjunction with low moisture levels may result in reduced photosynthesis, which predisposes plants to stalk rot later in the season.

Low-Temperature Sunscald and Cold Banding

Sunscald usually happens when cool (5–15°C), clear, dewy nights are followed by clear, sunny mornings. The symptoms of sunscald are often seen along tree lines in plants where dew accumulates in the whorl and the sun rises over the tree line and shines into the water-filled whorl. The condition commonly begins when plants are between 20 and 45 cm in height. Yellow or water-soaked bands may also appear across one or more leaves on young plants (Fig. 285) or mature plants (Fig. 286).

Fig. 282. Drought injury. (Courtesy L. E. Claflin)

Fig. 283. Drought injury. (Courtesy L. E. Claflin)

Fig. 284. Frost injury to a seedling. (Courtesy B. Wilson, DuPont Pioneer)

Fig. 285. Early symptoms of sunscald on a seedling. (Courtesy D. G. White)

The streaks on leaves usually eventually turn white (Fig. 287). Sunscald is not easily confused with any of the diseases caused by bacteria or fungi because the dead tissue is usually white. Plants usually recover rapidly. However, sunscald symptoms may appear on plants with stalk rot because of poor vascular function in combination with rapid transpiration.

Wind

Strong winds can cause several types of damage to corn plants that are 10–25 cm tall. Leaf tips and edges turn white and then brown. Driving winds may carry sand and soil particles that cause abrasions on the leaves or stalks on the windward side of seedlings. Plants may also break or be injured below the soil line, hampering growth if the growing point remains viable and killing the plant if not. A broken ring may form around the coleoptile just below the soil line. Such seedlings have purple or reddish purple leaves, similar to those with phosphorus deficiency. Shredding of old and young plants occurs in areas where high winds prevail. Secondary roots are broken off (stubby) and do not become established in the soil. Plants lodged before complete establishment of the brace roots (and growth) may exhibit a bowed stalk that tends to make harvesting and walking between the rows difficult. Plants lodged after full attainment of vegetative growth may not recover. Wind injury can predispose plants to common smut or Goss's bacterial wilt in areas where the disease is prevalent.

Fig. 286. Sunscald. (Courtesy D. G. White)

Fig. 287. Sunscald symptoms on a mature leaf several weeks after injury. (Courtesy D. G. White)

Fig. 288. Greensnap. (Courtesy R. Elmore)

Midseason, wind can cause breakage usually known as greensnap. Corn is most susceptible to greensnap between stage V10 until just prior to tasseling. Plants affected by greensnap can break anywhere along the stalk, but they usually break at a node below the primary ear (Fig. 288). Broken plants may still be able to produce a nubbin, a small nonproductive ear. Yield loss of snapped plants is directly related to the amount of stalk breakage. If 10% of plants are broken, this results in a 10% yield reduction.

Sometimes there is a pinching of the stalk rather than a clean break. Pinching is the crushing of stalk tissue on one side of the stalk, causing the plant to lean over but still remain attached and intact. Pinched and root-lodged plants try to reorient into a vertical position and produce an ear. Yield reduction occurs from both pinched and root-lodged plants. Yield loss from damage like this is difficult to predict. Expect harvesting to be more difficult in some areas.

Hybrids vary dramatically in their tolerance to greensnap. Stage of growth also affects breakage as mentioned above. Factors that increase early-season growth tend to increase breakage susceptibility, such as high nitrogen, phosphorus, and potassium rates; spring-applied nitrogen; tillage; and high levels of organic matter.

Selected References

Arny, D. C., Lindow, S. E., and Upper, C. D. 1976. Frost sensitivity of *Zea mays* increased by application of *Pseudomonas syringae*. Nature 262:282-284.

Hoppe, P. E. 1961. Sun scald of corn in Wisconsin in 1961. Plant Dis. Rep. 45:787.

Lindow, S. E., Arny, D. C., and Upper, C. D. 1982. Bacterial ice nucleation: A factor in frost injury to plants. Plant Physiol. 70:1084-1089.

(Prepared by E. L. Stromberg, E. S. Hagood, and D. G. White)

Air Pollutants

Air pollutants cause substantial annual losses by causing visible injury to leaves or by suppressing plant growth without visible injury. Pollutants may be either primary or secondary. Primary pollutants (e.g., sulfur dioxide and hydrogen fluoride) are emitted in a form toxic to plants. They are usually emitted from combustion of fossil fuel, ore, or fertilizer processing and other industrial sources. Secondary pollutants (e.g., ozone and peroxyacetyl nitrate) result from photochemical reactions between primary pollutants (waste hydrocarbons and nitrogen dioxide) in the air and oxygen. The energy for these reactions is provided by sunlight.

In general, the extent and nature of air pollution injury depend on genetic and meteorological factors and on the concentration and duration of the exposure to various pollutants alone or in combination. When plants are examined for air pollution damage, the possible source of pollutants, terrain, previous weather patterns, abnormal symptoms on other vegetation in the area, and time of symptom onset all have to be considered, and there is always the possibility that the symptoms are caused by microscopic pathogens, insects, abnormal nutrition, or various environmental stresses.

Sweet corn is much more sensitive to air pollutants than field corn. Symptoms of fluoride toxicity and ozone injury are often similar to those produced by potassium deficiency or normal senescence.

Ozone

Sweet corn exposed to ozone develops dull, gray-green, water-soaked lesions that become tan to white necrotic areas on both leaf surfaces. Sensitivity varies among hybrids. Small, silver gray to tan, necrotic areas develop on some cultivars in bands across one or both leaf surfaces. The oldest leaves may be injured at the bases, the next oldest leaves across the middle, and the youngest leaves at the tips. Leaf margins are most severely injured, but the midrib remains undamaged.

Ozone injury may be confused with insect or mite damage, chlorine or fluoride injury, or natural senescence.

Fluorides

Sweet corn exposed to fluorides shows a chlorotic mottle or "flecking" along the leaf margins and tips. Small, irregular, chlorotic spots develop between the veins and may form continuous chlorotic bands. More severe injury causes interveinal and marginal chlorosis, and some areas become necrotic.

Zinc or potassium deficiency, mite injury, genetic variation, or normal senescence cause symptoms resembling fluoride injury.

Chlorine

Chlorine injury to corn causes interveinal, tannish yellow, necrotic streaks. Middle-aged or older leaves are usually more susceptible than young leaves.

Sulfur Dioxide

Sulfur dioxide can injure corn plants of all ages. Acute injury appears as white, necrotic areas between the veins. Acute or chronic injury results in tip dieback and premature senescence of leaves.

Peroxyacetyl Nitrate

Corn is generally considered resistant to peroxyacetyl nitrate (PAN) injury. Distinct, bleached, chlorotic bands form across the leaf blade. Young leaf tips may die back. If injury is severe, leaf cells in the band may totally collapse, and the leaf folds downward at this point. Young, rapidly growing plants are more sensitive to peroxyacetyl nitrate than older plants.

Selected References

Heck, W. W. 1968. Factors influencing expression of oxidant damage to plants. Annu. Rev. Phytopathol. 6:165-188.

Heggestad, H. E. 1968. Diseases of crops and ornamental plants incited by air pollutants. Phytopathology 58:1089-1097.

Heggestad, H. E., and Heck, W. W. 1971. Nature, extent, and variation of plant response to air pollutants. Adv. Agron. 23:111-145.

Hicks, D. R., ed. 1976. Diagnosing Vegetation Injury Caused by Air Pollution. Environmental Protection Agency, Washington, DC.

Jacobson, J. S., and Hill, A. C., eds. 1970. Recognition of Air Pollution Injury to Vegetation: A Pictorial Atlas. Air Pollution Control Association, Pittsburgh.

(Prepared by E. L. Stromberg, E. S. Hagood,
A. G. Hager, and D. G. White)

Other Abnormalities

Buggy-Whip

The characteristic symptom of buggy-whip is the slight or complete fusion of the upper part of the plant into a tight tube that encloses the upper leaves and tassel. The tube develops above the soil line and is composed of thick, heavily lignified tissues that resemble a buggy whip or an onion leaf (Fig. 289). Buggy-whip may be caused by some herbicides, such as 2,4-dichlorophenoxyacetic acid (2,4-D). However, it is not always the result of herbicide injury. Certain hybrids may be more severely affected with this disorder, particularly under very warm growing conditions. Generally, hybrids with a B73-related inbred line are more likely to develop buggy-whip. Plants outgrow this condition. Buggy-whipped leaves usually unfurl in 1 or 2 weeks and appear crinkled in some areas and overall very yellow. The yellow disappears after the leaves are exposed to the sun for a day or so.

Rootless Corn Syndrome

Rootless corn usually appears in plants between stages V3 and V8 that lack most or all nodal roots. Existing nodal roots may appear stubby and blunt and are not anchored to the soil (Fig. 290). Because of a lack of a root system, the affected plants may lodge, wilt, have stunted growth, and die. Rootless corn is attributed to many things, including hot, dry surface soils; excessive rainfall that erodes soil near the base of the plant; or strong, abrasive winds that break upper roots or inhibit establishment of a strong root system. Management factors

Fig. 289. Buggy-whip symptom. (Courtesy R. Elmore)

Fig. 290. Rootless corn syndrome. (Courtesy R. Elmore)

that may inhibit nodal root formation include shallow planting depth, compact soils, and loose or cloddy soils. In loose soils or with rotary hoeing, shoots are exposed to light sooner than normal, causing nodal roots to form closer to the soil surface. Plants can recover if rainfall promotes nodal root development. Cultivation to move soil over exposed roots aids in recovery, yet this is extremely difficult if plants are lodged or in no-till.

Soil Crusting and Leafing Out Underground

Crusting is caused by planting being in wet soils, intensive tillage over the row, and heavy rains after planting. It worsens when these factors are coupled with fine-textured soils. Crusted soils:

- delay or inhibit seedling emergence, reducing plant population because of some plants not emerging
- increase variability of emergence
- limit nodal root growth
- reduce yield because of any or all of the above factors

Fields should be assessed by counting the number of emerged plants and determining whether struggling seedlings will reach the soil surface. Yield potential should be estimated from the remaining stand to determine whether replanting is warranted.

Fig. 291. Purple leaf sheath. (Courtesy D. G. White)

Fig. 292. Adaxial surface of a leaf sheath with purple leaf sheath symptoms. (Courtesy D. G. White)

Rotary hoeing may be an option for some situations, although it can often reduce stands by 5–10%.

Purple Leaf Sheath

Purple leaf sheath is often on stalks of hybrids and inbred lines late in the growing season. It is a harmless discoloration of the sheath of the leaf produced when saprophytic fungi, such as yeast and *Fusarium* spp., as well as bacteria grow on pollen or other particulate matter lodged between the stalk and

Fig. 293. Genetic striping. (Courtesy P. E. Lipps)

Fig. 294. White striping caused by genetic mutation in a breeding nursery. (Courtesy D. G. White)

Fig. 295. White leaf blade caused by genetic mutation. (Courtesy G. P. Munkvold)

the leaf sheath (Figs. 291 and 292). The stalk tissue under the sheath is not discolored or harmed. Purple leaf sheath may be confused with Physoderma brown spot; however, Physoderma brown spot also produces lesions on leaves.

Genetic Stripe, Leaf Spots, and Flecks

Numerous genes are known to cause abnormal morphology and color patterns in corn plants. Many genes are simple recessive; only a few are known to be dominant. Some alter or limit the ability of the corn plant to produce chlorophyll and result in albino plants or plants with white to yellow stripes and other variegations (Figs. 293–295). Some leaf spots and mosaic patterns mimic symptoms caused by infectious diseases (Figs. 296 and 297). When seen in a commercial production field, they often are rogue plants. Generally, plants with these genetic abnormalities are seen only in corn-breeding nurseries, but occasionally, they may be found on one parent in a hybrid seed production field or, rarely, in hybrid grain production fields. For example, the mosaic pattern on leaves of some C103- and Mo17-type inbred lines is easily confused with symptoms of a viral infection (Fig. 298).

Barren Stalks, Red Stalk, and Red Leaf

Low fertility, drought, dense plant populations, aphid or herbicide damage, chewed off silks or poor timing of silking, and poor pollen shed may all contribute to barren stalks or cobs without grain. Red leaf may also be caused by a broken midrib. A red or purple coloration is caused by the accumulation of sugars and other photosynthetic products in leaves, sheaths, or stems (Figs. 299 and 300). This can also be found in some cases when ear rot or other disease prevents pollination.

Fig. 296. Genetic ring spot. (Courtesy C. De Leon)

Fig. 297. Genetic leaf spotting of a corn inbred line, resembling infectious leaf spot symptoms. (Courtesy G. P. Munkvold)

Fig. 298. Genetic yellow flecking common on inbred line Mo17 (pictured) and related inbred lines that is often mistaken for a virus symptom. (Courtesy D. G. White)

Fig. 299. Discoloration caused by broken midribs; healthy leaf on the top. (Courtesy D. G. White)

Fig. 300. Red stalk caused by poor pollination (right) compared with a pollinated plant (left). (Courtesy D. G. White)

Kernel Red Streak

Red streaking of the kernel pericarp is a symptom of kernel red streak (Figs. 301 and 302). Streaked kernels are more common at the ear tips, particularly if the husks are loose and the kernels are exposed. Kernel red streak is most common on yellow dent corn and least common on white corn, but it may be found on sweet corn and popcorn. Streaks on the pericarp extend from the kernel base to the crown and vary from deep red in yellow kernels to pink-purple in white kernels. Kernel red streak is associated with a toxin produced during feeding by a strain of the wheat curl mite, *Eriophyes tulipae*, the vector of *Wheat streak mosaic virus*.

Exposed Ear Tips

Exposed ear tips are most common when extreme drought during kernel set is followed by abundant rainfall and good growing conditions. In extreme cases, the ears may outgrow the husk cover.

Popped Kernel and Silk-Cut

Popped kernel and silk-cut, although common in breeders' nurseries, are rarely seen on commercial hybrids. It is assumed that this characteristic is inherited, and it is usually eliminated during the breeding and selection process. Popped kernel is an irregular break in the seed coat over the kernel crown (Fig. 303). The kernel resembles a partially expanded popcorn kernel. This phenomenon is believed to result from irregular growth because it is most common during years with irregular rainfall, particularly when conditions are very hot and dry. Silk-cut is the embedment of silks in ruptured areas on tips of the kernel or occasionally in the sides of kernels between kernel rows (Figs. 304–306). The exact cause of silk-cut is unknown, but it may be caused by irregular growth of the pericarp around unpollinated silks. Even on an ear that appears to be fully pollinated, as many as 10% of the ovules are not fertilized. Pollinated silks die and dry up, but silks attached to unpollinated

Fig. 301. Kernel red streak. (Courtesy D. G. White)

Fig. 302. Kernel red streak. (Courtesy R. Nielsen)

Fig. 303. Popped kernel and silk-cut. (Courtesy D. G. White)

Fig. 304. Silk-cut. (Courtesy D. G. White)

Fig. 305. Silk-cut. (Courtesy D. G. White)

ovules remain viable for an extended period of time. The viable silks push against the developing pericarp of kernels, causing the pericarp to rupture. Both popped kernel and silk-cut result in the rupture of the pericarp and affected kernels are often heavily colonized by ear-rotting and saprophytic fungi.

Branched Ear

The lateral branch of the corn plant that bears the ear (shank) is a shortened structure somewhat similar to the main stem. The ear shank is composed of very short internodes with modified leaf sheaths and leaf blades (husks) that come off each internode and cover the ear. Rudimentary buds in the axils of the husk leaves may remain as meristematic rudiments and not be noticed, or they may develop fully into ear-bearing branches. In some primitive cultivars and modern hybrids of corn, these axillary branches produce secondary ears large enough to be worth harvesting. Usually, these small secondary ears in the axils of the husk produce little, if any, grain because the main plant has completed pollen shed (Fig. 307).

Blunt Ear Syndrome

Ears are markedly short with fewer kernels per row but with a normal number of rows per ear (Figs. 308–310). The ear tip is extremely small compared with the lower portion of the ear. Husk length also may be normal (Fig. 311). Multiple possible causes are proposed, yet temperature stress (brief cold shock) during stages V8 through V12 and applications of surfactants

Fig. 306. Silk-cut. (Courtesy G. Odvody)

Fig. 307. Branched ear. (Courtesy D. G. White)

Fig. 308. Blunt ear syndrome. (Courtesy G. P. Munkvold)

Fig. 309. Ears with blunt ear syndrome; healthy ear on the left. (Courtesy A. E. Robertson)

Fig. 310. Blunt ear syndrome. (Courtesy A. E. Robertson)

Fig. 311. Plant with a normal husk length but abnormally blunt ear. (Courtesy G. P. Munkvold)

Fig. 312. Tassel ear. (Courtesy D. G. White)

associated with pesticides before stage VT are often contributing factors.

Floral Abnormalities—Tassel Ear and Tassel Tip

Early in their development, tissues that will form tassels or ears are similar and contain initials of both male and female parts. During development, flower parts of one sex are usually suppressed so that the flower is ultimately imperfect (i.e., only one sex). Tassel ears often occur on tillers (suckers) and thus are associated with plants injured by hail, frost, flooding, herbicides, or mechanical damage and along field edges.

Functionally perfect flowers may develop. An occasional female flower (cob and kernel tissues) with its long silks may be found in the tassel of almost any corn hybrid (Fig. 312). Plants with certain genetic backgrounds have a tendency to do this more often than others. Male flowering parts may occasionally be found in the ear, usually at the tip. In some sweet corn with particular genetic backgrounds, this is common and is referred to as tassel tip. Floral abnormalities are common in plants with certain genetic backgrounds; but in most commercially used corn hybrids, their occurrence is triggered by environmental factors.

Selected References

Ashman, R. B., and Ullstrup, A. J. 1976. A heritable, noninfectious necrosis of maize leaves. J. Hered. 67:220-222.

Nault, L. R., Briones, M. L., Williams, L. E., and Barry, B. D. 1967. Relation of the wheat curl mite to kernel red streak of corn. Phytopathology 57:986-989.

Neuffer, M. G., and Calvert, O. H. 1975. Dominant disease lesion mimics in maize. J. Hered. 66:265-270.

Neuffer, M. G., Coe, E. H., and Wessler, S. R. 1997. Mutants of Maize. Cold Spring Harbor Laboratory Press, Cold Spring Harbor, MI.

Weatherwax, P. 1955. Structure and development of reproductive organs. Pages 89-121 in: Corn and Corn Improvement. G. F. Sprague, ed. Academic Press, New York.

(Prepared by E. L. Stromberg, E. S. Hagood, A. G. Hager, and D. G. White)

Appendix

Diseases of Corn (syn. Maize) (*Zea mays* L.)

BACTERIAL DISEASES

Bacterial leaf blight
Pseudomonas avenae Manns subsp. *avenae* Manns
(syns. *Acidovorax avenae* (Manns) Willems et al. subsp. *avenae* (Manns) Willems et al.; *Bacillus avenae* (Manns) Burgvits; *Pseudomonas alboprecipitans* Rosen)
Bacterial leaf streak
Xanthomonas vasicola pv. *vasculorum* (Cobb) comb. nov.
Bacterial stalk rot
Erwinia carotovora (Jones) Bergey et al. subsp. *carotovora* (Jones) Bergey et al.
(syns. *Erwinia carotovora* (Jones) Bergey et al. f. sp. *zeae* Sabet; *Erwinia chrysanthemi* Burkholder et al. pv. *zeae* (Sabet) Victoria et al.; *Pectobacterium carotovorum* (Jones) Waldee emend. Hauben et al. subsp. *carotovorum* (Jones) Waldee emend. Hauben et al.)
Bacterial stripe and leaf spot
Pseudomonas andropogonis (Smith) Stapp
(syn. *Burkholderia andropogonis* (Smith) Gillis et al.)
Chocolate spot
Pseudomonas syringae van Hall pv. *coronafaciens* (Elliott) Young et al.
(syn. *Pseudomonas coronafaciens* (ex Elliott) Schaad & Cunfer pv. *zeae* Ribeiro et al.)
Goss's bacterial wilt and blight
Clavibacter michiganensis (Smith) Davis et al. subsp. *nebraskensis* (Vidaver & Mandel) Davis et al.
(syn. *Corynebacterium michiganense* (Smith) Jensen subsp. *nebraskense* (Vidaver & Mandel) Carlson & Vidaver)
Holcus spot
Pseudomonas syringae van Hall pv. *syringae* van Hall
(syn. *Pseudomonas holci* Kendrick)
Stewart's bacterial wilt
Pantoea stewartii (Smith) Mergaert et al. subsp. *stewartii* (Smith) Mergaert et al.
(syns. *Erwinia stewartii* (Smith) Dye; *Xanthomonas stewartii* (Smith) Dowson)

FUNGAL AND OOMYCETE DISEASES

Downy Mildews

Brown stripe downy mildew
Sclerophthora rayssiae R. G. Kenneth, Koltin & I. Wahl var. *zeae* Payak & Renfro
Crazy top
Sclerophthora macrospora (Sacc.) Thirum., C. G. Shaw & Naras.
(syns. *Phytophthora macrospora* (Sacc.) S. Ito & I. Tanaka; *Sclerospora macrospora* Sacc.)
Graminicola downy mildew
Sclerospora graminicola (Sacc.) J. Schröt.
Java downy mildew
Peronosclerospora maydis (Racib.) C. G. Shaw
(syn. *Sclerospora maydis* (Racib.) E. J. Butler)
Leaf splitting downy mildew
Peronosclerospora miscanthi (T. Miyake) C. G. Shaw
(syn. *Sclerospora miscanthi* T. Miyake)
Philippine downy mildew
Peronosclerospora philippinensis (W. Weston) C. G. Shaw
(syn. *Sclerospora philippinensis* W. Weston)
Rajasthan downy mildew
Peronosclerospora heteropogoni Siradhana, Dange, Rathore & S. D. Singh
Sorghum downy mildew
Peronosclerospora sorghi (W. Weston & Uppal) C. G. Shaw
(syns. *Scleropora graminicola* (Sacc.) J. Schröt. var. *andropogonis-sorghi* Kulk.; *Scleropora sorghi* W. Weston & Uppal)
Spontaneum downy mildew
Peronosclerospora spontanea (W. Weston) C. G. Shaw
(syn. *Sclerospora spontanea* W. Weston)
Sugarcane downy mildew
Peronosclerospora sacchari (T. Miyake) C. G. Shaw
(syn. *Sclerospora sacchari* T. Miyake)

Ear and Kernel Rots

Aspergillus ear rot
Aspergillus flavus Link:Fr.
Aspergillus niger Tiegh.
Aspergillus oryzae (Ahlb.) Cohn
Aspergillus parasiticus Speare
Black kernel rot
Lasiodiplodia theobromae (Pat.) Griffon & Maubl.
(syn. *Botryodiplodia theobromae* Pat.)
Cladosporium rot
Cladosporium cladosporioides (Fresen.) G. A. de Vries
(syn. *Hormodendrum cladosporioides* (Fresen.) Sacc.)
Cladosporium herbarum (Pers.:Fr.) Link
Diplodia (Stenocarpella) ear rot
Stenocarpella macrospora (Earle) B. Sutton
(syn. *Diplodia macrospora* Earle)
Stenocarpella maydis (Berk.) B. Sutton
(syns. *Diplodia maydis* (Berk.) Sacc.; *Diplodia zeae* Lév.)
Fusarium ear rot or pink ear rot
Fusarium spp.
Fusarium oxysporum Schltdl.:Fr.
Fusarium poae (Peck) Wollenw.

Fusarium proliferatum (Matsush.) Nirenberg ex Gerlach & Nirenberg
(syns. *Gibberella fujikuroi* (Sawada) Ito, mating population D; *Gibberella fujikuroi* var. *intermedia* Kuhlman; *Gibberella intermedia* (Kuhlman) Samuels, Nirenberg & Seifert)
Fusarium sporotrichioides Sherb.
Fusarium subglutinans (Wollenw. & Reinking) P. E. Nelson, Toussoun & Marasas
(syns. *Gibberella fujikuroi* (Sawada) Ito, mating population E; *Gibberella subglutinans* (E. T. Edwards) P. E. Nelson, Toussoun & Marasas)
Fusarium temperatum Scaufl. & Munaut
Fusarium tricinctum (Corda) Sacc.
Fusarium verticillioides (Sacc.) Nirenberg
(syns. *Fusarium moniliforme* J. Sheld.; *Gibberella fujikuroi* (Sawada) Ito, mating population A; *Gibberella moniliformis* Wineland)

Gibberella ear rot or red ear rot
Fusarium boothii O'Donnell et al.
Fusarium culmorum (Wm. G. Sm.) Sacc.
Fusarium graminearum Schwabe
(syn. *Gibberella zeae* (Schwein.) Petch)
Fusarium meridionale T. Aoki et al.

Gray ear rot
Botryosphaeria zeae (G. L. Stout) Arx & E. Müll.
(syns. *Macrophoma zeae* Tehon & E. Y. Daniels; *Physalospora zeae* G. L. Stout)

Nigrospora ear and cob rot
Nigrospora oryzae (Berk. & Broome) Petch
(syns. *Basisporium gallarum* Molliard; *Khuskia oryzae* H. J. Huds.)

Penicillium ear rot and blue eye
Penicillium spp.
Penicillium chrysogenum Thom
Penicillium cyclopium Westling
Penicillium expansum Link
(syn. *Penicillium glaucum* Link)
Penicillium funiculosum Thom
Penicillium oxalicum Currie & Thom
Penicillium verrucosum Dierckx
Penicillium viridicatum Westling

Rhizoctonia ear rot
Waitea circinata Warcup & Talbot
(syn. *Rhizoctonia zeae* Voorhees)

Rhizopus ear rot
Rhizopus stolonifer (Ehrenb.:Fr.) Vuill.

Trichoderma ear rot
Trichoderma viride Pers.:Fr.

Ergot
Claviceps gigantea S. F. Fuentes, Isla, Ullstrup & A. E. Rodr.

Foliar Diseases

Alternaria leaf spot
Alternaria alternata (Fr.:Fr.) Keissl.

Anthracnose leaf blight
Colletotrichum graminicola (Ces.) G. W. Wilson
(syn. *Glomerella graminicola* D. J. Politis)

Ascochyta leaf spot
Ascochyta maydis G. L. Stout
Ascochyta zeae G. L. Stout
Ascochyta zeina Sacc.

Curvularia leaf spot
Curvularia spp.
Curvularia aeria (Bat., J. A. Lima & C. T. Vasconc.) Tsuda
(syn. *Curvularia lunata* (Wakker) Boedijn var. *aeria* (Bat., J. A. Lima & C. T. Vasconc.) M. B. Ellis)

Didymella leaf spot
Didymella exitialis (Morini) E. Müll.

Diplodia (Stenocarpella) leaf spot
Stenocarpella maydis (Berk.) B. Sutton
(syns. *Diplodia maydis* (Berk.) Sacc.; *Diplodia zeae* Lév.)

Diplodia (Stenocarpella) leaf streak
Stenocarpella macrospora (Earle) B. Sutton
(syns. *Diplodia macrospora* Earle; *Stenocarpella zeae* Syd. & P. Syd.)

Eyespot
Kabatiella zeae Narita & Y. Hirats.
(syn. *Aureobasidium zeae* (Narita & Y. Hirats.) Dingley)

Fusarium leaf scorch
Fusarium verticillioides (Sacc.) Nirenberg
(syns. *Fusarium moniliforme* J. Sheld.; *Gibberella fujikuroi* (Sawada) Ito, mating population A; *Gibberella moniliformis* Wineland)

Gray leaf spot
Cercospora spp.
Cercospora zeae-maydis Tehon & E. Y. Daniels
Cercospora zeina Crous & U. Braun

Hyalothyridium leaf spot
Astromella spp.
Hyalothyridium maydis Latterell & A. E. Rossi
Leptosphaerulina australis McAlpine

Leptosphaeria leaf spot
Paraphaeosphaeria michotii (Westend.) O. E. Erikss.
(syn. *Leptosphaeria zeae* G. L. Stout)
Phaeosphaeria variiseptata (G. L. Stout) Shoemaker & C. E. Babc.
(syn. *Leptosphaeria variiseptata* G. L. Stout)

Northern corn leaf blight
Exserohilum turcicum (Pass.) K. J. Leonard & Suggs
(syns. *Bipolaris turcica* (Pass.) Shoemaker; *Drechslera turcica* (Pass.) Subram. & P. C. Jain; *Helminthosporium turcicum* Pass.; *Setosphaeria turcica* (Luttr.) K. J. Leonard & Suggs; *Trichometasphaeria turcica* Luttr.)

Northern corn leaf spot
Bipolaris zeicola (G. L. Stout) Shoemaker
(syns. *Cochliobolus carbonum* R. R. Nelson; *Drechslera zeicola* (G. L. Stout) Subram. & P. C. Jain; *Helminthosporium carbonum* Ullstrup)

Phaeosphaeria leaf spot
Phaeosphaeria maydis (Henn) Rane, Payak & Renfro
Phoma sp.

Physoderma brown spot
Physoderma maydis (Miyabe) Miyabe
(syn. *Physoderma zeae-maydis* F. J. F. Shaw)

Rhizoctonia banded leaf and sheath blight
Rhizoctonia solani J. G. Kühn
(syn. *Thanatephorus cucumeris* (A. B. Frank) Donk)

Rostratum leaf spot
Exserohilum rostratum (Drechsler) K. J. Leonard & Suggs
(syns. *Bipolaris rostrata* (Drechsler) Shoemaker; *Drechslera rostrata* (Drechsler) M. J. Richardson & E. M. Fraser; *Exserohilum halodes* (Drechsler) K. J. Leonard & Suggs; *Helminthosporium rostratum* Drechsler; *Setosphaeria rostrata* K. J. Leonard)

Selenophoma leaf spot
Selenophoma spp.

Septoria leaf blotch
Septoria spp.

Southern corn leaf blight
Bipolaris maydis (Y. Nisik. & C. Miyake) Shoemaker
(syns. *Cochliobolus heterostrophus* (Drechsler) Drechsler; *Drechslera maydis* (Y. Nisik. & C. Miyake) Subram. & P. C. Jain; *Helminthosporium maydis* Y. Nisik. & C. Miyake; *Ophiobolus heterostrophus* Drechsler)

Tar spot complex
Coniothyrium phyllachorae Maubl.
Monographella maydis E. Müll. & Samuels
Phyllachora maydis Maubl.
Yellow leaf blight
Didymella maydis (Arny & R. R. Nelson) Qian Chen & O. Cai
(syn. *Phyllosticta maydis* Arny & R. R. Nelson)
Zonate leaf spot
Microdochium sorghi (D. C. Bain & Edgerton ex Deighton) U. Braun
(syn. *Gloeocercospora sorghi* D. C. Bain & Edgerton ex Deighton)

Fungal Stalk Rots

Anthracnose stalk rot
Colletotrichum graminicola (Ces.) G. W. Wilson
(syn. *Glomerella graminicola* D. J. Politis)
Black bundle disease
Sarocladium strictum (W. Gams) Summerb.
(syn. *Cephalosporium acremonium* Corda)
Charcoal rot
Macrophomina phaseolina (Tassi) Goid.
(syns. *Botryodiplodia phaseoli* (Maubl.) Thirum.; *Macrophomina phaseoli* (Maubl.) S. F. Ashby; *Sclerotium bataticola* Taubenh.)
Diplodia (Stenocarpella) stalk rot
Stenocarpella maydis (Berk.) B. Sutton
(syns. *Diplodia maydis* (Berk.) Sacc.; *Diplodia zeae* Lév.)
Fusarium stalk rot
Fusarium spp.
Fusarium proliferatum (Matsush.) Nirenberg ex Gerlach & Nirenberg
(syns. *Gibberella fujikuroi* (Sawada) Ito, mating population D; *Gibberella intermedia* (Kuhlman) Samuels, Nirenberg & Seifert)
Fusarium subglutinans (Wollenw. & Reinking) P. E. Nelson, Toussoun & Marasas
(syns. *Gibberella fujikuroi* (Sawada) Ito, mating population E; *Gibberella subglutinans* (E. T. Edwards) P. E. Nelson, Toussoun & Marasas)
Fusarium temperatum Scaufl. & Munaut
Fusarium verticillioides (Sacc.) Nirenberg
(syns. *Fusarium moniliforme* J. Sheld.; *Gibberella fujikuroi* (Sawada) Ito, mating population A; *Gibberella moniliformis* Wineland)
Gibberella stalk rot
Fusarium graminearum Schwabe
(syn. *Gibberella zeae* (Schwein.) Petch)
Nigrospora stalk rot
Nigrospora oryzae (Berk. & Broome) Petch
Other stalk rot
Phaeocytostroma ambiguum (Mont.) Petr.
(syn. *Phaeocytosporella zeae* G. L. Stout)
Pythium stalk rot
Pythium aphanidermatum (Edson) Fitzp.
(syn. *Pythium butleri* Subram.)
Red root rot
Setophoma terrestris (H. N. Hansen) Gruyter, Aveskamp & Verkley
(syns. *Phoma terrestris* H. N. Hansen; *Pyrenochaeta terrestris* (H. N. Hansen) Gorenz, J. C. Walker & Larson)

Late Wilt

Magnaporthiopsis maydis (Samra, Sabet & Hing.) Klaubauf, Lebrun & Crous
(syns. *Cephalosporium maydis* Samra, Sabet & Hing.; *Harpophora maydis* (Samra, Sabet & Hing.) W. Gams)

Mycotoxins and Mycotoxicoses

Aflatoxins
Aspergillus spp.
Aspergillus flavus Link:Fr.
Aspergillus parasiticus Speare
Deoxynivalenol and other trichothecenes
Fusarium spp.
Fusarium cerealis (Cooke) Sacc.
(syn. *Fusarium crookwellense* L. W. Burgess, P. E. Nelson & Toussoun)
Fusarium culmorum (Wm. G. Sm.) Sacc.
Fusarium equiseti (Corda) Sacc.
Fusarium graminearum Schwabe
(syn. *Gibberella zeae* (Schwein.) Petch)
Fusarium poae (Peck) Wollenw.
Fusarium sporotrichioides Sherb.
Diplodiatoxin
Stenocarpella maydis (Berk.) B. Sutton
(syns. *Diplodia maydis* (Berk.) Sacc.; *Diplodia zeae* Lév.)
Fumonisins
Fusarium spp.
Fusarium proliferatum (Matsush.) Gerlach & Nirenberg
(syn. *Gibberella intermedia* (Kuhlman) Samuels, Nirenberg & Seifert)
Fusarium verticillioides (Sacc.) Nirenberg
(syns. *Fusarium moniliforme* J. Sheld.; *Gibberella moniliformis* Wineland)
Ochratoxins
Aspergillus spp.
Aspergillus alutaceus Berk. & M. A. Curtis var. *alutaceus* Berk. & M. A. Curtis
(syn. *Aspergillus ochraceus* G. Wilh.)
Penicillium spp.
Penicillium verrucosum Dierckx
Other mycotoxins
Penicillium spp.
Aspergillus flavus Link:Fr.
Aspergillus parasiticus Speare
Fusarium proliferatum (Matsush.) Gerlach & Nirenberg
(syn. *Gibberella intermedia* (Kuhlman) Samuels, Nirenberg & Seifert)
Fusarium subglutinans (Wollenw. & Reinking) P. E. Nelson, Toussoun & Marasas
(syns. *Gibberella fujikuroi* (Sawada) Ito; *Gibberella subglutinans* (E. T. Edwards) P. E. Nelson, Toussoun & Marasas)
Zearalenone
Fusarium cerealis (Cooke) Sacc.
(syn. *Fusarium crookwellense* L. W. Burgess, P. E. Nelson & Toussoun)
Fusarium culmorum (Wm. G. Sm.) Sacc.
Fusarium graminearum Schwabe
(syn. *Gibberella zeae* (Schwein.) Petch)

Root Rots

Fusarium root rot
Fusarium spp.
Fusarium acuminatum Ellis & Everh.
Fusarium equiseti (Corda) Sacc.
Fusarium graminearum Schwabe
(syn. *Gibberella zeae* (Schwein.) Petch)
Fusarium oxysporum Schltdl.:Fr.
Fusarium proliferatum (Matsush.) Gerlach & Nirenberg
Fusarium solani (Mart.) Sacc.
Fusarium subglutinans (Wollenw. & Reinking) P. E. Nelson, Toussoun & Marasas
Fusarium verticillioides (Sacc.) Nirenberg
Pythium root rot (and feeder root necrosis)
Phytopythium helicoides (Drechsler) Abad et al.
(syn. *Pythium helicoides* Drechsler)

Phytopythium vexans (de Bary) Abad et al.
(syn. *Pythium vexans* de Bary)
Pythium spp.
Pythium acanthicum Drechsler
Pythium adhaerens Sparrow
Pythium angustatum Sparrow
Pythium aphanidermatum (Edson) Fitzp.
Pythium arrhenomanes Drechsler
Pythium graminicola Subram.
Pythium irregulare Buisman
Pythium lutarium Ali-Shtayeh
Pythium paroecandrum Drechsler
Pythium pulchrum Minden
Pythium rostratum E. J. Butler
Pythium splendens Hans Braun
Pythium sylvaticum W. A. Campb. & F. F. Hendrix
Pythium tardicrescens Vanterp.
Pythium torulosum Coker & P. Patt.
Pythium ultimum Trow
Red root rot
Setophoma terrestris (H. N. Hansen) Gruyter, Aveskamp & Verkley
(syns. *Phoma terrestris* H. N. Hansen; *Pyrenochaeta terrestris* (H. N. Hansen) Gorenz, J. C. Walker & Larson)
Rhizoctonia crown and brace root rot
Rhizoctonia solani J. G. Kühn
(syn. *Thanatephorus cucumeris* (A. B. Frank) Donk)

Rusts
Common rust
Puccinia sorghi Schwein.
Southern rust
Puccinia polysora Underw.
Tropical rust
Phakopsora zeae (Mains) Buriticá
(syns. *Angiopsora zeae* Mains; *Physopella zeae* (Mains) Cummins & Ramachar)

Seed Rots and Seedling Blights (Damping-Off)
Alternaria spp.
Aspergillus spp.
Aspergillus flavus Link:Fr.
Aspergillus tamarii Kita
Bipolaris spp.
Fusarium spp.
Fusarium graminearum Schwabe
(syn. *Gibberella zeae* (Schwein.) Petch)
Fusarium oxysporum Schltdl.:Fr.
Fusarium proliferatum (Matsush.) Nirenberg ex Gerlach & Nirenberg
Fusarium solani (Mart.) Sacc.
Fusarium subglutinans (Wollenw. & Reinking) P. E. Nelson, Toussoun & Marasas
Fusarium temperatum Scaufl. & Munaut
Fusarium verticillioides (Sacc.) Nirenberg
Nigrospora spp.
Penicillium spp.
Penicillium oxalicum Currie & Thom
Phytopythium vexans (de Bary) Abad et al.
(syn. *Pythium vexans* de Bary)
Pythium spp.
Pythium debaryanum R. Hesse
Pythium irregulare Buisman
Pythium paroecandrum Drechsler
Pythium rostratum E. J. Butler
Pythium ultimum Trow
Rhizoctonia spp.
Rhizoctonia solani J. G. Kühn
Rhizopus spp.
Stenocarpella maydis (Berk.) B. Sutton
Trichoderma spp.

Smuts
Common smut
Ustilago maydis (DC.) Corda
(syn. *Ustilago zeae* (Beckm.) Unger)
False smut
Ustilaginoidea virens (Cooke) Takah.
(syn. *Ustilaginoidea oryzae* (Pat.) Bref.)
Head smut
Sporisorium reilianum (J. G. Kühn) Langdon & Full.
(syns. *Sorosporium reilianum* (J. G. Kühn) McAlpine; *Sphacelotheca reiliana* (J. G. Kühn) G. P. Clinton; *Sporisorium holci-sorghi* (Rivolta) Vánky; *Sporisorium reilianum* (J. G. Kühn) Langdon & Full. f. sp. *zeae* (J. G. Kühn) Langdon & Full.; *Ustilago reiliana* J. G. Kühn)

Storage Molds
Alternaria spp.
Aspergillus spp.
Aspergillus candidus Link:Fr.
Aspergillus flavus Link:Fr.
Aspergillus fumigatus Fresen.
Aspergillus glaucus (L.) Link:Fr.
Aspergillus ochraceus G. Wilh.
Aspergillus restrictus G. Sm.
Aspergillus versicolor (Vuill.) Tirab.
Cladosporium spp.
Fusarium spp.
Fusarium graminearum Schwabe
(syn. *Gibberella zeae* (Schwein.) Petch)
Fusarium verticillioides (Sacc.) Nirenberg
Mucor spp.
Nigrospora oryzae (Berk. & Broome) Petch
Penicillium spp.
Penicillium brevi-compactum Dierckx
Penicillium chrysogenum Thom
Penicillium citrinum Thom
Penicillium cyclopium Westling
Penicillium funiculosum Thom
Penicillium oxalicum Currie & Thom
Penicillium viridicatum Westling
Rhizopus spp.
Rhizopus stolonifer (Ehrenb.:Fr.) Vuill.
Trichothecium roseum (Pers.) Link

MOLLICUTE DISEASES

Corn stunt
Spiroplasma kunkelii Whitcomb et al.
Maize bushy stunt
Maize bushy stunt phytoplasma: '*Candidatus* Phytoplasma asteris', 16S rDNA group I-B
Maize redness
Stolbur phytoplasma: '*Candidatus* Phytoplasma solani', 16S rDNA group XII-A

NEMATODES (PARASITIC)

Awl nematode
Dolichodorus heterocephalus Cobb
Cyst nematodes
Corn cyst nematode
Heterodera zeae Koshy, Swarup & Sethi
Mexican corn cyst nematode
Punctodera chalcoensis Stone, Sosa Moss & Mulvey
Goosegrass cyst nematode
Vittatidera zeaphila Bernard et al.
Dagger nematodes
Xiphinema americanum Cobb
Xiphinema chambersi Thorne
Xiphinema rivesi Dalmasso
Lance nematodes
Hoplolaimus columbus Sher

Hoplolaimus galeatus (Cobb) Thorne
Needle nematode
Longidorus breviannulatus Norton & Hoffmann
Ring nematodes
Criconemella ornata (Raski) Luc & Raski
Criconemella sphaerocephala (Taylor) Luc & Raski
Root-knot nematodes
Meloidogyne arenaria (Neal) Chitwood
Meloidogyne chitwoodi Golden et al.
Meloidogyne hapla Chitwood
Meloidogyne incognita (Kofoid & White) Chitwood
Meloidogyne javanica (Treub) Chitwood
Root lesion nematodes
Pratylenchus spp.
Pratylenchus brachyurus (Godfrey) Filipjev & Schuurmans-Stekhoven
Pratylenchus hexincisus Taylor & Jenkins
Pratylenchus neglectus (Rensch) Filipjev & Schuurmans-Stekhoven
Pratylenchus penetrans (Cobb) Chitwood & Oteifa
Pratylenchus scribneri Steiner
Pratylenchus zeae Graham
Spiral nematodes
Helicotylenchus digonicus Perry
Helicotylenchus dihystera (Cobb) Sher
Helicotylenchus pseudorobustus (Steiner) Golden
Scutellonema brachyurum (Steiner) Andrássy
Sting nematodes
Belonolaimus spp.
Belonolaimus longicaudatus Rau
Stubby-root nematodes
Paratrichodorus minor (Colbran) Siddiqi (syns. *Paratrichodorus christiei* (Allen) Siddiqi; *Trichodorus christiei* Allen; *Trichodorus minor* Colbran; *Trichodorus obesus* Razjivin & Penton)
Paratrichodorus porosus (Allen) Siddiqi
Stunt nematodes
Tylenchorhynchus claytoni Steiner
Tylenchorhynchus maximus Allen

NONINFECTIOUS OR ABIOTIC DISEASES AND DAMAGE

Air Pollutants
Chlorine
Fluorides
Ozone
Peroxyacetyl nitrate (PAN)
Sulfur dioxide

Environmental Factors
Drought damage
Excess moisture damage
Damage intensified by warmer weather; plants without nodal roots
Frost damage
Most severe in organic soil and low-lying areas; increases with density of bacterial cells
High-temperature scalding
Usually in association with low soil moisture and hot, dry winds
Low-temperature sunscald and cold banding
Happens with cool, clear, dewy nights followed by clear, sunny mornings
Wind damage
Abrasions, breakage, lodging, pinching, shredding
Greensnap

Herbicide Injury
Excess rate of use; use of herbicides to which corn has a low tolerence; carryover or drift

Nutrient Deficiencies
Boron
Rare; drought; high-pH soil;, and sandy, low-organic-matter soil
Calcium
Rare; very-low-pH soil; soil high in potassium and magnesium
Copper
Rare; high-organic-matter soil
Iron
Rare; high-pH soil; wet, compacted, poorly aerated soil; cool weather
Magnesium
Strongly acidic, sandy soil in regions of moderate to high rainfall; high-potassium soil; and soil treated with limestone low in magnesium
Manganese
High-pH soil; peat, muck, and sandy soils high in organic matter
Molybdenum
Rare; low-pH soil; heavily weathered soil
Nitrogen
Cold, wet, or flooded soil; drought (especially after mid-season); sandy, low-organic-matter soil; heavy, leaching rains; standing water in warm weather; intensive row cropping
Phosphorus
Cool, wet soil
Potassium
Sandy, organic, wet, or compacted, weathered soil; potassium removal by preceding crops
Sulfur
Acidic, sandy, low-organic-matter soil; cold, wet soil
Zinc
High-phosphorus soil or fertilizer; high-pH soil; cool, wet soil; low-organic-matter soil (exposed subsoil)

Nutrient Toxicities
Anhydrous ammonia or aqua-ammonia damage
Ammonia (NH_3) release from fertilizing practices; sandy, dry, compact, and low-organic-matter soil; uneven soil surface; deep planting; shallow placement of NH_3
Fertilizer burn
Nutrient toxicity (excessive soluble nitrogen, potassium, or boron fertilizers); sandy soil; dry weather; high-salt-index fertilizer
Other nutrient excesses
Accumulation of aluminum oxides, iron oxides, nickel, and chromium

Other Abnormalities
Barren stalks
Low fertility; drought; dense plant populations; aphid or herbicide damage
Blunt ear syndrome
Temperature stress (brief cold shock); surfactants application
Branched ear
Undetermined
Buggy-whip
Some herbicides (e.g., 2,4-D); warm growing conditions; genetic connection
Exposed ear tips
Extreme drought during kernel set followed by abundant rainfall and good growing conditions
Genetic stripe, leaf spots, and flecks
Genetic abnormality
Kernel red streak
Associated with a toxin produced by a strain of the wheat curl mite, *Eriophyes tulipae*

Popped kernel
Inherited; irregular rainfall, particularly with very hot, dry conditions
Purple leaf sheath
Genetic connection; particulate matter lodged between the stalk and the leaf sheath for saprophytic fungi bacteria to grow on
Red leaf
Accumulation of sugars and other photosynthetic products
Red stalk
Broken midrib
Rootless corn syndrome
Hot, dry surface soil; excessive rainfall; or strong, abrasive winds; shallow planting; compact soil; loose or cloddy soil
Silk-cut
Inherited; exact cause unknown
Soil crusting and leafing out underground
Wet soil; intensive tillage; heavy rains after planting; fine-textured soil
Tassel ear (floral abnormality)
Plants injured by hail, frost, flooding, herbicides, or mechanical damage
Tassel tip (floral abnormality)
Genetic connection

PARASITIC PLANT

Witchweed
Striga spp.
Striga asiatica (L.) Kuntze
(syn. *Striga lutea* Lour.)
Striga hermonthica (Del.) Benth.

VIRAL DISEASES

Barley yellow dwarf
Genus *Luteovirus*:
Barley yellow dwarf virus-MAV, strain (BYDV-MAV)
Barley yellow dwarf virus-PAV, strain (BYDV-PAV)
Genus *Polerovirus*:
Cereal yellow dwarf virus-GPV, strain (CYDV-GPV)
Cereal yellow dwarf virus-RPV, strain (CYDV-RPV)
Brome mosaic
Genus *Bromovirus*, *Brome mosaic virus* (BMV)
Cereal chlorotic mottle
Genus *Nucleorhabdovirus*, *Cereal chlorotic mottle virus* (CCMV)
Cucumber mosaic
Genus *Cucumovirus*, *Cucumber mosaic virus* (CMV)
High Plains disease
Genus *Emaravirus*, *Wheat mosaic virus* (WMoV)
(syns. *High Plains virus* [HPV], *Maize red stripe virus* [MRStV])
Maize chlorotic dwarf
Genus *Waikavirus*, *Maize chlorotic dwarf virus* (MCDV)
Maize chlorotic mottle
Genus *Machlomovirus*, *Maize chlorotic mottle virus* (MCMV)
Maize dwarf mosaic
Genus *Potyvirus*:
Johnsongrass mosaic virus (JGMV)
Maize dwarf mosaic virus (MDMV)
Pennisetum mosaic virus (PenMV)
Sorghum mosaic virus (SrMV)
Sugarcane mosaic virus (SCMV)
Zea mosaic virus (ZeMV)
Genus *Tritimovirus*, *Wheat streak mosaic virus* (WSMV)
Maize fine streak
Genus *Nucleorhabdovirus*; *Maize fine streak virus* (MFSV)
Maize leaf fleck
Genus has not been assigned, *Maize leaf fleck virus* (MLFV)
Maize lethal necrosis
Genus *Machlomovirus*, *Maize chlorotic mottle virus* (MCMV) combined with Potyviridae infection, e.g., *Maize dwarf mosaic virus* (MDMV) (*Potyvirus*); *Sugarcane mosaic virus* (SCMV) (*Potyvirus*); *Wheat streak mosaic virus* (WSMV) (*Tritimovirus*)
Maize mosaic
Genus *Nucleorhabdovirus*, *Maize mosaic virus* (MMV)
Maize necrotic streak
Genus *Zeavirus*, *Maize necrotic streak virus* (MNeSV)
Maize rayado fino
Genus *Marafivirus*, *Maize rayado fino virus* (MRFV)
Maize rough dwarf
Genus *Fijivirus*:
Maize rough dwarf virus (MRDV)
Rice black streak dwarf virus (RBSDV)
Maize streak
Genus *Mastrevirus*, *Maize streak virus* (MSV)
Maize stripe
Genus *Tenuivirus*, *Maize stripe virus* (MStV)
Maize white line mosaic
Genus *Aureusvirus*, *Maize white line mosaic virus* (MWLMV)
Mal de rio cuarto
Genus *Fijivirus*, *Mal de Rio Cuarto virus* (MRCV)
Rice stripe
Genus *Tenuivirus*, *Rice stripe virus* (RSV)
Wheat streak mosaic
Genus *Tritimovirus*, *Wheat streak mosaic virus* (WSMV)

Glossary

a.i.—active ingredient
C—Celsius or centigrade
cm—centimeter (1 cm = 0.01 m = 0.3937 in.)
F—Fahrenheit
g—gram (1 g = 0.03527 oz)
gal—gallon (1 gal liquid [U.S.] = 3.785 L)
h—hour
ha—hectare (1 ha = 2.471 acres)
in.—inch (1 in. = 2.540 cm)
kg—kilogram (1 kg = 2.205 lb)
L—liter (1 L = 1.057 quarts liquid [U.S.])
lb—pound (1 lb = 453.59 g)
m—meter (1 m = 39.37 in.)
mg—milligram (1 mg = 0.001 g)
min—minute
ml—milliliter (1 ml = 0.001 L)
mm—millimeter (1 mm = 0.001 m = 0.03937 in.)
µg—microgram (1 µg = 10^{-6} g)
µm—micrometer (1 µg = 10^{-6} m)
nm—nanometer (1 nm = 10^{-9} m)
oz—ounce (1 oz = 28.35 g); fluid ounce (1 fl oz [U.S.] = 29.57 ml)
ppm—parts per million
sec—second

abaxial—directed away from the axis; pertaining to the side of a leaf facing away from the stalk (*see* adaxial)
abiotic—pertaining to or caused by nonliving factors, e.g., abiotic disease
acervulus (pl. acervuli)—an erumpent, saucer-shaped, cushionlike fruiting body of a fungus bearing conidiophores, conidia, and sometimes setae
acute—pertaining to symptoms that develop suddenly (*see* chronic)
adaxial—directed toward the axis; pertaining to the side of a leaf facing toward the stalk (*see* abaxial)
additive gene effect—a gene action in which the effects on a genetic trait are enhanced by each additional gene (*see* dominant gene effect)
aeciospore—a binucleate, dikaryotic spore produced in an aecium of a rust fungus
aecium (pl. aecia; adj. aecial)—a cup-shaped fruiting body of a rust fungus; a "cluster cup" that produces chains of aeciospores
aerobic—living only in the presence of air or oxygen
agar—a gelatin derived from algae and used to solidify culture media; a growth medium containing agar
alternate host—one of two species of plants on which some parasitic fungus (e.g., a heteroecious rust) must develop to complete its life cycle
anaerobic—living in the absence of air or oxygen
anamorph (adj. anamorphic)—the imperfect state or asexual form in the life cycle of certain fungi (*see* teleomorph)
anastomosis (pl. anastomoses)—the fusion of branches of the same or different structures (e.g., hyphae) to make a network
annulation—the division of the cuticle of nematodes into transverse striae (appearing as lines) and raised annules (appearing as ridges)
antheridium (pl. antheridia)—the male sexual organ (male gametangium) of certain fungi
anthracnose—a leaf or stem disease caused by acervuli-forming fungi (order Melanconiales); characterized by sunken lesions and necrosis
antigen—any foreign chemical (normally, a protein) that induces antibody formation in warm-blooded animals
apex (pl. apices; adj. apical)—the tip of a root or shoot
aplerotic—pertaining to oospores (e.g., of *Pythium* spp.), not filling the oogonium
apothecium (pl. apothecia)—the open, cuplike or saucerlike, ascus-bearing fruiting body (ascocarp) of some ascomycetous fungi; often supported on a stalk
appressorium (pl. appressoria)—the swollen, flattened portion of a fungal filament (e.g., germ tube or hypha) that adheres to the surface of a higher plant (host), providing anchorage for invasion by the fungus
ascocarp—a sexual fruiting body (e.g., apothecium, cleistothecium, perithecium, pseudothecium) of ascomycetous fungi, producing asci and ascospores
ascomycete (adj. ascomycetous)—a member of a group of fungi in the phylum Ascomycota (kingdom Fungi) that produces sexual spores (ascospores) within an ascus
ascospore—a sexual spore borne in an ascus
ascus (pl. asci)—a saclike or clavate structure containing ascospores (typically eight) usually borne within a fungal fruiting body
aseptate—lacking cross-walls (septa); nonseptate; coenocytic
asexual—vegetative; lacking sex organs, gametes, or sexual spores (as in the imperfect or anamorphic state of a fungus)
avirulent—nonpathogenic; unable to cause disease (*see* virulent)

backcross—to cross (mate) an hybrid offspring with one of its parents or with a genetically equivalent organism
basidiocarp—a sexual fruiting body of basidiomyceteous fungi, producing basidia
basidiomycete (adj. basidiomycetous)—a member of a group of fungi in the class Basidiomycetes (phylum Basidiomycota; kingdom Fungi) that has septate mycelium, sometimes with clamp connections, and produces sexual spores (basidiospores or sporidia) on a basidium
basidiospore—a haploid (*n*) sexual spore produced on a basidium
basidium (pl. basidia; adj. basidial)—a short, club-shaped, haploid (*n*) promycelium produced by basidiomycetous fungi (including smut and rust fungi) on which basidiospores are produced externally
biotic—pertaining to or caused by living organisms
biotype—a subdivision of a species, subspecies, or race having a like genetic makeup (i.e., some identifiable physiological trait)
blade—the flat or expanded portion of a leaf or petal above the sheath
blight—any sudden, severe, and extensive spotting, discoloration, wilting, or destruction of leaves, flowers, stems, or entire plants that usually affects young, growing tissues (in disease names, it is often coupled with the name of the affected part of the host, e.g., leaf blight and tip blight)
blotch—an irregularly shaped, usually superficial, spot or blot
bursa—a lateral, cuticular extension or flap at the posterior of the sexual organ of certain male nematodes; used for orienting the body during mating

capsid—the protective protein coat surrounding the nucleic acid core of a virus
cephalic framework—a subcuticular framework that supports the lip region of plant-parasitic nematodes and to which are attached the stylet protractor muscles

chlamydospore—a thick-walled or double-walled, asexual resting spore formed from hyphal cells (terminal or intercalary) or by transformation of conidial cells

chlorosis (adj. chlorotic)—the failure of chlorophyll development caused by disease or nutritional disturbance (e.g., lack of iron, zinc, or magnesium); or fading of green plant parts to light green, yellow, or white

chronic—pertaining to symptoms that are slow developing, persistent, or recurring (*see* acute)

circulative—pertaining to viruses that circulate in the body or pass through the gut of an insect vector before being transmitted to a host

cleistothecium (pl. cleistothecia)—a usually spherical ascocarp that is closed at maturity, typical of the powdery mildew fungi

coenocytic—having multiple nuclei embedded in the cytoplasm and not separated by cross-walls; nonseptate

conidiophore—a simple or branched fertile hypha on which conidia are produced

conidium (pl. conidia)—an asexual, nonmotile fungal spore that develops externally at the tip or side of a specialized conidiophore; an asexual spore other than an intercalary chlamydospore or sporangiospore (syn. conidiospore)

cortex (adj. cortical)—the region of parenchyma tissue between the epidermis and phloem in stems and roots; the region beneath the rind of a sclerotium; a more or less thick outer covering; peridium

culm—the stem of a grass plant

cultivar—a cultivated variety of a plant species, resulting from deliberate genetic manipulation and having recognizable characteristics, such as color; shape of flowers, fruits, and seeds; height; and form (syn. variety)

cuticle (adj. cuticular)—the waxy, noncellular layer formed on the outer walls of plant epidermal cells; outer sheath or membrane of a nematode or plant

cyst—in fungi, a saclike covering formed around zoospores; in nematodes, the dead body of a female nematode that surrounds and protects her eggs

cytoplasmic sterility—the sterility of male flowers resulting from specific cytoplasmic genes

damping-off—the rapid, lethal decline of a germinating seed or seedling before or shortly after emergence caused by decomposition of the root or lower stem (it is common to distinguish between preemergence and postemergence damping-off)

dikaryon (adj. dikaryotic)—an organism having two sexually compatible haploid nuclei ($n + n$) per cell (common in rust and smut fungi)

diploid—having two complete sets of chromosomes ($2n$) per cell (*see* haploid)

dissemination—the spread of infectious material (inoculum) from a diseased to a healthy plant by wind, water, man, insects, mites, machinery, etc. (syn. dispersal)

dominant—pertaining to a gene of a phenotypic trait that expresses itself in a hybrid (a dominant gene needs to be present in only one of two parents of a single-cross hybrid in order to be expressed) (*see* recessive)

dominant gene effect—a gene action in which the offspring is more like one parent than the other (*see* additive gene effect)

dormant—resting; living in a state of reduced physiological activity

dorsal—situated toward the back or upper surface of the body (ant. ventral)

double cross—a cross between two single crosses (a double-cross corn hybrid has as a female parent a cross between two inbred lines and as a male parent a cross between two inbred lines)

echinulate—having small, pointed spines projecting from the cell wall (e.g., spores)

ELISA—*see* enzyme-linked immunosorbent assay

embryo—the seed "germ"; the rudimentary plant within a seed

endemic—native to a particular place, such as one country or geographical region

enzyme-linked immunosorbent assay (acronym ELISA)—a serological test in which the sensitivity of the antigen–antibody reaction is increased by attaching an enzyme that produces a colored product in response to one of the reactants

eradicate—to destroy or remove the pathogen or the plants that carry the pathogen after disease is established

etiology—the science or study of the causes or origins of a disease or abnormality

eukaryote (adj. eukaryotic)—an organism with a membrane-bound nucleus and other organelles (*see* prokaryote)

exudate—a substance excreted by or discharged from diseased tissues, roots, leaves, and fungi; an ooze

facultative—capable of changing lifestyle (e.g., from saprophytic to parasite or the reverse)

fascicle—a small group, bundle, or cluster (especially of fungi)

filamentous—threadlike (syn. filiform)

flagellum (pl. flagella)—hairlike, whiplike, or tinsellike appendage of a motile cell (e.g., bacterium or zoospore) that provides locomotion

Fungi Imperfecti—a group of fungi for which the sexual stage is not known; a group comprising the asexual stages of fungi in the groups ascomycetes and basidiomycetes (syn. deuteromycetes)

fusiform—spindle shaped; tapering at each end

fusoid—somewhat fusiform

gamete—a sex cell, especially a cell formed in a gametangium, that fuses with another sex cell in sexual reproduction

general resistance—a nonspecific host resistance; resistance to all or most races of a pathogen (*see* specific resistance)

genotype—the genetic constitution of an individual or group; a class or group of individuals sharing a specific genetic makeup

genus (pl. genera)—a taxonomic category that consists of species that are closely related structurally or phylogenetically (the genus or generic name is the first name in a Latin binomial)

germ tube—a hypha resulting from an outgrowth of the spore wall, cytoplasm, or both after germination

gram-negative—pertaining to bacteria, staining red or pink after treatment with Gram's stain

gram-positive—pertaining to bacteria, staining violet or purple after treatment with Gram's stain

haploid—having a single complete set of chromosomes (n) per cell (*see* diploid)

haustorium (pl. haustoria)—a specialized hyphal branch of a parasite formed within a living host cell for absorption of food, often associated with rusts, downy and powdery mildews, parasitic flowering plants, and other obligate parasites

helix (adj. helical)—a spiral shape or form

heterosis—the increased vigor of a hybrid compared with that of its parents, resulting from crossing genetically unlike parents (syn. hybrid vigor)

heterothallic—having a life cycle in which sexual reproduction can occur only in the presence of genetically different mycelia (e.g., sexes separated in different mycelia) (*see* homothallic)

homothallic—having a life cycle in which sexual reproduction occurs in a single, self-fertile thallus (e.g., both sexes in the same mycelium) (*see* heterothallic)

honeydew—an ooze or exudate of sweetish liquid

hybrid—the offspring of two individuals of different genetic character or genotype

hypha (pl. hyphae; adj. hyphal)—a tubular filament of a fungal thallus or mycelium; the basic structural unit of a fungi

immunity—the nonsusceptibility to infection by a specific pathogen

imperfect state—the anamorph; the asexual form in the life cycle of certain fungi, in which asexual spores (e.g., conidia) or no spores are produced (*see* perfect state)

inbred line—a product of inbreeding created by self-pollination for many generations

infect—to establish a parasitic relationship with a host plant

infest—to contaminate (as with organisms); to attack as a pest (used especially for insects and nematodes)

inoculate—to place inoculum (e.g., a microorganism or virus) on or in the healthy tissue of a plant

inoculum (pl. inocula)—a pathogen or its parts capable of establishing a live colony and producing disease when transferred to a favorable location (e.g., host)

intercellular—between or among cells

internode—the area between two adjacent nodes on a culm, stem, or stalk
intracellular—through or within a cell or cells

land race—an early cultivated form of a crop species evolved from the wild populations (e.g., a locally developed strain)
larva (pl. larvae)—a juvenile; the growth stage of an insect between the embryo and the adult
latent—present but not manifested or visible
lateral field—an interruption of the transverse striae in the cuticle of nematodes by longitudinal cuticular thickenings consisting of lines and ridges situated on top of the lateral chords
lesion—a wound or delimited diseased area
lodge—pertaining to hay or grain crops, to lie down or fall over, by which culms approach a horizontal rather than a vertical position

macroconidium (pl. macroconidia)—a long or large conidium (*see* microconidium)
mechanical transmission—the spread or introduction of inoculum to an infection court (especially a wound) by hand manipulation accompanied by physical disruption of host tissue (syn. mechanical inoculation)
meristem (adj. meristematic)—a plant tissue that functions principally in cell division and differentiation
microconidium (pl. microconidia)—a small conidium; a microspore (*see* macroconidium)
midrib—the central, thickened vein of a leaf
mollicute—a member of a group of prokaryotic organisms bound by a flexuous membrane and lacking a cell wall (order Mycoplasmatales, class Mollicutes) (*see* phytoplasma)
molt—to shed a cuticle or body encasement during a growth phase
monogenic—pertaining to the inheritance of a character controlled by a single gene (*see* polygenic)
mosaic—a disease symptom characterized by nonuniform coloration of foliage or variegated patterns of dark and light green to yellow caused by the disarrangement or unequal development of chlorophyll content; usually caused by a virus (often used interchangeably with "mottle")
mycelium (pl. mycelia; adj. mycelial)—the mass of hyphae constituting the thallus or body of a fungus

necrosis (adj. necrotic)—the death of plant cells or tissues, usually accompanied by blackening or browning, commonly a symptom of fungal, nematode, viral, or bacterial infection
node (adj. nodal)—a joint of a culm or stalk; the enlarged portion of a shoot in which leaves and spikelets are attached

obligate parasite—an organism that can survive only on or in living tissues as a parasite in association with its host plant and cannot be grown in artificial culture media
oogonium (pl. oogonia)—a one-celled female sex organ of oomycetes in which one or more eggs (oospores) are produced
oospore—the thick-walled, resting spore that develops from an oosphere produced by fertilization or parthenogenesis
ostiole (adj. ostiolate)—a pore; an opening in the papilla or neck of a perithecium, pseudothecium, or pycnidium through which spores are released
overwinter—to survive or persist through the winter

parasexual—pertaining to the recombination of genetic characters without a sexual process
parthenogenesis (adj. parthenogenetic)—the development of an egg (female gamete) into a new individual without fertilization by a male sperm (male gamete)
penetration—the initial invasion of a host by a pathogen
penetration peg—a specialized, thin strand of hypha located on the underside of an appressorium and involved in penetration of the host (syn. infection peg)
perfect state—the teleomorph; the sexual form in the life cycle of certain fungi, in which spores (e.g., ascospores or basidiospores) are formed after nuclear fission; the stage during which a fungus is capable of sexual reproduction (*see* imperfect state)
pericarp—the outer tissue layer of a ripened ovary (seed)
perithecium (pl. perithecia)—the flask-shaped or subglobose, thin-walled fruiting body (ascocarp) of an ascomycetous fungus containing asci and ascospores and having a pore (ostiole) at its apex or tip through which ascospores are expelled or released
persistent virus—a circulatory virus that remains infectious within its vector for a long period and is transmitted via salivary fluids
phage—a virus that attacks bacteria
physiologic race—a group of individuals belonging to the same species and differing from other members of the species in their behavior or other characteristics (such as virulence, symptom expression, biochemical and physiological properties, or in some cases, host range) but not in morphology
phytoplasma—a very small mollicute that is an obligate parasite of plants, is enveloped only by a single membrane (lacking cell walls), and is found in phloem tissue; spread by phloem-feeding insects
pleomorphic—able to assume various shapes
polygenic—pertaining to the inheritance of a character controlled by a group of three or more genes (*see* monogenic)
predispose—to make prone to infection and disease
primary inoculum—an inoculum that initiates rather than spreads or magnifies disease; the first inoculum produced after overwintering
prokaryote (adj. prokaryotic)—an organism lacking internal membrane-bound organelles and a distinct nucleus (*see* eukaryote)
pseudocoelom—a fluid-filled body cavity where the organs are held in place loosely by a partial lining of mesoderm-derived tissue
pseudothecium (pl. pseudothecia)—a perithecium-like ascocarp with a dispersed rather than an organized hymenium and asci in numerous unwalled locules
pycnidium (pl. pycnidia)—an asexual, flask-shaped or globose fruiting body of certain imperfect fungi, producing conidia

quantitative resistance—a resistance controlled by a group of genes at different loci that are cumulative in their effect

race—a subgroup or biotype within a species or variety distinguished from other races by behavior (e.g., differences in virulence, symptom expression, or to some extent, host range) but not by morphology (*see* physiologic race)
recessive—pertaining to a gene that is not expressed in a hybrid in the presence of a contrasting dominant allele (e.g., it is necessary to have recessive resistance genes in both parents of a single-cross hybrid in order for the resistance to be expressed) (*see* dominant)
resistant (n. resistance)—possessing properties that suppress, prevent, or impede disease development (e.g., a host plant resistant to the activity of a pathogen) (*see* susceptible)
resting spore—a temporarily dormant, usually thick-walled spore that germinates later and is capable of producing other spores or initiating infection

saprophyte (adj. saprophytic)—an organism that survives (obtains nutrients) on dead organic matter (syn. saprobe)
sclerotium (pl. sclerotia)—a hard, frequently rounded, and usually darkly pigmented resting body of a fungus composed of a mass of special hyphal cells that can remain dormant for long periods under unfavorable conditions (sclerotia germinate when conditions improve to produce a stroma, fruiting body, mycelium, or conidiophores)
sclerotized—characterized by hardened, refractive regions; used to describe certain arthropod and nematode tissues
scutellum (adj. scutellar)—the cotyledon of a grass embryo
secondary infection—infection resulting from the spread of infectious material produced after a primary infection (the first infection by a pathogen after a resting period) or from other secondary infections without an intervening inactive period
secondary inoculum—the inoculum produced by a primary infection
seminal root—a root derived from a seed
semipersistent virus—a virus that circulates in its insect vector for a short time
senesce (n. senescence)—to decline or degenerate, as with maturation, age, or disease stress
septum (pl. septa; adj. septate)—a cross-wall (e.g., in a hypha)
seta (pl. setae)—a stiff, hairlike appendage
sexual dimorphism—a phenotypic difference between males and females of the same species; in plant-parasitic nematodes, this condition can be manifested as swollen females and vermiform males or as a reduction in the feeding apparatus (stylet and cephalic framework) of males compared with that of females

sheath—the lower part of a grass (corn) leaf that clasps the culm; a membranous cover

sign—an indication of disease by the visible presence of a pathogen or its parts on a host

sorus (pl. sori)—a compact fruiting structure, especially with reference to the spore masses of rusts and smuts; occasionally a group of fruiting bodies

sp. (pl. spp.)—species; one kind of plant or animal life subordinate to a genus but above a race, strain, or variety (a genus name followed by sp. indicates that the particular species is undetermined and followed by spp. indicates that several species are grouped together without being named individually)

specific resistance—a host resistance to specific races of a pathogen (*see* general resistance)

spermagonium (pl. spermagonia)—a flask-shaped, walled fungal structure in which reproductive cells are produced

spermatium (pl. spermatia)—a nonmotile, uninucleate haploid (+ or –) sex cell; a haploid male gamete

sporangiophore—a differentiated hypha that bears a sporangium

sporangium (pl. sporangia)—a saclike or flasklike fungal structure whose entire contents are converted into an indefinite number of asexual spores (e.g., sporangiospores or zoospores)

spore—the one- to many-celled reproductive body of fungi and lower plants

sporidium (pl. sporidia)—a basidiospore

sporodochium (pl. sporodochia)—a conidial spore mass supported by a cushion-shaped mass of short conidiophores interwoven on a stroma

stele—the central, vascular cylinder of a plant

stigma—the portion of a flower that receives pollen

stroma (pl. stromata)—a compact mass of specialized fungal hyphae (with or without host tissue or substratum), sometimes sclerotium-like in form, in or on which fruiting bodies and/or spores are produced

stylet—the stiff, slender, hollow feeding organ of plant-parasitic nematodes and some insects

susceptible (n. susceptibility)—not immune; lacking resistance; prone to develop disease when infected by a pathogen (*see* resistant)

synergistic—pertaining to a host response to concurrent pathogens that exceeds the sum of the separate responses to each pathogen

teleomorph (adj. teleomorphic)—the perfect state or sexual form in the life cycle of certain fungi (*see* anamorph)

teliospore—a thick-walled resting spore produced by some fungi, notably rusts and smuts, that germinates to form a basidium

telium (pl. telia)—the fruiting body (sorus) of a rust fungus that produces teliospores

thallus—the vegetative body of a fungus

three-way cross—the progeny of a cross between a single cross and an inbred line (a three-way cross corn hybrid has a cross between two inbred lines as the female parent and an inbred line as the male parent)

tiller—a lateral shoot, culm, or stalk arising from a crown bud

tolerant—capable of sustaining disease, adverse conditions, or chemical injury without serious damage or yield loss

transmission—the spread of a virus or other pathogen from plant to plant or from one generation of host plant to the next

ultramicroscopic—too small to be seen with an ordinary laboratory (optical) microscope

urediniospore—the binucleate, dikaryotic ($n + n$), asexual, one-celled, repeating spore of a rust fungus borne in a uredinium (syns. uredospore, urediospore)

uredinium (pl. uredinia)—the fruiting body (sorus) of a rust fungus that produces urediniospores (syn. uredium [pl. uredia])

variety (adj. varietal)—a group of closely related plants of common origin within a species, resulting from deliberate genetic manipulation and differing from other members of the species in certain minor characteristics, such as form, color, flower, or fruit

vector—a living agent (e.g., insect, mite, animal, human) capable of transmitting a pathogen (e.g., virus, bacterium, fungus, mycoplasma, nematode)

vesicle—the thin, bladder-like sac or structure in which zoospores are differentiated or from which they are released; the bulbous head terminating in a conidiophore of *Aspergillus* spp.

virion—an individual virus particle

virulence—the degree or measure of pathogenicity; the capacity to cause disease

virulent—pathogenic; having the capacity to cause disease (*see* avirulent)

water-soaked—pertaining to plants or lesions that appear wet and dark (lesions are usually sunken and transparent)

whorl—three or more leaves, flowers, twigs, or other plant parts arranged in a circle and radiating from one point

zoospore—an asexually produced fungal spore with cilia or flagella capable of locomotion in water; a swarm spore

zygote—a diploid cell arising from the fusion of two haploid cells or gametes

Index